Probability, Random Variables, and Data Analytics
with Engineering Applications

P. Mohana Shankar

Probability, Random Variables, and Data Analytics with Engineering Applications

 Springer

P. Mohana Shankar
Electrical and Computer Engineering
Drexel University
Philadelphia, PA, USA

ISBN 978-3-030-56261-8   ISBN 978-3-030-56259-5   (eBook)
https://doi.org/10.1007/978-3-030-56259-5

The solutions and slides for this book can be found at https://www.springer.com/us/book/9783030562588

This Springer imprint is published by the registered company Springer Nature Switzerland AG
The registered company address is: Gewerbestrasse 11, 6330 Cham, Switzerland

Dedicated to my parents Padmanabharao and Kanakabai who were teachers.

Preface

This book presents the essential elements of probability and statistics and their applications relevant to engineering applications. The topics covered along with examples and demos are expected to meet the requirements of the undergraduate course in probability for students pursuing engineering education. A unique feature of the book has been the inclusion of computational aspects involving examples and exercises requiring random number simulations and data analysis. This paradigm shift in the contents of the book offers an opportunity to link concepts in statistics (including random variables) to practical applications in engineering, business and medicine. With Excel spreadsheets of data provided, the book offers a balanced mix of traditional topics and data analytics, expanding the scope, diversity and applications of engineering probability. The solution manual will be available to instructors. It contains solutions to both types of exercises namely the traditional ones and analytical ones and others based on data sets relevant to machine vision, machine learning and medical diagnostics.

The book took almost 3 years to complete and grew out of the notes, examples, demos and exercises created while teaching engineering probability in the Department of Electrical and Computer Engineering, Drexel University during the past several years. The timely completion of the book was possible only through the wholehearted support and active engagement of my wife Raja and daughter Raji. Their support was immeasurable and invaluable.

I also want to express my sincere thanks to Mary James, Brian Halm and Zoe Kennedy at Spinger (New York) and the Springer production team (Brinda Megasyamalan, Silembarasan Pannerselvam and Mario Gabriele) for their commitment and support to the book project.

Philadelphia, PA, USA   P. Mohana Shankar

Contents

Chapter 1
Introduction

The topics in probability and statistics are concept driven, and textbooks devoted to these require a different mode of presentation than the traditional ones seen in mathematics. The key difference is that the examples must be relevant to the students for whom the course in probability is a curricular requirement. While the students might have been exposed to some elements of probability, the idea of experiments with outcomes being described in statistical terms is novel to them. Thus the introduction of randomness and the notion of a random variable are abstract and require example driven methodology in presentation. Since engineering applications of probability include modeling of outcomes and testing the validity of the models, the book must be thematic with particular emphasis on engineering problems that students may encounter either in school or in their professional world after graduation. This book has been prepared keeping this central theme, namely, application-oriented presentation in terms of examples and exercises. Industrial, commercial, and medical applications involve collection, analysis, interpretation of data, etc., and therefore students must also learn computational techniques. This aspect adds another component to the presentation as well as the choice of examples and exercises.

Chapter 2 begins with the elementary aspects of probability by starting with sets and Venn diagrams. The concepts of probability follow with appropriate descriptions of marginal, joint, conditional, and total probabilities, Bayes' rule, Bernoulli trials, etc. Examples include those that examine the notion of continuous probability as a prelude to the presentation of random variables in the next chapter. Keeping with the theme of application oriented content, the chapter contains topics in data analytics such as the estimation of á priori, conditional and á posteriori probabilities associated with a given set of data collected from measurements. The presentation of the subject matter is organized to offer the reader the importance and relevance of the topics to present day engineering problems. The association between transition matrix and confusion matrix is introduced to illustrate the connection of the probability concepts to data science. Examples and exercises include conceptual and data analytics based ones.

© Springer Nature Switzerland AG 2021
P. M. Shankar, *Probability, Random Variables, and Data Analytics with Engineering Applications*, https://doi.org/10.1007/978-3-030-56259-5_1

The concept of a random variable and its importance in modeling the outcomes of experiments are presented in Chap. 3. While providing traditional coverage of discrete and continuous random variables, densities, distribution functions, conditional densities, readers are also introduced to mixed random variables as well as mixture densities to reflect the current trends in modeling of data from experiments. The examples of transformation of variables presented offer a peek into the various engineering applications. Data analytic techniques are presented again with the offering of receiver operating characteristics curves (ROC) and performance measures such as the area under the ROC curve (AUC), positive predictive values, etc., thereby relating the subject matter to the topics covered in Chap. 2. Chapter summary provided offers detailed descriptions of densities and their properties. Multiple examples of transformation of variables are also presented. Examples and exercises include traditional analytical ones alongside data based ones requiring computational approaches.

Chapter 4 is devoted to multiple random variables (mainly two variables). The transformation of two variables is presented initially by expanding on the notions of conditional densities in Chap. 3, before invoking the approaches requiring the use of Leibniz theorem and Jacobian. Modeling of outcomes in an experiment is presented as a two-stage experiment. Characteristic functions and Laplace transforms are offered for the determination of the densities of the sum and difference of variables. Mellin transforms (a topic not covered in textbooks) are presented as a means to obtain the densities of the products and ratios of two or more random variables. Meijer G functions are introduced to express the densities of products of random variables. The chapter offers detailed descriptions of the central limit theorem (sums and products) and order statistics. The examples and exercises are applications oriented and often involve the use of computational approaches.

Chapter 5 is exclusively devoted to data analytics applied to machine vision (target detection) tasks. The topics from Chaps. 2, 3, and 4 are invoked to make connections to hypothesis testing (chi-square tests), parameter estimation, ROC, performance analysis with the aim of developing data analytic approaches. Instead of presenting hypothesis testing and parameter estimation as theoretical topics, they are offered as tools in the context of data analytics and examples reflect this paradigm shift. ROC analysis is revisited to understand the statistics of the area under the ROC curve through the study of bootstrapping. Examples of bootstrapping are given which examine the method of comparing the areas under the ROC curves generated from two different sensors. A modeling example is presented by examining the statistics of signal fluctuations in wireless channels demonstrating the genesis of several densities. Diversity is introduced as a means to mitigate signal strength fluctuations in wireless channels. All the exercises involve data analytics.

While Chap. 5 examines the parametric method of hypothesis testing using the chi-square tests, Appendix A offers a look at other hypothesis tests such as the t-tests and z-tests. Estimation of confidence intervals and p-values are presented along with examples using data sets. Appendix B provides various examples of Matlab (www. mathworks.com) scripts to solve exercises and ways of expanding the concepts covered in the book. While Matlab scripts are provided within each chapter wherever necessary, Appendix B provides a broad overview of Matlab-based approach to engineering applications of probability. It is important to note that computational work expected during the course instruction may be undertaken using software other than Matlab.

Even though the chapters are organized to cover different topics, certain sections in Chap. 5 can be taught early. For example, sections in Chap. 5 devoted to Neyman-Pearson criterion and chi-square testing can be covered as soon as the topic of a single random variable is covered (Chap. 3). Similarly, following the completion of the discussion of two random variables, topics from Chap. 5 dealing with performance improvement may be taught.

Efforts to introduce data analytics into the undergraduate engineering probability course were initiated a few years ago. This book grew out of these efforts to bring a unique perspective emphasizing practical examples and computational approaches using data analytics bridging the gap between theory and practice. This is reflected in the substantial number of examples and exercises data analysis while offering traditional examples and exercises requiring analytical skills in all areas of mathematics. The electronic supplementary materials contain data sets for the exercises along with a number of videos. These videos (more than 60) provide step-by-step explanation of concepts, transformation of variables, and data analytics examples covering Chaps. 2–5.

Chapter 2
Sets, Venn Diagrams, Probability, and Bayes' Rule

2.1 Introduction

The study of the probability involves analysis of outcomes, observations, events, occurrences, etc. This suggests that a simple way to build a basic understanding of the concepts of probability is through the use of methods commonly used to group outcomes or events and analysis of these groups or sets. Thus the starting point of discussion will be the development of an understanding of the properties of sets, manipulation of sets, and outcomes. Once the algebraic manipulations of sets (addition, multiplication, subtraction) are understood, the transition to probability becomes easy and straightforward.

2.2 Algebra of Sets

A set is a collection of events or objects. For example, we can define a set α as a group consisting of the lowercase letters of English alphabet,

$$\alpha = \{a, b, \ldots, z\}. \tag{2.1}$$

In this case, α represents a countable set (elements are discrete and hence countable). We see that a belongs to α expressed as

$$a \in \alpha. \tag{2.2}$$

Electronic supplementary material: The online version of this chapter (https://doi.org/10.1007/978-3-030-56259-5_2) contains supplementary material, which is available to authorized users.

P. M. Shankar, *Probability, Random Variables, and Data Analytics with Engineering Applications*, https://doi.org/10.1007/978-3-030-56259-5_2

We note that the uppercase letter P does not belong to the set α and this is expressed as

$$P \notin \alpha. \tag{2.3}$$

The whole English alphabet can be referred to as a universal set S of English alphabet which means that

$$S = \{a, b, \cdots, z, A, B, \cdots Z\}. \tag{2.4}$$

Comparing Eqs. (2.1) and (2.4), we can see that α is a subset of S, expressed as

$$\alpha \subset S \text{ or } S \supset \alpha. \tag{2.5}$$

Let us define another set β consisting of just three letters, a, b, and c.

$$\beta = \{a, b, c\}. \tag{2.6}$$

We can state that β is a **subset** of α which itself is a subset of S expressed as

$$\beta \subset \alpha \subset S \text{ or } S \supset \alpha \supset \beta \tag{2.7}$$

In certain literature, the notation $\subseteq$ represents a subset (implying that the sets are also equal) and $\subset$ represents a **proper subset** (β is **contained** in α). We will always use the notation $\subset$ for a subset.

A set that has no elements is an empty or a null set represented as

$$\varnothing = \{\}. \tag{2.8}$$

It is easy to see that **null** set is a subset of **every** set.

It should be noted that we may also have an uncountable set. For example, if we define a set γ to consist of numbers between 2 and 4, the set has an uncountable number of elements since decimal values are allowed. The set γ is

$$\gamma = \{x | 2 < x < 4\}. \tag{2.9}$$

Equation (2.9) implies that γ contains elements x such that x lies between 2 and 4. Let us define a new set as

$$\kappa = \{x | x > 5\}. \tag{2.10}$$

Another property of the sets is evident through an examination of the elements of γ and κ. They contain no common elements and they are said to be disjoint. The commonality of two sets is generally expressed as the **intersection** using the symbol, $\cap$. This means that

$$\gamma \cap \kappa = \varnothing = \{\}. \tag{2.11}$$

Consider a set ψ consisting of a couple of upper- and lowercase English alphabets,

$$\psi = \{a, A, c, H\}. \tag{2.12}$$

If we compare the sets α and ψ, they have a few common elements, and these are identified as belonging to a new set, namely, the intersection of the two sets. The common elements are a and c. The intersection of α and ψ is expressed as

$$\alpha \cap \psi = \alpha \,\&\, \psi = \alpha \psi = \psi \cap \alpha = \{a, c\}\ . \tag{2.13}$$

If we now combine all the elements of α and ψ, we have the union of two sets consisting of the elements of α and ψ. The union of two sets is expressed as

$$\alpha \cup \psi = \alpha + \psi = \{a, b, c, \cdots, z, A, H\}. \tag{2.14}$$

Define another set μ as

$$\mu = \{A, B, \ldots, Z\} \tag{2.15}$$

Comparing α and μ, it is clear that they have no common elements and the union of α and μ is S. S represents the universal set, and its universal aspect (i.e., what is contained in it) is determined and defined by the user. In other words, the universal set is subjective. We can define the ***complement*** of α as

$$\overline{\alpha} = \alpha^C = \alpha' = \{\text{elements in } S \text{ that are not in } \alpha\} = \mu\ . \tag{2.16}$$

Note that

$$\alpha \cup \mu = S. \tag{2.17}$$

A few additional points of observation can be made from these properties. When we formed the union of α and ψ, and the intersection of α and ψ, we only used the common elements a and c only once. This means that

$$\begin{aligned} \alpha \cup \alpha &= \alpha \\ \alpha \cap \alpha &= \alpha. \end{aligned} \tag{2.18}$$

In other words, the union or the intersection of a set with itself is the set itself (and not two times the set).

Another way of representing the relationship among any two sets is through the **difference** of two sets. The differences of the two sets β and ψ is expressed as

$$\begin{aligned} \beta - \psi &= \{b\} \\ \psi - \beta &= \{A, H\}. \end{aligned} \tag{2.19}$$

Figure Ex.2.1

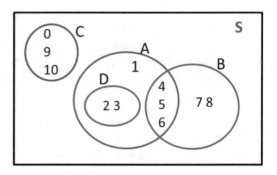

It is clear that the set $\beta - \psi$ is different from the set $\psi - \beta$. The difference $\beta - \psi$ results in a set that has elements in β that do not exist in ψ, and similarly, the difference $\psi - \beta$ results in a set that has elements of ψ that do not exist in β.

It is often easy to understand the relationships among the various sets and the universal set through the sample space representation in terms of Venn diagrams. In a Venn diagram, the rectangle represents the universal set, and subsets (or other contained sets) are represented by other shapes including rectangles.

Example 2.1 Let us define a universal set as integers from 0 to 30. Let us also define four sets, A, B, C, and D as follows:

$$A = \{1, 2, 3, 4, 5, 6\}$$
$$B = \{4, 5, 6, 7, 8\}$$
$$C = \{0, 9, 10\}$$
$$D = \{2, 3\}$$

These sets can be represented in a Venn diagram as shown in Figure Ex.2.1.

The Venn diagram in Figure Ex.2.1 shows that the sets A and C, D and C, B and D, and B and C are disjoint. D is a subset of A and the intersection of A and B is the set consisting of elements $\{4, 5, 6\}$. These observations may be summarized as

$$A \cap C = \{\} = \varnothing$$
$$B \cap C = \{\} = \varnothing$$
$$D \cap C = \{\} = \varnothing$$
$$B \cap D = \{\} = \varnothing$$

$$D \subset A$$

$$A \cap B = \{4, 5, 6\}$$

$$A \cup B \cup C \subset S$$

The equation above means that if we define P as the set that is complementary to $A \cup B \cup C$, we have

$$P \cup A \cup B \cup C = S$$

$$P = \overline{A \cup B \cup C}$$

From the subset D of A, we can also define a new set as the **difference** of two sets, A–D. If E is defined as the set A–D, we have E consisting of elements of A that are not present in D,

$$E = A - D = \{1, 4, 5, 6\}$$

Let us go back to the use of Venn diagrams to illustrate the relationships among sets. Let us consider two sets A and B (Figs. 2.1, 2.2, 2.3, and 2.4).

Fig. 2.1 Disjoint sets A and B

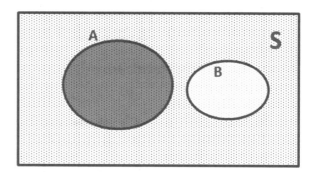

Fig. 2.2 A and B are not disjoint. $A \cup B$ is the **shaded region** and unshaded region is $\overline{A \cup B}$

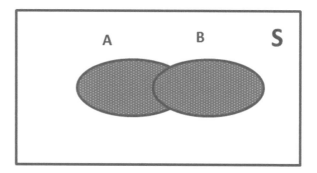

Fig. 2.3 Intersection of
A and B (overlap of A and
B): A and B **not** disjoint

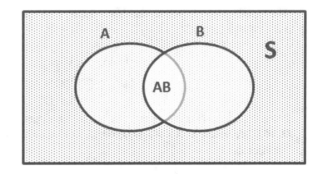

Fig. 2.4 $A \cup B$ expressed as
the union of three disjoint
sets, each of which is an
intersection of two sets

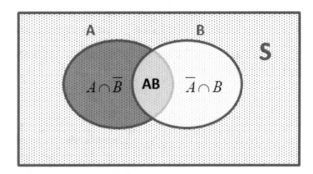

2.2.1   Summary of Relationships

$$S \Rightarrow \text{universal set}$$
$$\varnothing \Rightarrow \text{null or empty set}$$
$$\overline{A} \Rightarrow \text{complement of } A$$
$$A \cup \overline{A} = S$$
$$A \cap \overline{A} = \varnothing \Rightarrow A \text{ and } \overline{A} \text{ are disjoint}$$
$$C = A \cup B = A + B \Rightarrow \{\text{outcomes in either } A, B, \text{ or both}\}$$
$$D = A \cap B = AB \Rightarrow \{\text{outcomes that are in both } A \text{ and } B\}$$
$$\overline{A \cup B} = \overline{A} \cap \overline{B}$$
$$\overline{A \cup B} = \overline{A \cap B}$$

For any three sets, A, B, and C,

Fig. 2.5 Difference between A and B, $B \subset A$

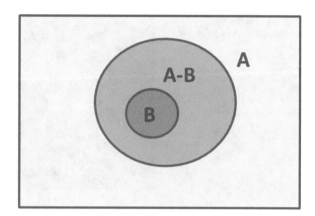

Fig. 2.6 Difference between B and A, $A \subset B$

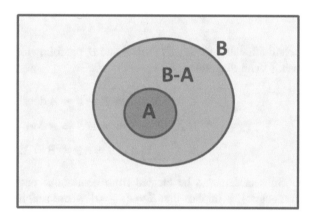

$$(A \cup B) \cup C = A \cup (B \cup C) = A \cup B \cup C \qquad (2.20)$$

$$(A \cap B) \cap C = A \cap (B \cap C) = A \cap B \cap C = ABC \qquad (2.21)$$

$$\begin{aligned} A \cup (B \cap C) = (A \cup B) \cap (A \cup C) \\ A \cap (B \cup C) = (A \cap B) \cup (A \cap C) \end{aligned} \qquad (2.22)$$

If B is a **subset** of A, it is expressed as $B \subset A$ or $A \supset B$ as shown in Fig. 2.5. This means that B is **contained** in A as seen below. If $B \subset A$, then B and A–B are disjoint.

$$\begin{aligned} A \cup B = A \\ A \cap B = B \end{aligned} \qquad (2.23)$$

$$\overline{A} \subset \overline{B} \qquad (2.24)$$

If $A \subset B$ (see Fig. 2.6), we see that A and $(B$–$A)$ are disjoint.

It is worthwhile to note that union of the event with itself and intersection of the event with itself is the event as seen in Eq. (2.18).

Fig. 2.7 Difference and
intersection: $\{A - B\} = A\bar{B}$

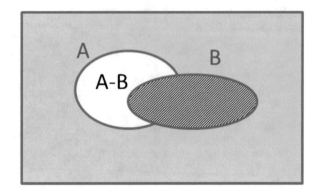

$$A \cup A \neq 2A, \quad A \cup A = A$$
$$A \cap A = AA = A \tag{2.25}$$

Additionally, for any two sets A and B, the following results can easily be proven
using Venn diagrams.

$$\text{If } A \cup B = \bar{A} \Rightarrow B \equiv S \Rightarrow A \text{ does not exist} \tag{2.26}$$

$$\text{If } A \cap B = \bar{A} \Rightarrow A \equiv S \Rightarrow B \text{ does not exist} \tag{2.27}$$

$$\text{If } A \cup B = AB \Rightarrow B = A \tag{2.28}$$

Subsets can also be created from continuous outcomes. If $S = \{$lifetimes of
computers$\}$, it implies that $S = \{t \geq 0\}$ where t represents the time (let us say in
years). If we now create two sets by observing the performance of systems over a
long period of time, we can create two events $A = \{0 \leq t \leq 5\}$ where computers
seem to last up to 5 years and $B = \{3 \leq t \leq 4\}$ where computers seem to operate
between 3 and 4 years. In this case, it is seen that $B \subset A$.

Let us examine the difference of two sets. The difference between A and B is the
set A–B. This means that A–B represents the elements of A that are not elements of B.
In this case, the events B and $(A$–$B)$ are disjoint.

$$A - B = A \cap \bar{B} = A\bar{B} \tag{2.29}$$

The set $(A$–$B)$ is the region of A excluding the overlap of A and B as shown in
Fig. 2.7. Therefore,

$$A \cap \bar{B} = A\bar{B} = A - (A \cap B) \tag{2.30}$$

$$A - A = \emptyset \tag{2.31}$$

If A and B are **disjoint**,

$$A - B = A, \quad B - A = B \tag{2.32}$$

If A and B are not disjoint,

$$\begin{aligned} A - B &= \{\text{Outcomes in } A \text{ that are not in } B\} \\ &= \{\text{Outcomes in } A \text{ that are not in } A \cap B\} \end{aligned} \tag{2.33}$$

Therefore, we have several ways of expressing the difference in two sets, A and B.

$$A - B = A\overline{B} \tag{2.34}$$

$$A\overline{B} = A - \{A \cap B\} \tag{2.35}$$

You will also notice that the events $(A–B)$ and AB are disjoint and A is the union of these two disjoint events,

$$\{A\} = \{A - B\} + \{AB\} = \{A\overline{B}\} + \{AB\} \tag{2.36}$$

Similarly, $B–A$ can be expressed as

$$B - A = \overline{A} \cap B = \overline{A}B . \tag{2.37}$$

Using Eq. (2.37), we can write the identity for B where B is expressed as the union of two disjoint sets $(B–A)$ and AB. This means that

$$\{B\} = \{B - A\} + \{AB\} = \{\overline{A}B\} + \{AB\} . \tag{2.38}$$

Example 2.2 If $A = \{1, 4, 5, 7, 8, 10\}$ and $B = \{2, 4, 5, 7, 9\}$, what is $A–B$? What is $B–A$?

Solution

$$\{A - B\} = \{1, 8, 10\}$$
$$\{B - A\} = \{2, 9\}$$
$$\{AB\} = \{4, 5, 7\}$$
$$A - \{AB\} = \{1, 8, 10\} = A - B$$
$$B - \{AB\} = \{2, 9\} = B - A$$

Example 2.3 If A is defined as the set of continuous numbers between 1 and 4 and B is defined as the set of continuous numbers between 3 and 6, find the $\{A + B\}$, $\{AB\}$ $\{(A + B)(AB)^C\}$.

Solution
$$A = \{1 \leq X \leq 4\}$$
$$B = \{3 \leq X \leq 6\}.$$

Since we are dealing with continuous numbers, the universal set is chosen as

$$S = \{-\infty \leq X \leq \infty\}$$

We could have chosen the universal set as numbers between 0 and 100 or -200 and 100, or any set of values that will contain A and B within the context of this problem.

$$\{A \cup B\} = \{1 \leq X \leq 6\}$$
$$\{AB\} = \{3 \leq X \leq 4\}$$
$$\{AB\}^C = \{-\infty \leq X \leq 3\} + \{4 \leq X \leq \infty\} = \{X \leq 3\} + \{X \geq 4\}$$
$$\left\{(A \cup B)(AB)^C\right\} = \{1 \leq X \leq 3\} + \{4 \leq X \leq 6\}$$

A note on the use of the symbols $<$ and $\leq$ is in order at this point. In this example, if X (a continuous variable) is taking two values E and F, $E \leq F$ is not different from $E < F$. Similarly, $E > F$ and $E \geq F$ represent identical sets. On the other hand, if X represented a discrete variable, the presence of the equal sign becomes important. We will revisit this issue in Chap. 3.

Example 2.4 If A and B are numbers such that each may take values randomly between 0 and 5, show the sample space of the experiment. If $C = \{A > 2\}$, $D = \{B > 2\}$, $E = \{A > B\}$, and $F = \{A > 3, B < 4\}$, show the sample space associated with these sets.

Solution (Figures Ex.2.4.1, Ex.2.4.2, and Ex.2.4.3)

2.2.2 Overview of Properties of Sets

A set is a collection of items, with each item identified as element of the set.

An empty set contains no elements and is expressed as Ø. An empty set is a NULL set.

The universal set contains all the elements (or outcomes in an experiment) in the context of the situation being described.

Two sets are equal, if they contain identical elements. Order of the elements is not important. Also, repetitions of elements have no effect. For example, the sets $A = \{2, 2, 3, 4, 5\}$ and $B = \{2,3,4,5\}$ are equal.

A set B is a subset of A if A always contains all the elements of B and B may or may not contain all the elements of A.

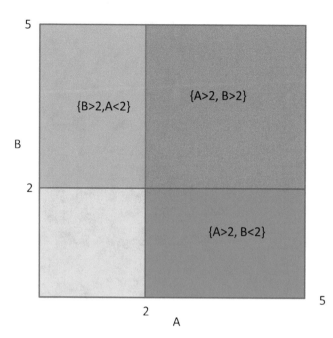

Figure Ex.2.4.1

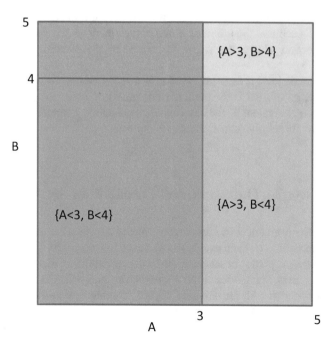

Figure Ex.2.4.2

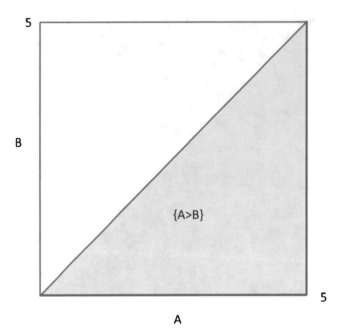

5

B

{A>B}

5

A

Figure Ex.2.4.3

The complement of a set A is the set formed by the remaining elements of the universal set S such that the union of A and complement of A is the universal set.

The intersection of two sets is the set formed by the common elements of the two sets.

Two sets are disjoint if their intersection is a NULL set.

The cardinality of a set is the number of elements of a set.

Two sets are equivalent, if their cardinalities are same. Therefore, if two sets are equal, they are also equivalent. Converse is not true.

2.3   Sets, Events, Outcomes, and Probabilities

We saw that events and occurrences can be grouped into sets. Since occurrences and events are statistical, i. e., each event or occurrence has a probability associated with it, we can use the properties of sets to describe the probabilities. Since S is a universal set, it implies that S includes all the events (or outcomes), meaning that the probability associated with the universal set is 1. Therefore, S constitutes the sample space (Leon-Garcia 1994; Ross 2006; Papoulis and Pillai 2002). The set S contains all the possible outcomes and therefore, S is a certain event with probability,

Fig. 2.8 Union of A and B shown in terms of three mutually exclusive events, $AB, A\bar{B}, \bar{A}B$

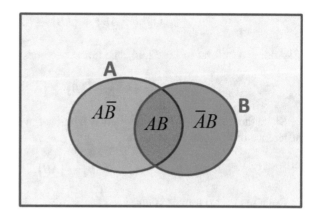

$$P(S) = 1. \tag{2.39}$$

For any subsets (sets comprising of selected outcomes from the same space), the probability associated with the subset will be less than 1. If A contains elements from the sample space S, we have

$$P(A) \le P(S). \tag{2.40}$$

The probability associated with the empty set or the null set is zero,

$$P(\varnothing) = 0. \tag{2.41}$$

We can therefore conclude that the probability p of any event or occurrence will always lie between 0 and 1,

$$0 \le p \le 1. \tag{2.42}$$

If A and B are two events, we can express the probability of the union of A and B (probability of the sum of A and B) as

$$P(A \cup B) = P(A) + P(B) - P(A \cap B) = P(B \cup A) = P(A) + P(B) - P(B \cap A). \tag{2.43}$$

Equation (2.43) follows directly from Fig. 2.3 because union of two events includes the joint space counted twice requiring the subtraction. Another way to explore the union is by writing the union of A and B as the union of three mutually exclusive (disjoint) events as shown in Fig. 2.8. Mutually exclusive events imply that the probability of the joint event is 0. Stated differently, mutually exclusive events imply that the occurrence of any one precludes the occurrence of others. This means that

$$A \cup B = \{A\overline{B}\} + \{AB\} + \{\overline{A}B\}. \tag{2.44}$$

Adding a null set to Eq. (2.44), the union becomes

$$A \cup B = \{A\overline{B}\} + \{AB\} + \{\overline{A}B\} + \{\}. \tag{2.45}$$

Since the difference of any two sets is a null set, we replace the null set with the difference of the joint event AB as

$$\{\} = \{AB\} - \{AB\}. \tag{2.46}$$

Equation (2.45) now becomes

$$A \cup B = \underbrace{\{A\overline{B}\} + \{AB\}}_{A} + \underbrace{\{\overline{A}B\} + \{AB\}}_{B} - \{AB\} \tag{2.47}$$

It can be seen that Eq. (2.47) leads directly to Eq. (2.43). If A and B are mutually exclusive, the total probability is

$$P(A \cup B) = P(A) + P(B). \tag{2.48}$$

While "mutually exclusive" events suggest that the presence or occurrence of one event precludes the presence of the other, "independence" of events implies that the occurrence of one has no influence on the occurrence of the other. For example, if we toss a coin, the two outcomes, heads (H) or tails (T), are mutually exclusive. Consider now the case of having a coin and a six-face die. We throw the pair up and watch the outcomes. The outcomes will be in pairs, $\{1,H\}$, $\{2,H\}$, $\{3,H\}$, $\{4,H\}$, $\{5,H\}$, $\{6,H\}$, $\{1,T\}$, $\{2,T\}$, $\{3,T\}$, $\{4,T\}$, $\{5,T\}$, and $\{6,T\}$. Any 1 of these 12 outcomes (occurring in pairs) is such that the 2 events, 1 from the coin and the other from the die, have no relationship and, therefore, the events are ***independent***. In general, if the events A and B are independent, the joint probability is the product of the marginal probabilities,

$$P(AB) = P(A)P(B). \tag{2.49}$$

In the experiment involving the coin and the die, this implies that

$$P(1, H) = P(1H) = P(H1) = P(1)P(H) = \frac{1}{6}\frac{1}{2} = \frac{1}{12}. \tag{2.50}$$

Equation (2.50) is obtained assuming that we have a fair coin and a fair die. You will also notice that the order is not important; $\{1,H\}$ and $\{H,1\}$ constitute the same event.

We may also arrive at the result in Eq. (2.50) by knowing that the 12 possible outcomes are mutually exclusive and, if the coin and die are unbiased, each outcome is equally likely. Therefore, we can invoke the frequency interpretation of the probability and express the probability of any one of the 12 outcomes as equal to 1/12.

The frequency interpretation of the probability is stated as follows (Ziemer 1997; Papoulis and Pillai 2002). If we conduct a large number of experiments (N) and the number of favorable outcomes (or outcomes of choice) is n, the probability of that event (specific outcome) p is the relative frequency expressed as the ratio n/N. This means that

$$p = \left(\frac{n}{N}\right)_{N \to \infty} \tag{2.51}$$

Independent vs. mutually exclusive:

If A and B are **mutually exclusive**, $P(AB) = 0$ because $(AB) = \varnothing$.
If A and B are **independent**, $P(AB) = P(A)P(B)$.
If A and B are **mutually exclusive and independent**, $P(AB) = 0 = P(A)P(B)$, one of the events must not occur.

Example 2.5 *If events A and B are independent*, show that A and $\overline{B}$ are also independent.

Solution

$$P(AB) = P(A)P(B)$$

Furthermore,

$$P(A) = P(AB) + P(A\overline{B})$$

Rewriting the expression for $P(A)$, we have

$$P(A\overline{B}) = P(A) - P(AB) \Rightarrow P(A) - P(A)P(B) = P(A)[1 - P(B)]$$

Note that

$$[1 - P(B)] = P(\overline{B})$$

This means that

$$P(A\overline{B}) = P(A)[1 - P(B)] = P(A)P(\overline{B}) \Rightarrow A \text{ and } \overline{B} \text{ are independent}$$

It can be shown similarly that events B and $\overline{A}$ are also independent. We have

$$P(B) = P(AB) + P(\overline{A}B).$$

Since A and B are independent, equation above becomes

$$P(B) = P(A)P(B) + P(\overline{A}B)$$

Rewriting,

$$P(B) - P(A)P(B) = P(\overline{A}B) \Rightarrow P(B)[1 - P(A)] = P(B)P(\overline{A}).$$

Equation above shows that B and $\overline{A}$ are independent. We can write an expression for $P(\overline{A})$ as

$$P(\overline{A}) = P(\overline{A}B) + P(\overline{A}\overline{B}).$$

If A and B are independent, B and $\overline{A}$ are also independent. Therefore, the equation above becomes

$$P(\overline{A}) = P(\overline{A})P(B) + P(\overline{A}\overline{B}).$$

Rewriting the equation above, we have

$$P(\overline{A}) - P(\overline{A})P(B) = P(\overline{A}\ \overline{B}) \Rightarrow P(\overline{A})[1 - P(B)] = P(\overline{A})P(\overline{B})$$

The equation above shows that if A and B are independent, then $\overline{A}$ and $\overline{B}$ are also independent.

If *events A, B, and C are independent*, they all must be independent taken two at a time as well as three at a time. *All four equalities must hold* for the events A, B, and C to be independent (Papoulis and Pillai 2002).

$$
\begin{aligned}
P(AB) &= P(A)P(B) \\
P(BC) &= P(C)P(B) \\
P(AC) &= P(A)P(C) \\
P(ABC) &= P(A)P(B)P(C)
\end{aligned}
\tag{2.52}
$$

Total Probability Revisited

$$P(A + B) = P(A) + P(B) - P(AB) \tag{2.53}$$

Note that for any two events A and B, the events A, $\overline{A}B$ and $\overline{A \cup B}$ constitute the complete sample space (see Fig. 2.8),

Fig. 2.9 Three events and mutually exclusive groupings

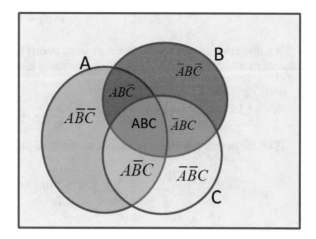

$$P(S) = 1 = P(A) + P(\overline{A}B) + P(\overline{A \cup B}) \tag{2.54}$$

We may now extend the concept of total probability to the case when the sample space contains more than two events. First consider the case when we have three events, A, B, and C. The total probability can be obtained by treating

$$\{A + B + C\} = \{A + D\} \tag{2.55}$$

$$\{D\} = \{B + C\} \tag{2.56}$$

By applying the total probability concept in Eq. (2.43) to Eq. (2.55), we get

$$P(A + B + C) = P(A) + P(B) + P(C) - P(AB) - P(BC) - P(AC) + P(ABC) \tag{2.57}$$

Just as in the case of two events expressed as the sum of three mutually exclusive events in Eq. (2.44), the union of three events can be expressed as the sum of seven mutually exclusive events shown in Fig. 2.9 as

$$A \cup B \cup C = \{A\overline{B}\overline{C}\} + \{ABC\} + \{AB\overline{C}\} + \{\overline{A}B\overline{C}\} + \{\overline{A}BC\} + \{\overline{A}\overline{B}C\} \\ + \{A\overline{B}C\} \tag{2.58}$$

If A, B, and C are mutually exclusive,

$$P(A + B + C) = P(A) + P(B) + P(C). \tag{2.59}$$

Furthermore, if the sample space is populated only by events A, B, and C, we have

$$P(A + B + C) = P(A) + P(B) + P(C) = 1 \tag{2.60}$$

We may now generalize the union of three events expressed in Eq. (2.57) to the case of a union of n events, $A_1, A_2, \ldots, A_n$ resulting in

$$P\left(\bigcup_{i=1}^{n} A_i\right) = \sum_{i=1}^{n} P(A_i) - \sum_{i=1}^{n-1}\sum_{j=i+1}^{n} P(A_i A_j) + \sum_{i=1}^{n-2}\sum_{j=i+1}^{n-1}\sum_{k=j+1}^{n} P(A_i A_j A_k) - \cdots \tag{2.61}$$

If all the events are mutually exclusive, the total probability becomes

$$P\left(\bigcup_{i=1}^{n} A_i\right) = \sum_{i=1}^{n} P(A_i) \tag{2.62}$$

Using Eq. (2.62), we can rewrite Eq. (2.61) as

$$P\left(\bigcup_{i=1}^{n} A_i\right) \leq \sum_{i=1}^{n} P(A_i). \tag{2.63}$$

The equality sign holds when the events are mutually exclusive. Equation (2.63) is known as the "union bound" or "Boole's inequality."

2.3.1   OR, AND, at Least, and at Most

If we have two events A and B (for the time being, we treat them as being **NOT** mutually exclusive), the probability associated with the union of two events implies the following, A or B or both. Therefore, **OR** is all **inclusive**.

$$P(A \text{ or } B \text{ or Both}) = P(A) + P(B) - P(A \text{ and } B) \tag{2.64}$$

Total probability implies events individually and collectively happening. On the other hand, **AND** implies the simultaneous occurrence (joint occurrence) of two or more events. For the case of two events A and B, the joint probability means

$$P(AB) = P(A \text{ and } B) = P(A \& B) \tag{2.65}$$

If five events are occurring, the term **at least** two implies two or more events (2, 3, 4, or 5). The term **at most** 3 implies 1, 2, or 3 events taking place. If you conduct an experiment with successes and failures,

$$P\{\text{at least one success}\} = 1 - P\{\text{no successes}\} = 1 - P\{\text{all fail}\} \tag{2.66}$$

If A and B are two events that are not mutually exclusive, $P(A + B)$ gives the probability that either A, B, or both occur and is given by Eq. (2.53) or Eq. (2.64). Therefore, the probability that neither event occurs is

$$P(\text{Neither event occurs}) = 1 - [P(A + B)] = 1 - P(A) - P(B) + P(AB). \quad (2.67)$$

While $P(A + B)$ gives the probability that either A, B, or both occur, the probability that *exactly one* of the events occurs is the probability associated with the union of $(A–B)$ and $(B–A)$.

$$P(\text{exactly one of the events}) = P[(A - B) \cup (B - A)]. \quad (2.68)$$

Using Eq. (2.34)

$$\begin{aligned} P(\text{exactly one of the events}) &= P[A - AB + B - AB] \\ &= P(A) + P(B) - 2P(AB) \end{aligned} \quad (2.69)$$

We may get the result in Eq. (2.69) by noting that taking out the joint event from the union of two events, we also get exactly one event. Therefore,

$$\begin{aligned} P(\text{exactly one of the events}) &= P(AUB) - P(AB) \\ &= P(A) + P(B) - P(AB) - P(AB). \end{aligned} \quad (2.70)$$

Note that Eqs. (2.69) and (2.70) are the same.

If A, B, and C are not mutually exclusive, the probability that exactly two of the three events occur is

$$P(AB) + P(BC) + P(CA) - 3P(ABC) \quad (2.71)$$

If $P(AB) = 1$, then we have certain events that are identical,

$$P(AB) = 1 \Rightarrow P(A) = 1, \;\; P(B) = 1, \;\; P(A + B) = 1 \quad (2.72)$$

In this case, the sample space only contains A and B that overlap 100%. Similarly, if $P(ABC) = 1$, we have

$$\begin{aligned} P(ABC) &= 1 \Rightarrow P(A) = P(B) = P(C) = P(AB) = P(BC) = P(AC) \\ &= P(A + B + C) = 1 \end{aligned} \quad (2.73)$$

In this case, the sample space contains only A, B, and C and they overlap 100%.

Example 2.6 In the linear algebra course (event B), the probability of any student receiving an A is 0.35, while the probability of receiving an A in the course in differential equations (event C) the following semester is 0.4. The probability that a student receives A in both is 0.2. (i) What is the probability that a student receives at least one A after completing both courses? (ii) What is the probability that the student receives exactly one A and (iii) the student receives an A in linear algebra, but not an A in differential equations?

Solution   Prob (a student receives at least one A)

$$P(C + B) = P(B) + P(C) - P(BC) = 0.35 + 0.4 - 0.2 = 0.55$$

Prob (a student receives exactly one A)

$$P(\overline{C}B) + P(C\overline{B}) = \underbrace{P(C) - P(BC)}_{P(C\overline{B})} + \underbrace{P(B) - P(BC)}_{P(\overline{C}B)} = P(B) + P(C) - 2P(BC)$$

$$= 0.35$$

Prob (a student receives A in linear algebra and different grade in differential equations)

$$P(B\overline{C}) = P(B) - P(BC) = 0.35 - 0.2 = 0.15$$

Example 2.7   We are given the following information: A, B, and C are events with the following probabilities: $P(A) = 0.6, P(AC) = 0.25, P(ABC) = 0.04, P(A\overline{B}\,\overline{C}) = 0.1$. What is $P(AB)$? (Figure Ex.2.7)

Figure Ex.2.7

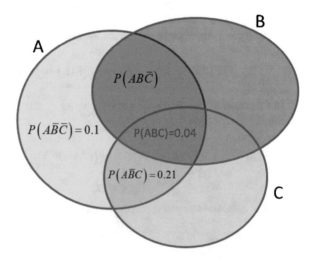

Solution   Note that the following events are mutually exclusive: (ABC), $(AB\overline{C})$, and $(A\overline{B}C)$. This means that

$$P(A\overline{B}C) = P(AC) - P(ABC) = 0.25 - .04 = 0.21$$

$$P(AB) = P(A) - P(A\overline{B}C) - P(A\overline{B}C) = 0.6 - 0.1 - 0.21 = 0.29$$

Example 2.8   Three sets A, B, and C are such that they constitute the universal set. Additionally, the probabilities of the events A, B, and C, respectively, are 0.4, 0.6, and 0.5. The probability of the event formed by the union of A and B is 0.8, union of B and C is 0.7, and the union of A and C is 0.8. What are the joint probabilities? Are the events independent? Are they mutually exclusive?

Solution   We are given the following:

$$P(A \cup B \cup C) = P(S) = 1$$

$$P(A) = 0.4, \quad P(B) = 0.6, \quad P(C) = 0.5$$
$$P(A \cup B) = 0.8, \quad P(B \cup C) = 0.7, \quad P(A \cup C) = 0.8$$
$$P(A) + P(B) + P(C) = 1.5 > P(S)$$

The marginal probabilities add to more than unity. Therefore, the events are not mutually exclusive because the sample space only consists of A, B, and C. If the events were to be mutually exclusive, the sum of the marginal probabilities would be exactly equal to 1 (neither larger than 1 nor smaller than 1).

$$P(AB) = P(A) + P(B) - P(A \cup B) = 0.2 \neq P(A)P(B)$$
$$P(BC) = P(B) + P(C) - P(C \cup B) = 0.4$$
$$P(AC) = P(A) + P(C) - P(A \cup C) = 0.1$$
$$P(ABC) = P(A \cup B \cup C) - P(A) - P(B) - P(C) + P(AB) + P(BC) + P(AC)$$
$$P(ABC) = 1 - 0.4 - 0.6 - 0.5 + 0.2 + 0.4 + 0.1 = 0.2$$

A, B, and C are not independent. There is only a need to show that the probability of any one of the joint events is not equal to the product of the respective marginal probabilities.

2.3.2   *Probability Expressed as a Continuous Function*

So far, we treated the probability as a form of measure taking a value between 0 and 1, generally expressed as a specific value such as 0.2 or 0.9. On the other hand, there is no rule barring us from using a continuous function to represent the probability. If the outcomes of an experiment X are continuous (e.g., voltage measured, power measured, etc.), let us express the probability associated with the set of outcomes specified by the event A

$$A = \{X|X > y\}. \tag{2.74}$$

Equation (2.74) implies that outcomes exceed a value of y. Let us assume that all the measurements are positive. In this case, the complementary event of A will be

$$A^C = \{X|0 < X < y\}. \tag{2.75}$$

Since A is an event, the probability of that event must lie between 0 and 1. One simple function that fits this requirement is

$$P(A) = P\{X|X > y\} = \exp\left(-\frac{y}{b}\right), \quad 0 < y < \infty, \quad b > 0 \tag{2.76}$$

The probability of the event is such that if $y = 0$, A includes all the outcomes from 0 to ∞ and, therefore, $P(A)$ must be unity. It is clear that the probability expressed in Eq. (2.76) meets this requirement. If $y = \infty$ (let us remember that the outcomes are continuous), the event A has no outcomes and the $P(A) = 0$. The probability expressed in functional form in Eq. (2.76) meets this requirement as well. In other words, it is possible to express probability in a functional form while retaining all its properties and characteristics. From this discussion, it is clear that

$$P\{X > y\} = \exp\left(-\frac{y}{b}\right) \tag{2.77}$$

$$P\{0 < X < y\} = 1 - P\{X > y\} = 1 - \exp\left(-\frac{y}{b}\right) \tag{2.78}$$

Another property that can be observed is that the right-hand side of Eq. (2.78) is a monotonically increasing function of y as we expect because as y goes up, the outcomes in the set $\{0 < X < y\}$ go up.

Example 2.9 Long-term observations of wait times at a teller window indicated that the probability a person has to wait for more than 10 minutes to access the window is 0.2. If X represents the wait time, long-term modeling has shown that

$$P\{X > y\} = \exp\left(-\frac{y}{b}\right), \quad y > 0.$$

(a) Is this a valid probability?
(b) What value of b will fit the model?

(c) What is the probability that the waiting time is less than 5 minutes? What is the probability that the wait time is between 5 and 10 minutes?

Solution

(a) If we put $y = 0$, $P(X > y\} = 1$. This is true since the probability that the wait time exceeds 0 is a certain event with a probability of 1. If we substitute $y = \infty$, we get $P(X > y\} = 0$. This is true since the probability that anyone has to wait beyond ∞ is zero. Furthermore, $P\{0 \le X \le y\} = 1 - \exp(-y/b)$. This probability is a monotonically increasing function of y. Therefore, the probability expression is valid.

(b) We are given that

$$P\{X > 10\} = 0.2 = \exp\left(-\frac{10}{b}\right).$$

Solving for b, we have

$$b = 6.2133$$

(c) Probability that the waiting time is less than 5 minutes is

$$P(X < 5) = 1 - \exp\left(-\frac{5}{b}\right) = 0.5528$$

Probability that the waiting time between 5 and 10 minutes is

$$P(5 < X < 10) = P(X > 5) - P(X > 10) = \exp\left(-\frac{5}{b}\right) - \exp\left(-\frac{10}{b}\right)$$

$$= 0.4472 - 0.2 = 0.2472$$

Probability that the waiting time between 5 and 10 minutes can also be expressed as

$$P(5 < X < 10) = P(X < 10) - P(X < 5) = \exp\left(-\frac{5}{b}\right) - \exp\left(-\frac{10}{b}\right)$$

$$= 0.4472 - 0.2 = 0.2472$$

It is worthwhile to note that X is continuous, and therefore,

$$P(5 < X < 10) = P(5 \le X \le 10) = P(5 < X \le 10) = P(5 \le X < 10)$$

In other words, the equality sign is redundant. We will revisit this issue in Chap. 3.

Example 2.10 A person is picking numbers between 0 and 5. The numbers also form a continuum (they are not discrete). What is the probability that (i) the number picked X will lie between 3 and 3.2, (ii) the number picked will lie between 4.5 and 4.7, and (iii) the number is exactly equal to 4?

Solution If x is a the number lying between 0 and 5, the probability that X lies between x and $x + \Delta x$ is easily observed as

$$P(\text{number lies between } x \text{ and } x + \Delta x) = \frac{\Delta x}{5}.$$

It is clear that it does not matter where the center of the range is located. This means that the probability that the number lies between 1 and 2 is equal to the probability that it lies between 2 and 3 and so on.

$$P(\text{number lies between 3 and 3.2}) = P(\text{number lies between 4.5 and 4.7}) = \frac{0.2}{5}$$

The equation above shows that the probability that the number picked will be exactly equal to a specific value is 0 [See Example 2.11 also. Read the discussion of uniform density in Chap. 3].

The possibility of being able to express the probability in a functional form allows us to expand the simple concepts to modeling systems, applications, and observations.

While the experiment of the coin toss was described as a discrete one having two outcomes with probabilities of $P(\text{Heads})$ and $P(\text{Tails})$, the presumption was that the probabilities are also discrete. It is also possible to introduce a continuous measure of probability associated with the experiment of coin toss as well. If we describe a variable X that takes values continuously in the range 0 to 1, it is possible to define the probabilities associated with the two complementary outcomes as

$$\begin{aligned} P(H) &= P(X < y), \quad 0 < y < 1 \\ P(T) &= P(X > y), \quad 0 < y < 1. \end{aligned} \tag{2.79}$$

As the value of y is varied, the probabilities of the two outcomes vary, and $y = 0.5$, the experiment results in unbiased outcomes.

This can be illustrated by picking numbers randomly between 0 and 1 followed by separating the numbers by setting a threshold and counting them. Figure 2.10 shows the scatter plot of 200 numbers generated indicating that the numbers are continuous.

Following Eq. (2.79), y is treated as a threshold and allowed to vary from 0 to 1. The resulting probabilities are estimated and plotted in Fig. 2.11.

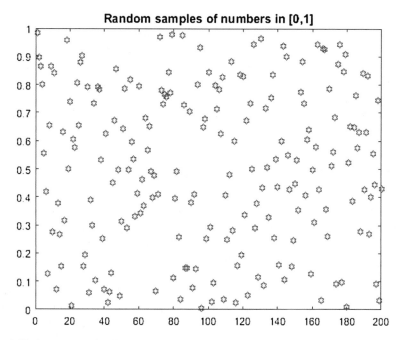

Fig. 2.10 Scatter plot of 200 random numbers in [0,1]

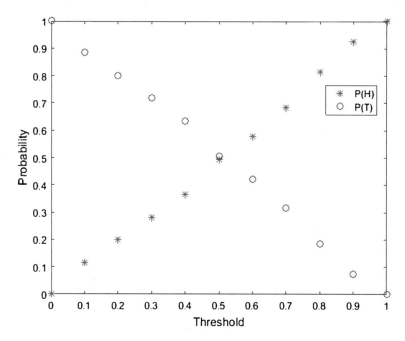

Fig. 2.11 Continuous probabilities associated with the experiment of a coin toss

It can be seen that the coin toss event can lead to a set of biased outcomes when the threshold is set to any value other than ½. The other usefulness of the approach is the possibility offered by a continuous function of the probability to create an experiment with a pair of discrete outcomes. It is also clear that the experiment described here can be modified and used to create discrete events such as the roll of a die. This aspect is discussed in Sect. 3.6 where the difference between a uniform continuous random variable and a uniform discrete random variable is presented.

2.3.3   Probability as a Geometric Measure

The continuous function approach to probability can be extended to geometric measures. Consider the experiment where a string of L units is likely to be cut along the length to produce two pieces. Since the cut can be made anywhere along the length, we can associate a probability with the event that one of the pieces is at least twice as long as the other or any such combination. Another experiment is one where a person is throwing a dart at a flat circular target of unit radius. It is possible to envision a wagering scheme and estimate the probability of a win if a person is able to have the dart land at a location 1/20th of a unit close to the center of the target. Similarly, we can extend this "geometrically derived measure" of probability to set of outcomes confined to one dimensional (a string), two dimensional (circle or a rectangle), three dimensional (a cube or a cylinder), etc. The geometric measure of probability can therefore be applied to cases where the domain has a geometric shape, with the firm basis of the outcomes being continuous such that the sample space S is

$$S = S\{X|0 \leq X \leq q\}. \tag{2.80}$$

Note that the upper limit of X is finite (i.e., q is finite) while the lower limit has been chosen to be 0. It may be taken as any number so long as the value is less than q.

If G_a represents the geometric measure of the event, the probability associated with the event is

$$P(G_a) = \frac{G_a}{G_A}. \tag{2.81}$$

In Eq. (2.81), G_A is the maximum measure associated with the experiment.

Example 2.11 A string of length 2 units is cut into two pieces. What is the probability that the shorter piece is less than half the dimension of the other piece? [See Example 2.10 also. Read the discussion of uniform density in Chap. 3]

Solution Since the string is 2 units long, if y is the length of the piece measured from initial point, the probability of having the piece to have a length up to y can be expressed as

$$P(\text{length} < y) = \frac{G_a}{G_A} = \frac{y}{2}.$$

If x is the length of the shorter piece, our event is such that

$$x \leq \frac{2-x}{2} \Rightarrow x \leq \frac{2}{3}.$$

Therefore, the probability associated with the required type of the cut is

$$P\left(\text{length} < \frac{2}{3}\right) = \frac{\left(\frac{2}{3}\right)}{2} = \frac{1}{3}.$$

If x is the longer piece, the event leads to

$$2 - x \leq \frac{x}{2} \quad \Rightarrow \quad x \geq \frac{4}{3}$$

$$P\left(\text{length} > \frac{4}{3}\right) = 1 - P\left(\text{length} < \frac{4}{3}\right) = 1 - \frac{\left(\frac{4}{3}\right)}{2} = 1 - \frac{2}{3} = \frac{1}{3}$$

Since there are two possibilities, the probability that the shorter piece is less than half the length of the longer piece will be $(1/3) + (1/3) = 2/3$.

Example 2.12 A dart is being thrown at a circle of radius 4 units. The event is that the dart lands at a location measured at a radius of 1 unit or less. What is the probability of this event?

Solution The probability that the dart falls within a circle of 1 radius w.r.t the center of the target is

$$P(r < 1) = \frac{\pi 1^2}{\pi 4^2} = \frac{1}{16}$$

The concepts associated with the representation of continuous functions will be revisited in Chap. 3 when we introduce the notion of random variables.

2.4 Conditional Probability and Bayes' Rule

Let us consider an experiment with event A with a non-zero probability. If we now examine the probability of seeing an outcome consisting of event A and another event B, the ratio of the probability of the joint event to the probability of the marginal event is termed as the conditional probability. Since we have two events A and B, each with non-zero probabilities, we can define the conditional probability of A given B as (Ghahramani 2000; Stark and Woods 1986; Peebles 2001; Ziemer 1997)

$$P(A|B) = \frac{P(AB)}{P(B)} = \frac{P(BA)}{P(B)}, \quad P(B) \neq 0. \tag{2.82}$$

The conditional probability of B *given* A is

$$P(B|A) = \frac{P(AB)}{P(A)} = \frac{P(BA)}{P(A)}, \quad P(A) \neq 0. \tag{2.83}$$

Equations (2.82) and (2.83) are example of Bayes' rule. Conditionality implies the existence of the conditioning event, B in Eq. (2.82) and A in Eq. (2.83), and therefore, we have more information a priori (ahead of time). If A and B are independent, Eq. (2.82) becomes

$$P(A|B) = \frac{P(AB)}{P(B)} = \frac{P(A)P(B)}{P(B)} = P(A). \tag{2.84}$$

2.4.1 Conditional Probability, Subsets, and Difference of Two Sets

If one of the events is a subset of another event, for example, $B \subset A$, we have (see Fig. 2.5)

$$P(B) \leq P(A) \tag{2.85}$$

$$P(A|B) = \frac{P(AB)}{P(B)} = \frac{P(B)}{P(B)} = 1 \tag{2.86}$$

$$P(B|A) = \frac{P(AB)}{P(A)} = \frac{P(B)}{P(A)} > P(B) \tag{2.87}$$

$$A - B = A \cap \overline{B} = A\overline{B} \tag{2.88}$$

$$P(A) = P(B) + P(A \cap \overline{B}) = P(B) + P(A\overline{B}), \quad B \subset A \tag{2.89}$$

If $A \subset B$, we have (see Fig. 2.6)

$$B - A = \overline{A} \cap B = \overline{A}B \tag{2.90}$$

$$P(B) = P(A) + P(\overline{A} \cap B) = P(A) + P(\overline{A}B) \tag{2.91}$$

The equations above lead to another identity when $A \subset B$,

$$P(B - A) = P(\overline{A}B) = P(B) - P(A) \tag{2.92}$$

Example 2.13 To understand the implications of prior knowledge, consider an experiment of randomly picking an integer between 1 and 10 (including both). A is the event of picking 4,

$$P(4) = P(A) = \frac{1}{10}.$$

What is the probability of picking a 4 given that the number picked is even?

Solution In this case, let $B = \{\text{even}\}$. We have

$$P(B) = \frac{5}{10} = \frac{1}{2}.$$

If we now simply use the frequency concept, we know the number of ways in which we can pick an even number is 5. The number of ways in which we can pick 4 out of the five even numbers is 1. Therefore, knowing that we pick an even number, the probability of picking a 4 is 1/5. Using Eq. (2.82), we have

$$P(A|B) = \frac{P(AB)}{P(B)} = \frac{P(A)}{P(B)} = \frac{\left(\frac{1}{10}\right)}{\left(\frac{1}{2}\right)} = \left(\frac{1}{5}\right).$$

We used the concept of subsets to find the joint probability $P(AB)$. Note that the event $A = \{4\}$ is a subset of the event $B = \{\text{even}\}$. Therefore,

$$P(AB) = P\{4, \text{even}\} = P\{4\} = P(A) = \frac{1}{10}, \quad A \subset B.$$

Let us explore the notion of independence, mutual exclusiveness, and conditionality by looking sample spaces that are continuous. Let the sample space consist of numbers picked randomly between 0 and 8. This means the event of picking a number in the range x_1 to x_2 is

$$A = \{x_1 \leq X \leq x_2\}. \tag{2.93}$$

The universal set is

$$S = \{0 \leq X < 8\}. \tag{2.94}$$

The probability that we have all the outcomes is unity, i.e.,

$$P(S) = 1. \tag{2.95}$$

The event and probability space can therefore be represented by a rectangle of sides 8 and (1/8) so that the area is unity as shown in Fig. 2.12.

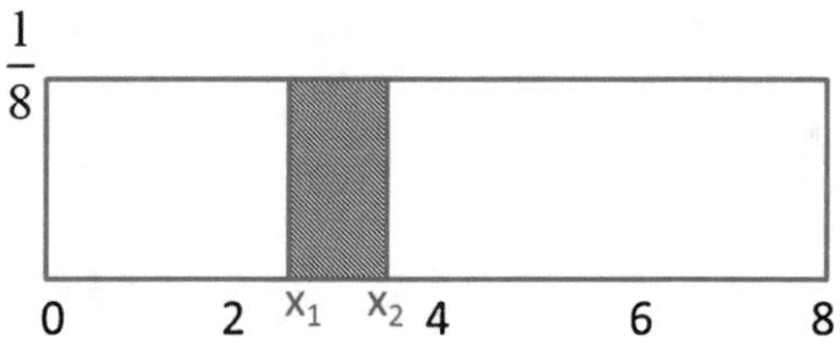

Fig. 2.12 The event and
sample space. Area
occupied by the slice shown
is $(x_2 - x_1)/8$

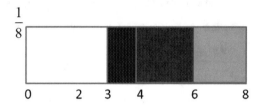

Fig. 2.13 The joint event
BC is the region between
4 and 6

From the probability space, we can easily see that

$$P(A) = P\{x_1 \le X \le x_2\} = \frac{x_2 - x_1}{8}. \tag{2.96}$$

Consider now two events, B and C,

$$B = \{3 \le X \le 6\} \tag{2.97}$$

$$C = \{4 \le X \le 8\}. \tag{2.98}$$

If past experiments indicate that the number picked seems to lie between 4 and
8, what is the probability that any number picked will lie between 3 and 6? This
problem translates to the interest in finding the conditional probability of B given
C (Fig. 2.13). This is expressed as

$$P(B|C) = \frac{P(BC)}{P(C)}. \tag{2.99}$$

Note that the joint event BC is the region between 4 and 6 as shown in Fig. 2.13.
We have

$$P(B) = \frac{3}{8}$$
$$P(C) = \frac{4}{8} \tag{2.100}$$
$$P(BC) = \frac{2}{8}$$

Equation (2.99) now becomes

$$P(B|C) = \frac{P(BC)}{P(C)} = \frac{P(4 < X < 6)}{P(4 < X < 8)} = \frac{\left(\frac{2}{8}\right)}{\left(\frac{4}{8}\right)} = \frac{1}{2} > P(B) = \frac{3}{8} \qquad (2.101)$$

We see that access to prior information increases the probability of the event B. Stated differently, *conditional probability will be greater than or equal to the marginal probability because the conditioning event C was favorable to the marginal event B. This means that*

$$P(B|C) > P(B) \qquad (2.102)$$

Let us see what happens if we find

$$P(C|B) = \frac{P(BC)}{P(B)} = \frac{P(4 < X < 6)}{P(3 < X < 6)} = \frac{\left(\frac{2}{8}\right)}{\left(\frac{3}{8}\right)} = \frac{2}{3} > P(C) = \frac{4}{8} \qquad (2.103)$$

Equation (2.103) implies

$$P(C|B) > P(C) \qquad (2.104)$$

Even though we see that conditional probabilities exceed marginal probabilities, it may not be the case always. Consider an event E such that

$$E = \{X < 5\} \qquad (2.105)$$

$$P(E) = \frac{5}{8} \qquad (2.106)$$

Let us find the conditional probability,

$$P(E|C) = \frac{P(EC)}{P(C)} = \frac{P(X < 5, 4 < X < 8)}{P(4 < X < 8)} = \frac{P(4 < X < 5)}{P(4 < X < 8)} = \frac{\left(\frac{1}{8}\right)}{\left(\frac{4}{8}\right)} = \frac{1}{4} < P(E) = \frac{5}{8}$$

$$(2.107)$$

In this case, prior knowledge leads to the *conditional probability being smaller than the marginal probability*. We may say that in the case of B, C carried positive information about B while in the case of the event E, C carried negative information about E. Let us now find the conditional probability of C given E,

$$P(C \mid E) = \frac{P(EC)}{P(E)} = \frac{P(4 < X < 5)}{P(X < 5)} = \frac{\left(\frac{1}{8}\right)}{\left(\frac{5}{8}\right)} = \frac{1}{5} < P(C) = \frac{4}{8} \qquad (2.108)$$

We see that

$$\begin{aligned} P(E \mid C) &< P(E) \\ P(C \mid E) &< P(C) \end{aligned} \qquad (2.109)$$

Example 2.14 Let us consider an experiment which involves choosing two numbers, each one being randomly between 0 and 8. Let X represent the outcomes in one and Y in the other one. The sample space will be a cube of size $8 \times 8 \times 1/64$, having a volume of unity (total probability). The sample space for X will be a rectangle of size $8 \times 1/8$. The sample space for Y will be of the same dimension.

Solution We have

$$P\{0 \leq X < 8\} = P\{0 \leq Y < 8\} = 1$$

Let us define two events A and B as

$$\begin{aligned} A &= \{0 \leq X < 6, \quad 4 < Y < 8\} \\ B &= \{4 \leq X < 8, \quad 0 < Y < 8\} \end{aligned}$$

Let us understand the relationships among A, B, $A+B$, AB, $\{A|B\}$, and $\{B|A\}$.

In Figure Ex.2.14 A is represented by the rectangle [$abcd$] with diagonal lines, and B is represented by the rectangle [$fghi$]. The rectangle [$ecbf$] represents the joint event $\{AB\}$.

From the figure, we can see that

$$P(A) = \frac{\text{area}[abcd]}{\text{total sample space}} = \frac{24}{64} = \frac{3}{8}$$

$$P(B) = \frac{\text{area}[fghi]}{\text{total sample space}} = \frac{32}{64} = \frac{1}{2}$$

$$P(AB) = \frac{\text{area}[fbec]}{\text{total sample space}} = \frac{8}{64} = \frac{1}{8}$$

You will note that A and B are not independent,

$$P(AB) \neq P(A)P(B)$$

The total probability,

Figure Ex.2.14

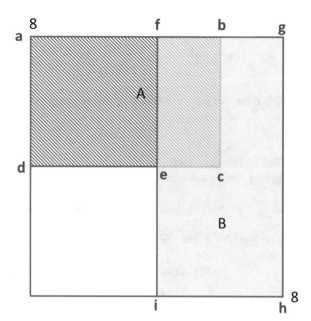

$$P(A + B) = P(A) + P(B) - P(AB) = \frac{1}{2} + \frac{3}{8} - \frac{1}{8} = \frac{3}{4}$$

The union of A and B can also be seen from the figure, and the probability can be directly calculated from the area as

$$P(A \cup B) = P(A + B) = \frac{\text{area}[adef] + \text{area}[fghi]}{\text{total sample space}} = \frac{16 + 32}{64} = \frac{48}{64} = \frac{3}{4}$$

$$P(A|B) = \frac{P(AB)}{P(B)} = \frac{\left(\frac{1}{8}\right)}{\left(\frac{1}{2}\right)} = \frac{1}{4} \neq P(A)$$

$$P(B|A) = \frac{P(AB)}{P(A)} = \frac{\left(\frac{1}{8}\right)}{\left(\frac{3}{8}\right)} = \frac{1}{3} \neq P(B)$$

Example 2.15 The manufacturer determined that the product suffers from two defects. Defect # 1 occurs with a probability of 0.01, and defect # 2 occurs with a probability of 0.015. The probability that both defects are present is 0.001. What is the probability that a randomly chosen product has no defect? If quality control person sees defect # 1, what is the probability that defect # 2 exists?

Solution
$$P(\#1) = 0.01$$
$$P(\#2) = 0.015$$
$$P(\#1\#2) = 0.001$$

It is clear that defects are not occurring independently since

$$P(\#1\#2) = 0.001 \neq P(\#1)P(\#2)$$

The probability that the product has at least one defect is the union of the events defect # 1 and defect # 2.

$$P(\#1 + \#2) = P(\#1) + P(\#2) - P(\#1\#2) = 0.025 - 0.001 = 0.024$$

The probability that there is no defect is

$$P(\text{No-Defect}) = 1 - P(\#1 + \#2) = 1 - 0.024 = 0.976$$

To determine the probability of having defect # 2, given that defect # 1 exists, we need to find the conditional probability,

$$P(\#2 \mid \#1) = \frac{P(\#1\#2)}{P(\#1)} = \frac{0.001}{0.01} = 0.1 > P(\#2)$$

Example 2.16 Consider a system having two components in series and in parallel. If each component is operating independently and each has a probability of failure of p, what is the probability that the system is ***functioning***?

Solution System in series (Figure Ex.2.16.1): In this case, the system fails when either one or both of them fail.

$$P(\text{Fail}) = P(\#1\text{fail}) + P(\#2\text{fail}) - P(\#1\text{fail})P(\#2\text{fail}) = 2p - p^2$$
$$P(\text{works}) = 1 - P(\text{Fail}) = 1 - 2p + p^2$$

We can also get the answer by examining each system in terms of it ***NOT*** failing. If $P(\#1W)$ and $P(\#2W)$ are the probabilities that the components are functioning,

$$P(\#1W) = P(\#2W) = 1 - p$$

The system in series can only function if both components function at the same time. Note that the components operate independently. Therefore,

Figure Ex.2.16.1

Figure Ex.2.16.2

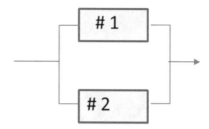

$$P(\text{works}) = P(\#1W)P(\#2W) = (1-p)(1-p) = 1 - 2p + p^2$$

System with components in parallel (Figure Ex.2.16.2):
When they are in parallel, the system fails when both components fail. Therefore,

$$P(\text{Fail}) = P(\#1\text{fail})P(\#2\text{fail}) = p^2. \tag{2.110}$$

The probability that the system works is

$$P(\text{works}) = 1 - P(\text{Fail}) = 1 - p^2 \tag{2.111}$$

Using the probabilities of NOT failing, the system works when either one or both of the systems are working. Therefore, the probability that the system with components in parallel works is

$$P(\text{works}) = P(\#1\text{works}) + P(\#2\text{works}) - P(\#1\text{works})P(\#2\text{works})$$
$$= (1-p) + (1-p) - (1-p)^2$$
$$P(\text{works}) = (1-p) + (1-p) - (1-p)^2 = 1 - p^2$$

2.4.2   *Conditional Probability and Multiplication Theorem of Probability*

Consider the joint probability of four events, A, B, C, and D expressed as $P(ABCD)$ when they are not *independent*. Let us start by examining two events first, three events next, and so on.
For *two* events A and B, we have

$$P(A + B) = P(A) + P(B) - P(AB) \tag{2.112}$$

$$P(AB) = P(A)P(B|A) \tag{2.113}$$

Furthermore, we may now rewrite the marginal probability representation given below in terms of conditional probabilities

$$P(A) = P(AB) + P(A\overline{B}) \tag{2.114}$$

Equation (2.114) becomes

$$P(A) = P(B|A)P(A) + P(\overline{B}|A)P(A) \tag{2.115}$$

Rewriting Eq. (2.115), we have

$$P(B|A) + P(\overline{B}|A) = 1, \quad P(A) \neq 0 \tag{2.116}$$

Similarly,

$$P(A|B) + P(\overline{A}|B) = 1, \quad P(B) \neq 0 \tag{2.117}$$

For *three* events A, B, and C, we have

$$P[(B \cup C)|A] = P(B|A) + P(C|A) - P(BC|A) \tag{2.118}$$

We have used Eq. (2.112) to obtain Eq. (2.118). Using Eq. (2.113), we can express the joint probability of the three events as

$$P(ABC) = P(C|AB)P(AB) = P(AB)P(C|AB) = \underbrace{P(A)P(B|A)}_{P(AB)}P(C|AB)$$

$$\tag{2.119}$$

We may now extend Eq. (2.119) to the case of four events A, B, C, and D. The joint probability of these four events is

$$P(ABCD) = P(ABC)P(D|ABC) = \underbrace{P(A)P(B|A)P(C|AB)}_{P(ABC)}P(D|ABC) \tag{2.120}$$

This concept can be extended to the product of several dependent random variables.

$$P(A_1 A_2 .. A_n) = P(A_1)P(A_2|A_1)P(A_3|A_1 A_2) .. P(A_n|A_1 A_2 .. A_{n-1}) \tag{2.121}$$

Another set of results can be expressed as

$$P(AB|C) = \frac{P(ABC)}{P(C)} = P(B|AC)P(AC)\frac{1}{P(C)}$$

$$= P(B|AC)P(A|C)P(C)\frac{1}{P(C)} \tag{2.122}$$

Simplifying further, Eq. (2.122) becomes

$$P(AB|C) = P(A|C)P(B|AC). \tag{2.123}$$

Equation (2.123) may now be extended to give

$$P(A_1A_2..A_n|C) = P(A_1|C)P(A_2|A_1C)P(A_3|A_1A_2C)..P(A_n|A_1A_2..A_{n-1}C) \tag{2.124}$$

We have seen the concepts of conditionality and independence being applied to practical problems of interest. It is possible to extend the conditioning process to multiple events and establish the relationship between the probability of a single event and probabilities of several conditioning events. This aspect is discussed in the next section.

2.4.3   Bayes' Rule of Total Probability: a priori and a posteriori Probabilities

Consider the case of an event A and a number of mutually exclusive events B_1, B_2, B_3, and B_4. Note that the events B_1, B_2, B_3 and B_4 constitute the universal set. In other words, B_1, B_2, B_3, and B_4 constitute a set of mutually exclusive and exhaustive set. The Venn diagram in Fig. 2.14 illustrates these events and their overlaps if any.

$$B = \{B_1, B_2, B_3, B_4\} \tag{2.125}$$

It can be seen that the event A can be written as the sum of the intersections of A with members of the group B as

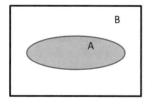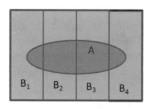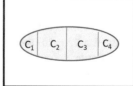

Fig. 2.14 Events A and B and their relationship. B is the universal set

$$\{A\} = \{AB_1\} + \{AB_2\} + \{AB_3\} + \{AB_4\} \qquad (2.126)$$

Note that AB_1, AB_2, AB_3, and AB_4 are also mutually exclusive. Rewriting in terms of sets C's, we have

$$C_k = A \cap B_k = (AB_k), \quad k = 1, 2, 3, 4 \qquad (2.127)$$

It can be seen that C_1, C_2, C_3, and C_4 are mutually exclusive and, hence,

$$\{A\} = \{C_1 \cup C_2 \cup C_3 \cup C_4\} = \{C_1\} + \{C_2\} + \{C_3\} + \{C_4\} \qquad (2.128)$$

Equation (2.128) is **Bayes' total probability rule** and can be written as

$$P(A) = P(AB_1) + P(AB_2) + P(AB_3) + P(AB_4) \qquad (2.129)$$

Note that Eq. (2.82) is Bayes' rule or **Bayes' product rule** to separate it from Eq. (2.129) which is **Bayes' total probability rule**. Rewriting in terms of conditional probabilities, Eq. (2.129) becomes

$$P(A) = P(A|B_1)P(B_1) + P(A|B_2)P(B_2) + P(A|B_3)P(B_3)$$
$$+ P(A|B_4)P(B_4) \qquad (2.130)$$

In the context of Bayes' rule, $P(B_k)$, $k = 1, 2, 3, 4$ are known as **a priori** probabilities. The **a posteriori** probabilities are defined as

$$P(B_k|A) = \frac{P(A|B_k)P(B_k)}{P(A)}, \quad k = 1, 2, 3, 4 \qquad (2.131)$$

$$\underbrace{P(B_k|A)}_{\text{a posteriori}} =$$

$$\frac{P(A|B_k)P(B_k)}{\underbrace{P(A|B_1)P(B_1) + P(A|B_2)P(B_2) + P(A|B_3)P(B_3) + P(A|B_4)P(B_4)}_{\sum \text{conditional probability * a priori probability}}}, k = 1, 2, 3, 4$$

$$(2.132)$$

It is clear that **posterior** probabilities are still a form of **conditional** probabilities.

It should be noted that **a priori** probability is known ahead of the experiment (such as what are the probabilities of transmitting a "0" or a "1" in communication systems; no conditions attached) while **a posteriori** is known only after the experience or after the experiment is over [such as having known that a "1" was received, the probability that it was originally transmitted as a "0," **P(Transmit 0\receive 1)**, or that it was originally transmitted as a "1," **P(Transmit 1\receive 1)**].

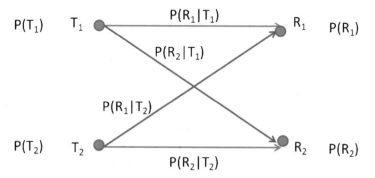

Fig. 2.15 Transition chart (input → output)

The Bayes' total probability rule in Eq. (2.130) can be written in general terms as

$$P(A) = \sum_{k=1}^{n} P(A|B_k)P(B_k).\qquad(2.133)$$

In Eq. (2.133), we have n *a priori* probabilities. Let us now explore the use of the Bayes' rule for total probability by looking at binary signal transmission in communication and computer networks.

Consider two mutually exclusive signals T_1 and T_2 representing bits (*0* and *1*). This means that $P(T_1) + P(T_2) = 1$. T_1 and T_2 constitute complementary events (mutually exclusive and exhaustive). Mutually exclusive and exhaustive is also known as collectively exhaustive. In communications/computer/networking terminology, $P(T_1)$ and $P(T_2)$ represent the *a priori* probabilities associated with the data being transmitted. Once the bits are transmitted through the channel (wired or wireless), because of noise, *0* might be received as a *0* or *1*. Similarly, *1* might be received as a *0* or *1*. Let R_1 and R_2 represent the received bits matching the levels corresponding to T_1 and T_2. Figure 2.15 displays the transition from input to output, and it represents the **flow chart** or the **transition chart** of the channel. The transition from input to output can also be represented in matrix form, with the **transition matrix** consisting of the conditional probabilities.

Using Bayes' rule of total probability, we can write

$$\begin{aligned}P(R_1) &= P(R_1|T_1)P(T_1) + P(R_1|T_2)P(T_2)\\ P(R_2) &= P(R_2|T_1)P(T_1) + P(R_2|T_2)P(T_2)\end{aligned}\qquad(2.134)$$

In matrix notation,

$$\begin{bmatrix} P(R_1) \\ P(R_2) \end{bmatrix} = \overset{\text{Transition Matrix}}{\begin{bmatrix} P(R_1|T_1) & P(R_1|T_2) \\ P(R_2|T_1) & P(R_2|T_2) \end{bmatrix}} \begin{bmatrix} P(T_1) \\ P(T_2) \end{bmatrix}\qquad(2.135)$$

If T_X represents the transition matrix, we have

$$T_X = \begin{bmatrix} P(R_1|T_1) & P(R_1|T_2) \\ P(R_2|T_1) & P(R_2|T_2) \end{bmatrix}. \tag{2.136}$$

We will notice that the sum of the elements of every column in the transition matrix is unity. *In other words, the elements of the column represent complementary or a set of mutually exclusive and exhaustive events or outcomes*. The transition matrix completely characterizes the transmission channel (and its properties), and it essentially encompasses the rules used for the classification of the output into two categories. In other words, the transition matrix represents the classifier.

If one notices R_1 at the receiver, one can ask the question: having received R_1, was it generated by T_1? This is gives the notion of posterior probability $P(T_1|R_1)$. Using Bayes' rule, the posterior (*a posteriori*) probability becomes

$$P(T_1|R_1) = \frac{P(R_1|T_1)P(T_1)}{P(R_1)} \tag{2.137}$$

Additional discussion on a posteriori probability appears in Sect. 2.4.4.

If we identify T_1 as "0" and T_2 as "1," using radar terminology, the *probability of false alarm* (P_F) and *probability of miss* (P_M) are

$$p_F = P(R_2|T_1) = \text{Probability of receiving 1 when 0 is transmitted}$$
$$p_M = P(R_1|T_2) = \text{Probability of receiving 0 when 1 is transmitted} \tag{2.138}$$

The probability of error (sometimes simply referred to as the error rate or error), p_e, is the **weighted** sum of the probabilities of miss and false alarm (we invoke the Bayes' rule of total probability). The error rate is

$$p_e = P(R_2|T_1)P(T_1) + P(R_1|T_2)P(T_2)$$
$$= p(\text{error}|T_1)P(T_1) + p(\text{error}|T_2)P(T_2). \tag{2.139}$$

In terms of probabilities of miss and false alarm, the error rate in Eq. (2.139) is

$$p_e = P_F P(T_1) + P_M P(T_2). \tag{2.140}$$

The complementary event of the error is accuracy, and accuracy rate (sometimes simply referred to as accuracy) is defined as

$$p_a = P(R_1|T_1)P(T_1) + P(R_2|T_2)P(T_2)$$
$$= p(\text{accuracy}|T_1)P(T_1) + p(\text{accuracy}|T_2)P(T_2). \tag{2.141}$$

In terms of the probabilities of miss and false alarm, the accuracy in Eq. (2.141) becomes

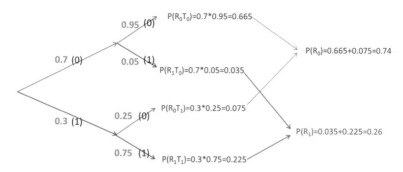

Figure Ex.2.17 TREE chart

$$p_a = [1 - P_F]P(T_1) + [1 - P_M]P(T_2). \tag{2.142}$$

The complementary nature of the error and accuracy allows us to express the rates in matrix form as

$$\begin{bmatrix} P_a \\ P_e \end{bmatrix} = \begin{bmatrix} P(R_1|T_1) & P(R_2|T_2) \\ P(R_2|T_1) & P(R_1|T_2) \end{bmatrix} \begin{bmatrix} P(T_1) \\ P(T_2) \end{bmatrix} \tag{2.143}$$

The square matrices in Eqs. (2.135) and (2.143) are different.

Note that probability of false alarm (P_F) is also known as **type I error** and probability of miss (P_M) is also called **type II error** in medical diagnostics (Greiner et al. 2000; Fawcett 2006; Eng 2012). Type I error occurs when a clinical test determines the person has a disease when the person is disease-free. Type II error occurs when test concludes the person is disease-free when the person actually has the disease. While in communication systems, the two a priori probabilities $p(T_1)$ and $p(T_2)$ may be equal or close, in medical diagnostics, the values will differ substantially. If $T_1 = \{$incidence of cancer in general population$\}$ and $T_2 = \{$absence of cancer in general population$\}$, $p(T_1)$ may be very small such as 10^{-5}, and $p(T_2)$ will be $1 - 10^{-5}$!

Instead of the *matrix-based approach* described above, one can also use a *tree diagram* to model the experiments described above (See Example 2.17 and Figure Ex.2.17).

Example 2.17 A binary communications channel carries data as either "0" or "1." Because of the noisy channel, "0" is received correctly only 95% of the time, and "1" is received correctly only 75% of the time. Assuming 70% of the time "0"s are transmitted, what is the probability that (*a*) "1" was received, (*b*) "0" was received, and (*c*) an error occurred. The procedure for estimating $p(R_0)$ and $p(R_1)$ is shown below (*follow the arrows*) (Figure Ex.2.17).

$P(T_1) = 0.3, \quad P(T_0) = 0.7$
$P(R_1/T_1) = 0.75, \quad P(R_0/T_1) = 0.25$
$P(R_0/T_0) = 0.95, \quad P(R_1/T_0) = 0.05$

Applying Bayes' total probability rule

$$P(R_1) = P(R_1/T_0)P(T_0) + P(R_1/T_1)P(T_1)$$
$$P(R_0) = P(R_0/T_0)P(T_0) + P(R_0/T_1)P(T_1) = 1 - P(R_1)$$
$$P(E) = P(R_1/T_0)P(T_0) + P(R_0/T_1)P(T_1)$$

Using the notion of the transition matrix

$$\begin{bmatrix} P(R_0) \\ P(R_1) \end{bmatrix} = \begin{bmatrix} 0.95 & 0.25 \\ 0.05 & 0.75 \end{bmatrix} \begin{bmatrix} 0.7 \\ 0.3 \end{bmatrix} = \begin{bmatrix} 0.95*0.7+0.25*0.3 \\ 0.05*0.7+0.75*0.3 \end{bmatrix} = \begin{bmatrix} 0.74 \\ 0.26 \end{bmatrix}$$

As you can see, both approaches lead to identical results.

Example 2.18 We have a binary channel corrupted by noise and the following information is provided: $P(0) = 0.6$ and $P(1|1) = P(0|0) = 0.8$. If we have cascaded N channels, what is the probability of receiving a "1" at the output? If at the output of first channel "1" was received, what is the probability that it was transmitted as a "0"? What is the probability that it was actually transmitted as a "1"?
What is the probability of error?

Solution Information given to us is summarized (in the nomenclature of probability) below.

$$P(T_0) = 0.6 \Rightarrow P(T_1) = 0.4$$
$$P(R_0|T_0) = P(R_1|T_1) = 0.8 \Rightarrow P(R_1|T_0) = P(R_0|T_1) = 0.2$$
$$\begin{bmatrix} P(R_0) \\ P(R_1) \end{bmatrix} = \begin{bmatrix} P(R_0|T_0) & P(R_0|T_1) \\ P(R_1|T_0) & P(R_1|T_1) \end{bmatrix}^N \begin{bmatrix} P(T_0) \\ P(T_1) \end{bmatrix}$$

$$\begin{bmatrix} P(R_0) \\ P(R_1) \end{bmatrix} = \begin{bmatrix} 0.8 & 0.2 \\ 0.2 & 0.8 \end{bmatrix}^N \begin{bmatrix} 0.6 \\ 0.4 \end{bmatrix}$$

$N = 1$

$$\begin{bmatrix} P(R_0) \\ P(R_1) \end{bmatrix} = \begin{bmatrix} 0.8 & 0.2 \\ 0.2 & 0.8 \end{bmatrix} \begin{bmatrix} 0.6 \\ 0.4 \end{bmatrix} = \begin{bmatrix} 0.8*0.6+0.2*0.4 \\ 0.2*0.6+0.8*0.4 \end{bmatrix} = \begin{bmatrix} 0.56 \\ 0.44 \end{bmatrix}$$

Directly,

$$P(R_1)=P(R_1|T_0)P(T_0)+P(R_1|T_1)P(T_1)=0.2*0.6+0.8*0.4=0.44$$

$$P(R_0)=1-P(R_1)=0.56$$

$$P(R_0)=P(R_0|T_0)P(T_0)+P(R_0|T_1)P(T_1)=0.8*0.6+0.2*0.4=0.56$$

$N=2$

$$\begin{bmatrix} P(R_0) \\ P(R_1) \end{bmatrix} = \begin{bmatrix} 0.8 & 0.2 \\ 0.2 & 0.8 \end{bmatrix}^2 \begin{bmatrix} 0.6 \\ 0.4 \end{bmatrix} = \begin{bmatrix} 0.68 & 0.32 \\ 0.32 & 0.68 \end{bmatrix} \begin{bmatrix} 0.6 \\ 0.4 \end{bmatrix} = \begin{bmatrix} 0.536 \\ 0.464 \end{bmatrix}$$

If we are doing the computation channel by channel,

$$\begin{bmatrix} P(R_0) \\ P(R_1) \end{bmatrix} = \begin{bmatrix} 0.8 & 0.2 \\ 0.2 & 0.8 \end{bmatrix} \begin{bmatrix} 0.56 \\ 0.44 \end{bmatrix} = \begin{bmatrix} 0.536 \\ 0.464 \end{bmatrix}$$

$N=3$

$$\begin{bmatrix} P(R_0) \\ P(R_1) \end{bmatrix} = \begin{bmatrix} 0.8 & 0.2 \\ 0.2 & 0.8 \end{bmatrix}^3 \begin{bmatrix} 0.6 \\ 0.4 \end{bmatrix} = \begin{bmatrix} 0.608 & 0.392 \\ 0.392 & 0.608 \end{bmatrix} \begin{bmatrix} 0.6 \\ 0.4 \end{bmatrix} = \begin{bmatrix} 0.5216 \\ 0.4784 \end{bmatrix}$$

We are looking for the *a posteriori probability*,

$$P(T_0|R_1) = \frac{P(R_1|T_0)P(T_0)}{P(R_1)} = \frac{0.2*0.6}{0.44} = \frac{0.12}{0.44} = 0.2727$$

Similarly, upon receiving "1," the probability that was correctly transmitted as a "1" is (*a posteriori probability*)

$$P(T_1|R_1) = \frac{P(R_1|T_1)P(T_1)}{P(R_1)} = \frac{0.8*0.4}{0.44} = \frac{0.32}{0.44} = 0.7273 = 1 - P(T_0|R_1)$$

Error occurs when *a transmitted 0 is received as a 1* and when *a transmitted 1 is received as a 0*.

$$p(\text{error}) = p(\text{error}|T_0)P(T_0) + p(\text{error}|T_1)P(T_1)$$
$$= p(R_1|T_0)P(T_0) + p(R_0|T_1)P(T_1)$$

$$p(error) = p(R_1|T_0)P(T_0)+p(R_0|T_1)P(T_1)=0.2*0.6+0.2*0.4=0.20$$

Example 2.19 In medical diagnostics, a company claims that it has developed a new technique to detect the presence of cancer with an accuracy of 96% when cancer is present and accuracy of 95% when cancer is absent. If the incidence of cancer among the general population is 0.01%, is this test as good as it appears to be?

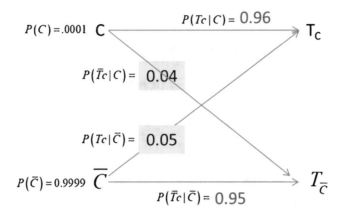

Figure Ex.2.19

Solution Let C represent cancer and Tc represent a positive test for cancer. The complementary events are $\overline{C}$ (absence of cancer) and $\overline{T}_C$ (negative test for cancer). We are given that

$$
\begin{aligned}
P(C) &= .0001 & P(\overline{C}) &= 0.9999 \\
P(Tc|C) &= 0.96 & P(\overline{T}c|C) &= 0.04 \\
P(Tc|\overline{C}) &= 0.05 & P(\overline{T}c|\overline{C}) &= 0.95
\end{aligned}
\tag{2.144}
$$

The transition chart is (Figure Ex.2.19)

$$
\begin{bmatrix} P(Tc) \\ P(\overline{T}c) \end{bmatrix} = \begin{bmatrix} P(Tc|C) & P(Tc|\overline{C}) \\ P(\overline{T}c|C) & P(\overline{T}c|\overline{C}) \end{bmatrix} \begin{bmatrix} P(C) \\ P(\overline{C}) \end{bmatrix}
$$

$$
= \begin{bmatrix} 0.96 & 0.05 \\ 0.04 & 0.95 \end{bmatrix} \begin{bmatrix} 0.0001 \\ 0.9999 \end{bmatrix}
\tag{2.145}
$$

$$
\begin{bmatrix} P(Tc) \\ P(\overline{T}c) \end{bmatrix} = \begin{bmatrix} 0.0501 \\ 0.9499 \end{bmatrix}
\tag{2.146}
$$

Let us now determine the posterior probability that cancer is present, given a positive test.

$$
P(C|Tc) = \frac{P(Tc|C)P(C)}{P(Tc)} = \frac{0.96 * 0.0001}{0.0501} = 0.0019
\tag{2.147}
$$

The probability of having the disease given that the test is positive is called ***positive predictive value*** (PPV). ***The probability that the cancer is absent given that the test is positive is***

$$P(\bar{C}|Tc)=1-P(C|Tc)=0.9981 \tag{2.148}$$

The probability of 0.9981 is unacceptably high pointing out that the test is unacceptable.

Let us now explore what happens if the accuracy of negative result is increased to 0.999.

$$\begin{bmatrix} P(Tc) \\ P(\bar{T}c) \end{bmatrix} = \begin{bmatrix} 0.96 & 0.001 \\ 0.04 & 0.999 \end{bmatrix} \begin{bmatrix} 0.0001 \\ 0.9999 \end{bmatrix} = \begin{bmatrix} 0.0011 \\ 0.9989 \end{bmatrix} \tag{2.149}$$

$$P(C|Tc) = \frac{P(Tc|C)P(C)}{P(Tc)} = \frac{0.96 * 0.0001}{0.0011} = 0.0873 \tag{2.150}$$

$$P(\bar{C}|Tc)=1-P(C|Tc)=0.9127 \tag{2.151}$$

The probability of a false positive test is still very high.

Example 2.20 Consider a pair of dice: one is fair, and the other is a "loaded" one with a probability of (1/3) associated with the outcome 2. A die is picked at random and rolled twice. If the value observed each time is a 2, what is the probability that the fair die was used both times?

Solution If L represents the biased or "loaded" die and F represents the fair one, we have

$$P(2|L) = \frac{2}{6}$$
$$P(2|F) = \frac{1}{6}$$

The probability of observing 2 both times is (loaded die)

$$P(2.2|L) = P(2|L)P(2|L) = \frac{4}{36}$$

The probability of observing 2 both times is (fair die)

$$P(2.2|F) = P(2|F)P(2|F) = \frac{1}{36}$$

Note that

$$P(F) = P(L) = \frac{1}{2}$$

The probability of having 2 and 2 is

$$P(2.2) = P(2.2|L)P(L) + P(2.2|F)P(F) = \frac{5}{72}$$

The probability that the fair die was used on both attempts is

$$P(F|2.2) = \frac{P(2.2|F)P(F)}{P(2.2)} = \frac{\left(\frac{1}{36}\right)}{\left(\frac{5}{72}\right)} = \frac{2}{5}$$

Figure Ex.2.21

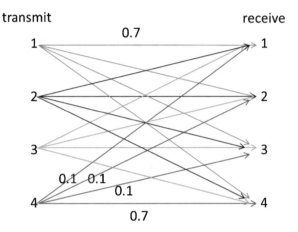

Example 2.21 In a quaternary communication system, a "4" is transmitted 4 times as frequently as a "1," a "3" is transmitted 3 times as frequently as a "1," and a "2" is transmitted twice as frequently as a "1." The values are received as they are 70% of the time, and in the remaining time, they may be received as any of the other three. If a "4" is observed, what is the probability that it was transmitted as a "2"?

Solution Based on the information provided, if p is the transmit probability of "1," we have (Figure Ex.2.21)

$$p + 2p + 3p + 4p = 1 \Rightarrow p = \frac{1}{10}$$

IN terms of the transition matrix, we have

$$\begin{bmatrix} P(R_1) \\ P(R_2) \\ P(R_3) \\ P(R_4) \end{bmatrix} = \begin{bmatrix} P(R_1|T_1) & P(R_1|T_2) & P(R_1|T_3) & P(R_1|T_4) \\ P(R_2|T_1) & P(R_2|T_2) & P(R_2|T_3) & P(R_2|T_4) \\ P(R_3|T_1) & P(R_3|T_2) & P(R_3|T_3) & P(R_3|T_4) \\ P(R_4|T_1) & P(R_4|T_2) & P(R_4|T_3) & P(R_4|T_4) \end{bmatrix} \begin{bmatrix} P(T_1) \\ P(T_2) \\ P(T_3) \\ P(T_4) \end{bmatrix}$$

$$\begin{bmatrix} P(R_1) \\ P(R_2) \\ P(R_3) \\ P(R_4) \end{bmatrix} = \begin{bmatrix} 0.7 & 0.1 & 0.1 & 0.1 \\ 0.1 & 0.7 & 0.1 & 0.1 \\ 0.1 & 0.1 & 0.7 & 0.1 \\ 0.1 & 0.1 & 0.1 & 0.7 \end{bmatrix} \begin{bmatrix} 0.1 \\ 0.2 \\ 0.3 \\ 0.4 \end{bmatrix} = \begin{bmatrix} 0.16 \\ 0.22 \\ 0.28 \\ 0.34 \end{bmatrix}$$

The probability that a 4 received was transmitted as a 2 is

$$P(T_2|R_4) = \frac{P(R_4|T_2)P(T_2)}{P(R_4)} = \frac{0.1 * 0.2}{0.34} = 0.0588$$

The probability that a received 1 was transmitted as a 1 is

$$P(T_1|R_1) = \frac{P(R_1|T_1)P(T_1)}{P(R_1)} = \frac{0.7 * 0.1}{0.16} = 0.5813$$

2.4.4 Conditional Probability Revisited: Conditionality and Correlation

Equations (2.101), (2.102), (2.103), (2.104), (2.105), (2.106), (2.107), (2.108), and (2.109) show that the conditional probability may be larger or smaller than the marginal probability depending on the relationship between the two events. The conditional probability is equal to the marginal probability if the two events are independent. We will now explore this aspect by writing down the expression for the probability of the event B in terms of the event A and its complementary event, $\overline{A}$. Following the representation in Eq. (2.114), we may write

$$P(B) = P(AB) + P(\overline{A}B) \tag{2.152}$$

Using Bayes' rule, Eq. (2.152) becomes

$$P(B) = P(A|B)P(B) + P(\overline{A}|B)P(B) \tag{2.153}$$

Equation (2.153) may be simplified by dividing it by $P(B)$; assuming that $P(B) \neq 0$, we get

$$1 = P(A|B) + P(\overline{A}|B), \quad P(B) > 0 \tag{2.154}$$

Rewriting Eq. (2.154), we have

$$P(A|B) = 1 - P(\overline{A}|B) \tag{2.155}$$

Equation (2.155) proves that the conditional events also form a set of complementary events just as A and its complement, $\overline{A}$,

$$P(A) + P(\overline{A}) = 1 \tag{2.156}$$

Equation (2.155) may be interpreted as follows. If "prior information" is beneficial to the event A, we expect

$$P(A|B) \geq P(A) \tag{2.157}$$

If Eq. (2.157) is true, then (from Eq. (2.155)), we have

$$P(\overline{A}|B) \leq P(\overline{A}) \tag{2.158}$$

This means that the conditional probability exceeds the marginal probability when prior information is favorable and the conditional probability will be less than the marginal probability when prior information is unfavorable.

We may also explain these properties of conditional probabilities in terms of correlation (we will define it shortly). We have

$$P(A|B) = P(A), \quad A \text{ and } B \text{ are independent} \tag{2.159}$$

Note that if A and B are independent, $\overline{A}$ and B are also independent implying that

$$P(\overline{A}|B) = P(\overline{A}), \quad \overline{A} \text{ and } B \text{ are independent} \tag{2.160}$$

On the other hand, if Eq. (2.157) is true, it suggests that A and B have **positive correlation**. Equation (2.158) suggests that if A and B have positive correlation, $\overline{A}$ and B have **negative correlation**. If Eq. (2.159) is true, A and B are **uncorrelated** (Ghahramani 2000).

As an example, consider the case where a number is picked randomly between 0 and 2. This means that the probability of picking a number between x and $x + \Delta x$ is $\Delta x/2$. Consider an event $A = \{X < 1.2\}$ and another event $B = \{X > 0.75\}$.

We have

$$P(A) = \frac{1.2}{2} = 0.6, \quad A = \{X < 1.2\} \tag{2.161}$$

$$P(\overline{A}) = 0.4, \quad \overline{A} = \{X > 1.2\} \tag{2.162}$$

$$P(B) = 1 - P(X < 0.75) = 1 - \frac{0.75}{2} = 0.625 \tag{2.163}$$

We may now estimate the conditional probabilities $P(A|B)$ and $P(\overline{A}|B)$.

$$P(A|B) = \frac{P(X < 1.2, X > 0.75)}{P(X > 0.75)} = \frac{P(0.75 < X < 1.2)}{P(X > 0.75)} = \frac{0.45/2}{0.625}$$

$$= 0.36 \tag{2.164}$$

This means that

$$P(A|B) < P(A) \tag{2.165}$$

If Eq. (2.165) is true (by virtue of Eq. (2.155)),

$$P(\overline{A}|B) > P(\overline{A}) \tag{2.166}$$

Let us see if Eq. (2.166) is true directly. Applying Bayes' rule,

$$P(\overline{A}|B) = \frac{P(\overline{A}B)}{P(B)} = \frac{P(X > 1.2, X > 0.75)}{P(X > 0.75)} = \frac{P(X > 1.2)}{P(X > 0.75)} = \frac{0.8/2}{1.25/2} = 0.64 > P(\overline{A}) \tag{2.167}$$

We may also obtain $P(\overline{A}|B)$ as 0.64 noting that $P(A|B)$ is 0.36.

Equation (2.167) shows that events A and B have **negative** correlation while events $\overline{A}$ and B have **positive** correlation. In other words, B is biased against A.

We may define correlation as

$$\rho = \frac{P(AB) - P(A)P(B)}{\sqrt{P(A)[1 - P(A)]}\sqrt{P(B)[1 - P(B)]}} = \frac{P(AB) - P(A)P(B)}{\sqrt{P(A)P(\overline{A})}\sqrt{P(B)P(\overline{B})}} \tag{2.168}$$

In terms of the Bayes' rule, we may write Eq. (2.168) as

$$\rho = \frac{P(A|B)P(B) - P(A)P(B)}{\sqrt{P(A)P(\overline{A})}\sqrt{P(B)P(\overline{B})}} = \frac{P(B)[P(A|B) - P(A)]}{\sqrt{P(A)P(\overline{A})}\sqrt{P(B)P(\overline{B})}} \tag{2.169}$$

Equation (2.169) suggests that

$$\begin{array}{lll} \rho > 0, & P(A|B) > P(A) & \text{positive correlation} \\ \rho < 0, & P(A|B) < P(A) & \text{negative correlation} \\ \rho = 0, & P(A|B) = P(A) & \text{uncorrelated} \end{array} \tag{2.170}$$

The association between correlation and conditionality is explored in detail in Chap. 4.

2.4.5 Data Analytics, Confusion Matrix, and Positive Predictive Value

While the examples shown earlier dealt with a priori probabilities and conditional probabilities that were already provided, only the data from measurements may be available in a number of experiments. These include experiments done to test the ability to detect the presence of a target in the field of view, data collected from subjects undergoing tests to see if a specific illness is present, data collected from subjects to test the efficacy of a new drug, etc. The data sets can be analyzed to see how much trust we can place on the measurements or tests.

Often a positive test for a specific illness needs to be analyzed further to see how much trust we can place on the test. As seen in Example 2.19, we are interested in finding out the following probability: the probability that the subject indeed had the illness, given that the test was positive. This probability is the a posteriori probability described earlier and is defined as the **positive predictive value** in this context. Positive predictive value is also sometimes referred to as the **precision**.

One of the critical aspects of data analytics with regard to obtaining the error rates (require probabilities of miss and false alarm), accuracy, and positive predictive value is that no information on the conditional probabilities seen in previous examples is given. This means that we have no knowledge of P(target detected| target present), P(target not detected|target absent), similar to $P(R_1|T_1)$ or $P(R_0|T_0)$ earlier. In other words, we have little information on the **transition matrix** defined earlier in Sect. 2.4.3 that will constitute the **classifier** which will make decisions on whether the target is present or absent (Kohavi and Provost 1998; Shankar 2019).

The absence of these transitional probabilities (i.e., conditional probabilities) makes it necessary to choose an appropriate threshold to estimate these conditional probabilities. The choice of the threshold is important because it impacts all the subsequent calculations. Indeed, once a threshold is fixed, the transition matrix or the characteristics of the classifier are set.

While advanced methods for choosing a threshold are presented in Chaps. 3 and 5, we will use simple methods to fix a threshold in Example 2.22 next. We will utilize Bayes' rule in the analysis of data collected from a driverless car experiment conducted as illustrated below.

Example 2.22 Data collected from a driverless car experiment is given. There are 40 measurements when there is *NO* target present and 25 measurements when *a target* is present. It can be seen that values in the two columns overlap because of the errors in measurement, and this leads to error in interpretation. A simple way to use the data is to set a threshold and analyze the data. Obtain the error, accuracy, and positive predictive value. State the method used to obtain the threshold (Table Ex.2.22.1).

Solution The first aspect we notice is that we have not been provided with a priori probabilities. If we are undertaking data analytics to test the efficacy of cancer detection techniques, we have the a priori probabilities in terms of the prevalence

Table Ex.2.22.1   The raw data sets on the left and the sorted ones (individual columns) are on the right

No target	Target
0.0844	2.7745
2.402	4.8782
0.8511	3.3295
0.4339	1.5498
1.4184	3.1265
1.6574	3.2108
1.7679	5.7838
0.8563	3.697
1.5297	1.4577
0.4089	4.7559
0.8165	5.1139
0.8395	3.545
0.9911	4.147
2.0259	2.2652
2.6641	2.7498
2.1169	6.0546
0.1475	1.4308
1.4065	3.7463
0.2849	1.1524
1.9604	3.1692
0.5707	3.6494
1.6152	1.8
0.4977	2.2059
1.7161	0.8309
0.9526	4.0036
2.0082	
0.3273	
0.6853	
0.5991	
2.8675	
1.1428	
3.2194	
0.3575	
1.9844	
1.3647	
1.3645	
0.1479	
2.7086	
0.5522	
1.8621	

No target	Target
0.0844	0.8309
0.1475	1.1524
0.1479	1.4308
0.2849	1.4577
0.3273	1.5498
0.3575	1.8
0.4089	2.2059
0.4339	2.2652
0.4977	2.7498
0.5522	2.7745
0.5707	3.1265
0.5991	3.1692
0.6853	3.2108
0.8165	3.3295
0.8395	3.545
0.8511	3.6494
0.8563	3.697
0.9526	3.7463
0.9911	4.0036
1.1428	4.147
1.3645	4.7559
1.3647	4.8782
1.4065	5.1139
1.4184	5.7838
1.5297	6.0546
1.6152	
1.6574	
1.7161	
1.7679	
1.8621	
1.9604	
1.9844	
2.0082	
2.0259	
2.1169	
2.402	
2.6641	
2.7086	
2.8675	
3.2194	

of the diseases in the general population and the probability of the complementary event, absence of cancer. These cases will be studied in Chap. 5. For the time being, we will limit ourselves to experiments in machine vision.

The a priori probabilities are obtained directly from the data counts and they are given below.

Number of TARGET absent samples, $N_0 = 40$ Number of TARGET present samples, $N_1 = 25$

Total number of samples from the experiment $N = N_0 + N_1 = 65$

The *a priori* probabilities are

$P(\text{target absent}) = N_0/N = \mathbf{40/65}$ $P(\text{target present}) = N_1/N == \mathbf{25/65}$

If the threshold value is chosen as T, we may estimate the transition probabilities (i.e., conditional probabilities) by noting that the presence of the target implies higher values of the data. The histograms of the two data sets are shown in Figure Ex.2.22.1. It can be seen that values of data from the target absent cohort exceeding the threshold constitute errors. Similarly, the values of the data from the target present cohort exceeding the threshold represent those values representing the correct identification of the target present.

We may sort the two sets of data either in ascending or descending order so that it is easy to count. In each group (TARGET and NO TARGET), count the number of samples **above** the threshold.

Prob. False Alarm $= P(\text{TARGET detected}|\text{Target absent})$

Prob. of False Alarm $=$ (Number of samples **above** the threshold in TARGET ABSENT group)/N_0

Prob. Miss $= P(\text{TARGET Not detected}|\text{Target present})$

Prob. of Miss $= 1 -$ Prob. Detection $= 1 -$ (Number of samples **above** the threshold in the TARGET present group)/N_1

Note that probability of detection (P_D) refers to the case of correctly detected cases among the samples of target present. Once these probabilities are calculated, it is possible to proceed with the calculations of $P(\text{target detected})$, error rate, accuracy, positive predictive value, etc.

Matlab function ksdensity(.) is used to obtain the approximate continuous histograms of the two sets, and the point of intersection is seen at 2.4. The threshold is chosen at 2.4.

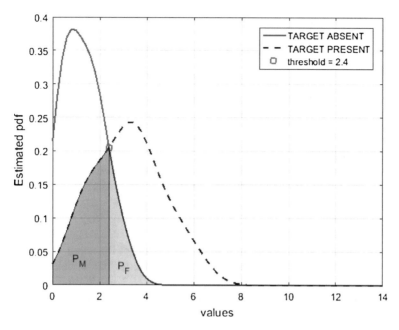

Figure Ex.2.22.1 Histograms generated using ksdensity(.). The threshold is chosen at the intersection of the two histograms

sorted data: red color values exceed the threshold

	Target Absent					Target Present			
3.219	1.984	1.418	0.851	0.434	6.055	4.147	3.545	2.775	1.55
2.868	1.96	1.407	0.84	0.409	5.784	4.004	3.33	2.75	1.458
2.709	1.862	1.365	0.817	0.358	5.114	3.746	3.211	2.265	1.431
2.664	1.768	1.365	0.685	0.327	4.878	3.697	3.169	2.206	1.152
2.402	1.716	1.143	0.599	0.285	4.756	3.649	3.127	1.8	0.831
2.117	1.657	0.991	0.571	0.148					
2.026	1.615	0.953	0.552	0.148					
2.008	1.53	0.856	0.498	0.084					

Number of samples of NO TARGET data above (exceeding) the threshold of 2.4 = 5

$P_F = P$(Target detected|target absent) $= (40\text{–}35)/N_0 = 5/40 = 0.125 =$ Prob. False alarm

P(No target detected|target absent) $= 35/N_0 = 35/40 = 0.875$

Number of samples of TARGET data (above) exceeding the threshold of 2.4 = 17

$P_M = P$(NO target detected|target present) $= 8/N_0 = 0.32 =$ Prob. Miss

$P_D = P$(target detected|target present) $= 17/N_1 = 0.68 =$ Prob. Detection

P(target detected) $= P$(target detected|target present)P(target present) $+ P$(target detected|target absent)P(target absent)

P(target detected) $= (17/25) * (25/65) + (5/40) * (40/65) = 0.3385$

P(Target not detected) $= P$(target not detected|target absent) P(target absent) $+$ P(target not detected|target present) P(target present)

P(Target not detected) $= (35/40) * (40/65) + (8/25) * (25/65) = 0.6615 = 1 - P$(target detected)

Positive Predictive value (PPV) $= P$(target present|target Detected)

$PPV = P$(Target detected|target present)P(target present)/P(target detected)

$\quad = (17/25) * (25/65)/0.3385 = 17/(17 + 5) = 17/22 = 0.7726$

This PPV is far higher than the PPV obtained in the example on cancer detection.

Error rate $= 0.125 * 40/65 + 0.32 * 25/65 = 0.2$

Accuracy $= 1 - 0.2 = 0.8$.

We may also calculate these probabilities simply by counting as

$$P(\text{Target Detected}) = \frac{\text{number of times target detected}}{\text{Total number of samples}} = \frac{17+5}{65} = \frac{22}{65} = 0.3385$$

$$P(\text{Target Not Detected}) = 1 - \frac{\text{number of times target detected}}{\text{Total number of samples}} = \frac{8+35}{65} = \frac{43}{65}$$
$$= 0.6615$$

$$\text{Error rate} = \frac{\text{number of times error made}}{\text{Total number of samples}} = \frac{8+5}{65} = \frac{13}{65} = \frac{1}{5} = 0.2$$

$$\text{Accuracy} = \frac{\text{number of correct decisions}}{\text{Total number of samples}} = \frac{17+35}{65} = \frac{52}{65} = \frac{4}{5} = 0.8$$

$$PPV = \frac{\text{number of times target detected when target was present}}{\text{number of times the target was detected}} = \frac{17}{5+17}$$
$$= \frac{17}{22} = 0.7727$$

We may write the transition matrix of this classifier (this means that the threshold has been decided).

$$T_X = \begin{bmatrix} P(\text{Target Not Detected}|\text{Target Absent}) & P(\text{Target Not Detected}|\text{Target Present}) \\ P(\text{Target Detected}|\text{Target Absent}) & P(\text{Target Detected}|\text{Target Present}) \end{bmatrix}$$

Based on the counts, we have the transition matrix as

$$T_X = \begin{bmatrix} \dfrac{35}{40} & \dfrac{8}{25} \\ \dfrac{5}{40} & \dfrac{17}{25} \end{bmatrix}$$

The decision strategy can now be expressed in matrix form as

$$\begin{bmatrix} P(\text{Target Not Detected}) \\ P(\text{Target Detected}) \end{bmatrix} = T_X \begin{bmatrix} P(\text{Target Absent}) \\ P(\text{Target Present}) \end{bmatrix}$$

Making appropriate substitutions, we have

$$\begin{bmatrix} P(\text{Target Not Detected}) \\ P(\text{Target Detected}) \end{bmatrix} = \begin{bmatrix} \dfrac{35}{40} & \dfrac{8}{25} \\ \dfrac{5}{40} & \dfrac{17}{25} \end{bmatrix} \begin{bmatrix} \dfrac{40}{65} \\ \dfrac{25}{65} \end{bmatrix} = \begin{bmatrix} \dfrac{43}{65} \\ \dfrac{22}{65} \end{bmatrix}$$

Often these counts are described in terms of what is defined as the **confusion matrix**, a term generally used by researchers in data science (Kohavi and Provost 1998). All the appropriate probabilities can easily be obtained from the elements of the matrix. While the elements of the transition matrix are probabilities, the elements of the confusion matrix are counts (Table Ex.2.22.2).

PPV $= 17/22$
Error rate $= 13/65$
P(TargetDetected) $= 22/65$
P(TargetNOTDetected) $= 43/65$

As it can be seen, it is a simple task to calculate the PPV, error rate, and the output probabilities directly from the confusion matrix.

In the example above, the threshold for separating the two cases was chosen on the basis of where the histograms overlapped. Such a choice is not the best way to optimize the performance. We will examine better (optimal) ways of determining the threshold once we study the densities and distributions in Chap. 3. One way of obtaining the threshold relies on the likelihood ratio tests which involve examination of the densities of measured quantities and minimizing the error. Yet another method uses the notion of minimizing the probability of false alarm while maximizing the probability of detection. The topic will also be explored in Chap. 5 which is devoted to data analytics.

It is also possible to choose the threshold values arbitrarily and explore how the PPV changes with the threshold. This is demonstrated in the next example.

Table Ex.2.22.2 Confusion matrix

Actual cases	Target detected	Target not detected	Total
Target absent	5	35	40
Target present	17	8	25
Total	22	43	65

Note that the confusion matrix is the 2 x 2 matrix of counts with rows of [5 35] and [17 8].

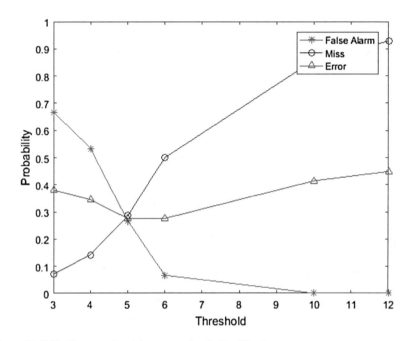

Figure Ex.2.23 Error rate is minimum at a threshold of 5 units

Example 2.23 You are given two data sets: X represents the data collected from an infrared (IR) receiver when there is no target in front of the transmitter, and Y represents the data collected from an IR receiver when a target is present in the field of view of the transmitter. With these data sets, **obtain** a priori probabilities. If threshold is set at 5 units (target detected when the measured value exceeds the threshold), **obtain** the probabilities of false alarm and miss. **What** is the error rate? Obtain the probabilities of false alarm, miss, and error rates of the threshold set at 3, 4, 6, 10, and 12. Obtain a plot of false alarm, miss, and error rates vs. threshold. Comment on the results. What value of the threshold gives the lowest error? Obtain the PPV at this value.

X = [2.6667, 3.4766, 3.7187, 4.0810, 9.4389, 5.0332, 1.0028, 4.2891, 5.2970, 5.1346, 4.6081   2.4199,   1.4691,   1.1571,   4.2389]
Y = [6.6541, 7.0132, 7.6296, 6.0201, 7.7718, 1.0703, 10.2948, 3.9876, 5.3993, 14.1822, 5.0649,   5.1110   4.2848   4.6071]

Solution There are 15 samples of X and 14 samples of Y. Therefore, the a priori probabilities are $P(X) = 15/29$ and $P(Y) = 14/29$ (Figure Ex.2.23).

The data sets are now sorted in descending order to make it convenient to count.

Threshold at 5:

Xsorted   =   [***9.4389   5.2970   5.1346   5.0332*** 4.6081   4.2891   4.2389
  4.0810   3.7187   3.4766 2.6667   2.4199   1.4691   1.1571   1.0028];
Ysorted = [***14.1822   10.2948   7.7718   7.6296   7.0132   6.6541   6.0201
  5.3993   5.1110   5.064*** 4.6071   4.2848   3.9876   1.0703];
Noting that X is the data collected when no target was present, the four samples in
  X above the threshold of 5 represent the number of times the presence of the target
  is assumed when there was no target. Therefore, the probability of false alarm is P
  (targetDetected|X) $= 4/15$.
Noting that Y is the data collected when target was present, the four samples of
  Y below the threshold represent the number of times the target is missed when
  there was a target present in the field of view. Therefore, the probability of miss is
  P(targetMissed|Y) $= 4/14$.
$P_F = 4/15$. This is the error given data X.
$P_M = 4/14$. This is the error given data Y.
Using Bayes' rule of total probability,
P(error) $=$ P(error|X)$P(X)$ $+$ P(error|Y)$P(Y)$ $=$ (4/15) $*$ 15/29 $+$ (4/14) $*$
  (14/29) $= 8/29 = 0.2759$.

Threshold at 3:

Xsorted   =   [***9.4389   5.2970   5.1346   5.0332*** 4.6081   4.2891   4.2389
  4.0810   3.7187   3.4766 2.6667   2.4199   1.4691   1.1571   1.0028];
Ysorted = [***14.1822   10.2948   7.7718   7.6296   7.0132   6.6541   6.0201
  5.3993   5.1110   5.064*** 4.6071   4.2848   3.9876   1.0703];
$P_F = 10/15$. This is the error given data X.
$P_M = 1/14$. This is the error given data Y.
P(error) $= (10/15) * 15/29 + (1/14) * (14/29) = 0.3793$.

Threshold at 4:

Xsorted   =   [***9.4389   5.2970   5.1346   5.0332*** 4.6081   4.2891   4.2389
  4.0810   3.7187   3.4766 2.6667   2.4199   1.4691   1.1571   1.0028];
Ysorted = [***14.1822   10.2948   7.7718   7.6296   7.0132   6.6541   6.0201
  5.3993   5.1110   5.064*** 4.6071   4.2848   3.9876   1.0703];
$P_F = 8/15$. This is the error given data X.
$P_M = 2/14$. This is the error given data Y.
P(error) $= (8/15) * 15/29 + (2/14) * (14/29) = 0.3448$.

Threshold at 6:

Xsorted   =   [**9.4389**   5.2970   5.1346   5.0332   4.6081   4.2891   4.2389
    4.0810   3.7187   3.4766 2.6667   2.4199   1.4691   1.1571   1.0028];
Ysorted   =   [**14.1822   10.2948   7.7718   7.6296   7.0132   6.6541   6.0201**
    5.3993   5.1110   5.064 4.6071   4.2848   3.9876   1.0703];
$P_F = 1/15$. This is the error given data X.
$P_M = 7/14$. This is the error given data Y.
$P(error) = (1/15) * 15/29 + (7/14) * (14/29) = 0.2759$.

Threshold at 10:

Xsorted   =   [9.4389   5.2970   5.1346   5.0332   4.6081   4.2891   4.2389
    4.0810   3.7187   3.4766 2.6667   2.4199   1.4691   1.1571   1.0028];
Ysorted   =   [**14.1822   10.2948**   7.7718   7.6296   7.0132   6.6541   6.0201
    5.3993   5.1110   5.064 4.6071   4.2848   3.9876   1.0703];
$P_F = 0/15$. This is the error given data X.
$P_M = 12/14$. This is the error given data Y.
$P(error) = (0/15) * 15/29 + (12/14) * (14/29) = 0.4138$.

Threshold at 12:

Xsorted   =   [9.4389   5.2970   5.1346   5.0332   4.6081   4.2891   4.2389
    4.0810   3.7187   3.4766 2.6667   2.4199   1.4691   1.1571   1.0028];
Ysorted   =   [**14.1822**   10.2948   7.7718   7.6296   7.0132   6.6541   6.0201
    5.3993   5.1110   5.064 4.6071   4.2848   3.9876   1.0703];
$P_F = 0/15$. This is the error given data X.
$P_M = 13/14$. This is the error given data Y.
$P(error) = (0/15) * 15/29 + (13/14) * (14/29) = 0.4483$.

It can be seen that with increasing threshold values, the false alarm rate goes down while the miss rates go up. From the figure, it is reasonable to conclude that the threshold of 5 appears to give the lowest value of the error rates.

PPV(at threshold of 5) = P(targetDetected|TargetPresent)/P(targetDetected)
P(targetDetected) = P(targetDetected|Y)$P(Y)$ + P(targetDetected|X)$P(X)$
P(targetDetected) = (10/14) * (14/29) + (4/15) * (15/29) = 0.4828
PPV = 10/(4 + 10) = 10/14 = 0.7142

A few points are in order. Probability of false alarm calculations only include the samples exceeding the threshold (excludes the threshold) from the target absent data. Similarly, the probability of detection only includes the samples exceeding the threshold (excludes the threshold) from the target present data. The reasons for this will become clear when the transition matrix is expressed in terms of the properties of random variables in Chap. 3. In Example 2.22 the intersection was not one of the sample points. The next example shows the case where the intersection is isolated as the data value closest to the intersection of the two densities.

Example 2.24 We have two sets of data collected, 70 samples of target absent and 30 samples of target present. Undertake the analysis.

The a priori probabilities are represented in terms of probabilities of the hypotheses, target absent (H_0) and target present (H_1).

The histograms of the data and the corresponding densities are shown in Figure Ex.2.24.1. The intersection point is 3.112, one of the sample values (in the present case, it is from the target absent cohort). Depending on the proximity of the intersection, the intersection point may come from either of the two cohorts. The data sets are shown in Figure Ex.2.24.2. It is seen that the threshold sample is not included in the P_F calculation as explained above. The confusion matrix and the transition matrix are shown in Figure Ex.2.24.3.

2.5   Permutations, Combinations, and Bernoulli Trials

When conducting experiments, it becomes necessary to arrange objects, events, or outcomes so that we may count objects, events, or outcomes. These counts allow us to estimate the probability of a specific set of outcomes and events using the concept of frequency interpretation of probability. Consider the case of a student taking a test consisting of a number of questions each requiring a binary (yes/no) answer. If there are 20 questions, student will receive an A if at least 17 questions are answered correctly. If we can determine the number of ways the student can have 17 correct ones, we can find out the probability that the student receives an A. In a slightly

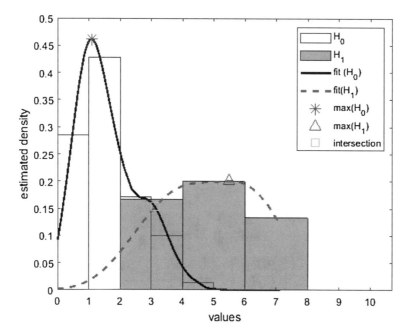

Figure Ex.2.24.1 Histograms, densities, peaks, and the intersection

Target Absent (H$_0$)							Target Present (H$_1$)		
0.118	0.691	1.007	1.229	1.58	2.005	2.861	2.13	4.078	5.834
0.323	0.714	1.02	1.255	1.594	2.011	2.99	2.394	4.127	5.979
0.369	0.844	1.044	1.256	1.6	2.059	3.086	2.709	4.29	6.007
0.401	0.853	1.075	1.263	1.616	2.152	3.112	2.979	4.749	6.086
0.501	0.861	1.099	1.355	1.743	2.368	3.197	3.06	4.862	6.399
0.538	0.861	1.1	1.374	1.861	2.61	3.33	3.455	4.896	6.746
0.547	0.865	1.113	1.403	1.921	2.66	3.33	3.709	4.896	6.754
0.557	0.913	1.118	1.416	1.974	2.674	3.522	3.733	5.493	6.77
0.636	0.961	1.122	1.489	1.988	2.748	3.728	3.826	5.687	6.814
0.651	0.985	1.148	1.553	1.993	2.797	4.136	3.988	5.818	7.144

Threshold (intersection): 3.112

ERROR (samples) in red

Figure Ex.2.24.2 Data sets

Confusion Matrix / Transition Matrix

Threshold (v_T) = 3.112 (intersection)

	Target Not Detected (D$_n$)	Target Detected (D$_p$)	Total Samples
Target Absent	64	False Alarm →6 ← N$_F$	70 ← N$_0$
Target Present	5 ← Miss	25 ← N$_C$	30 ← N$_1$
Total Decisions	69	31	100 ← N

$$\text{ERROR RATE} = \frac{N_F + (N_1 - N_C)}{N} = \frac{11}{100} \qquad \text{PPV} = \frac{N_C}{N_F + N_C} = \frac{25}{31}$$

$$T_X = \begin{bmatrix} P(D_n|H_0) & P(D_n|H_1) \\ P(D_p|H_0) & P(D_p|H_1) \end{bmatrix}$$

$$T_X = \begin{bmatrix} 1 - P_F & P_M \\ P_F & 1 - P_M \end{bmatrix} = \begin{bmatrix} 64 & 5 \\ 6 & 25 \end{bmatrix} \begin{bmatrix} \frac{1}{70} & 0 \\ 0 & \frac{1}{30} \end{bmatrix} = \begin{bmatrix} \frac{64}{70} & \frac{5}{30} \\ \frac{6}{70} & \frac{25}{30} \end{bmatrix}$$

$$\text{a priori prob.} \rightarrow \begin{bmatrix} P(H_0) \\ P(H_1) \end{bmatrix} = \frac{1}{N} \begin{bmatrix} N_0 \\ N_1 \end{bmatrix} = \frac{1}{100} \begin{bmatrix} 70 \\ 30 \end{bmatrix}$$

$$\begin{bmatrix} P(D_n) \\ P(D_p) \end{bmatrix} = T_X \begin{bmatrix} P(H_0) \\ P(H_1) \end{bmatrix} = \frac{1}{100} \begin{bmatrix} 69 \\ 31 \end{bmatrix}$$

Figure Ex.2.24.3 Confusion matrix and transition matrix along with all the probabilities needed estimated

different scenario, lotteries operate in two ways. If the lottery requires the picking of six correct numbers, in *one format*, the win is based on the six numbers matching the ones picked by the machine, while in the *other format*, it requires that the six numbers must appear in a certain order. Simply put, the counting technique associated with choosing at least 17 correct questions is such that the order in which the student gets them correct does not matter. Similarly, when the lottery requires six specific numbers in no specific order, the counting methodology is not tied to the ordered selection. On the other hand, if the six numbers have to be in a specific pattern, the order does matter. These two forms of counting, one where order matters and the one where order does not matter, give us the techniques of **permutation** and **combination**. It is assumed that repetitions are not allowed in either case.

Permutation If out of **n** items, **r** items (**n** $\geq$ **r**) are taken at a time, the number of ways doing this (when **order is important**) is **permutation** of n things taken r at a time and is expressed as

$$_nP_r = {}_r^nP = \frac{n!}{(n-r)!}. \tag{2.171}$$

If out of **n** items, **n** items are taken at a time, the number of ways of doing this is expressed as

$$_nP_n = \frac{n!}{0!} = n! \tag{2.172}$$

Note that permutation is like a lineup in a police station where the location of an individual (such as third person from the left) is important.

Combination If the **order is unimportant**, the number of ways of taking **r** items from **n** is expressed in terms of the combinatorial,

$$_nC_r = {}_r^nC = \binom{n}{r} = \frac{n!}{r!(n-r)!} \tag{2.173}$$

It is easy to see that combinations lead to lower counts than permutations because if we consider the order unimportant, $r!$ counts are identical leading to Eq. (2.173) from Eq. (2.171).

Bernoulli Trials
Consider the case of an experiment consisting of two complementary outcomes (A and B) that have probabilities of p and q, i. e.,

$$p + q = 1. \tag{2.174}$$

If the experiment is repeated N times, the probability of seeing the outcome (A) r times implies that one also observes $(N-r)$ times the outcome (B) at the same time. Thus, the probability of seeing r outcomes of A is

$$P\{A\ r\ \text{times}\} = \binom{N}{r} p^r q^{N-r} = \binom{N}{r} p^r (1-p)^{N-r} \tag{2.175}$$

In this experiment (i.e., Bernoulli trial), it is assumed that N experiments are conducted **independently** (hence we are able to multiply the probabilities). The presence of $\binom{N}{r}$ implies that the order in which outcomes appear is **not important** as long as there are r occurrences of A (Peebles 2001; Yates and Goodman 2005).

An **example** will be the tossing of a coin N times and determining the probability of observing r heads. This experiment (toss of a single coin N times) is identical to the tossing of N coins at the same time and determining the probability of r heads which implies $N-r$ tails as well.

Extension to more than two mutually exclusive events (Papoulis and Pillai 2002). In some experiments, it is possible to have more than 2 outcomes such as the roll of a die where the mutually exclusive and exhaustive outcomes will be 1,2,3,4,5,6 (these outcomes also constitute the sample space). Consider the case of an experiment with r mutually exclusive outcomes (constituting the sample space). We can now extend the notion of repeated trials to this experiment. If the experiment is repeated N times and k_j, $j = 1,2,...,r$ represents the number of outcomes in each category, we have

$$k_1 + k_2 + \cdots + k_r = N. \tag{2.176}$$

We identify A_1, A_2, .., A_r as the r mutually exclusive outcomes, each with probability $p(A_j) = p_j, j = 1,2,...,r$. Taking advantage of the independence of events and the order of occurrence being unimportant, the probability of the sequence is

$$P_N\{k_1, k_2, .., k_r\} = \frac{N!}{k_1! k_2! .. k_r!} p_1^{k_1} p_2^{k_2} .. p_r^{k_r} \tag{2.177}$$

The probabilities p_1, p_2, .., p_r add to unity because the mutually exclusive outcomes constitute the sample space. For $r = 2$, we get the traditional Bernoulli trial,

$$P_N\{k_1, k_2\} = \frac{N!}{k_1! k_2!} p_1^{k_1} p_2^{k_2} = C_r^N p^r (1-p)^{N-r}. \tag{2.178}$$

Equation (2.178) represents the probability of observing $k_1 = r$ events, each with a probability of p when N trials are conducted. The number of complementary events is $k_2 = (N-r)$. Note that Eqs. (2.178) and (2.175) are identical.

Example 2.25 Numbers 1 through 6 are given. Find the total number of ways of picking two numbers (a) without replacement where the order does not matter (b) without replacement where order does matter (c) repetition is allowed where order does matter and (d) repetition is allowed where order does not matter.

Solution We have n equal to 6 and r equal to 2.

(a) There are $C_r^n = \dfrac{n!}{r!(n-r)!} = C_2^6 = \dfrac{6!}{2!(6-2)!} = 15$ ways. The arrangements are shown below. IN this case, the order is not important.

12	23	34	45	56
13	24	35	46	
14	25	36		
15	26			
16				

Notice that the repetitions (11, 22, 33, 44, 55, 66) are not present.

(b) There are $P_r^n = \dfrac{n!}{(n-r)!} = P_2^6 = \dfrac{6!}{(6-2)!} = 30$ ways. The arrangements are shown below. In this case, the order *is* important.

12	21	31	41	51	61
13	23	32	42	52	62
14	24	34	43	53	63
15	25	35	45	54	64
16	26	36	46	56	65

Notice that the repetitions (11, 22, 33, 44, 55, 66) are not present.

(c) Now, the arrangement is such that order still matters; but repetitions are allowed (see the diagonal elements). In this case, the count is $n^r = 6^2 = 36$.

11	21	31	41	51	61
12	*22*	32	42	52	62
13	23	*33*	43	53	63
14	24	34	*44*	54	64
15	25	35	45	*55*	65
16	26	36	46	56	*66*

(d) Now the arrangement is one where repetition is allowed with order being unimportant. In this case, the count is $C_r^{n+1} = \dfrac{(n+1)!}{r!(n+1-r)!} = C_2^7 = \dfrac{7!}{2!(7-2)!} = 21$.

11					
12	*22*				
13	23	*33*			
14	24	34	*44*		
15	25	35	45	*55*	
16	26	36	46	56	*66*

Example 2.26 At an archery training center, archers make attempts to hit the bull's-eye or very close to it. Expert archers succeed to hit the bull's-eye 80% of the time and 20% of the time they miss. If 20 experts are there, what is the probability that none of them hit the bull's-eye. What is the probability that at least one succeeds. What is the probability that at least 70% of the experts succeed to hit the bull's-eye?

Solution We have an experiment that is described using the Bernoulli trial with the probability of success being 0.8 and the probability of failure of 0.2.

Probability that none of them succeed $= 0.2^{20}$.

Probability that at least one is successful $= 1 - 0.2^{20}$.

The probability that at least 70% succeed is

$$C_{14}^{20}0.8^{14}0.2^6 + C_{15}^{20}0.8^{15}0.2^5 + C_{16}^{20}0.8^{16}0.2^4 + C_{17}^{20}0.8^{17}0.2^3 + C_{18}^{20}0.8^{18}0.2^2$$
$$+ C_{19}^{20}0.8^{19}0.2^1 + 0.8^{20}$$

Example 2.27 How many passwords can be created with English alphabet if upper- and lowercase letters are used (no repetition allowed) if each password contains eight letters.

Solution There are 52 letters, and since order is important for passwords, the number of passwords is the permutation of 52 taken 8 at a time.

$$\text{Number of passwords} = \frac{52!}{(52-8)!} = \frac{52!}{44!}$$

Example 2.28 If there are ten true/false questions with students picking the answer randomly, what is the probability that a student gets at least eight correct?

Solution

$$P\{\text{at least 8 correct}\} = P\{8 \text{ or } 9 \text{ or } 10\} = C_8^{10}\left(\frac{1}{2}\right)^{10} + C_9^{10}\left(\frac{1}{2}\right)^{10} + \left(\frac{1}{2}\right)^{10}$$
$$= \left(\frac{1}{2}\right)^{10}\left[C_8^{10} + C_9^{10} + 1\right]$$

Example 2.29 We may now revisit Example 2.16 pertaining to the two circuits that were operating in parallel.

Solution In this case, we may apply Bernoulli principle and conclude that the system will work when at least one circuit does not fail. This means that

$$P(\text{system works}) = C_1^2(1-p)p + C_2^2(1-p)^2 = 2(1-p)p + (1-p)^2$$
$$= 2p - 2p^2 + 1 - 2p + p^2 = 1 - p^2$$

2.6 Summary

Basic elements of probability and statistics that are essential for the understanding of engineering problems have been presented. The topics of Bayes' rule and related topics of conditional probability will be seen repeatedly used in the upcoming chapters. The use of continuous measures of probability introduced in this chapter offers an early start to the topics of random variables that constitute the major portion of this book. We also saw the beginnings of data analytics undertaken using simple probabilistic concepts. Two specific entities, namely, the confusion matrix and the transition matrix, offer views of Bayes' rule from a practical viewpoint with regard to applications in data analytics. These two are summarized in Figs. 2.16 and 2.17.

Exercises

1. A random noise measurement shows that the values lie between -15 and 15 V. (a) Define the universal set for this noise voltage. (b) What is the set to describe the voltages if the noise is passing through a half-wave rectifier?
2. A school offers athletics (#A), music (#M), and quiz (#Q) clubs. #A participation is 25%, #M participation is 15%, and #Q participation is 20%. 2% participate in all three, 8% participate in #A and #M, 5% participate in #M and #Q, and 4% participate in #A and #Q. What is the probability that a randomly chosen student does not participate in any of these activities?
3. A random number generator is providing numbers between 5 and 20, expressed as $S = \{x|5 \le x < 20\}$. If two new sets are created as $A = \{x|5 \le x < 10\}$ and $B = \{x|x > 8\}$, obtain the following sets: $A \cup B$, $A \cap B$, $\bar{A} \cap \bar{B}$.
4. Four circuits in an engineering system are connected in parallel. The probability that anyone of them fails is 0.05. Assuming that they operate independently, what is the probability that the system **works**.

Data Analytics: Error rate, Accuracy, PPV

Number of samples (TARGET ABSENT) = N_0

Number of samples (TARGET PRESENT) = N_1

Total number of samples = $N = N_0 + N_1$

a' priori' probability (TARGET ABSENT) $p(T_a) = N_0/N$

a' priori' probability (TARGET PRESENT) $p(T_p) = N_1/N$

Number of samples (TARGET PRESENT) above threshold = N_C

Number of samples (TARGET ABSENT) above threshold = N_F

Probability of Detection $(P_D) = N_C/N_1$ $\Rightarrow$ **Sensitivity**

Probability of Miss $(P_M) = 1 - P_D$

Probability of False Alarm $(P_F) = N_F/N_0 \Rightarrow$ **1- Specificity**

Probability of Error $(P_e) = P_M\, p(T_p) + P_F\, p(T_a) = \dfrac{(N_1 - N_C) + N_F}{N}$

Accuracy (ACC) = $P_D\, p(T_p) + (1 - P_F)\, p(T_a) = \dfrac{N_C + (N_0 - N_F)}{N}$

P(TARGET Detected) = $P_{TP} = P_D\, p(T_p) + P_F\, p(T_a) = \dfrac{N_C + N_F}{N}$

Positive Predictive Value: PPV = $\dfrac{P_D\, p(T_p)}{P_{TP}} = \dfrac{N_C}{N_C + N_F}$

Fig. 2.16 Summary of terms and relationships

5. A system with four switches is shown below. The probability that any of the switches *operate* is p_k, $k = 1,2,3,4$. The switches function independently. **Find the probability that the system works.**

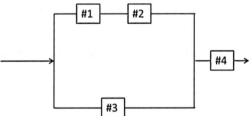

6. A circuit has four components connected as shown. If each component can fail independently with a probability of p, what is the probability that the system does not fail?

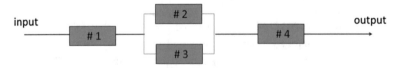

7. Companies do not test every component they produce. A company that manu-factures microphones takes two microphones from a batch consisting of 50 (in a

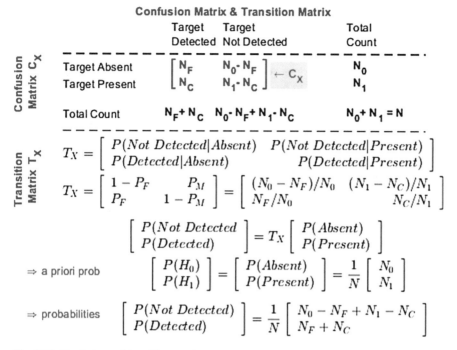

Fig. 2.17 Confusion and transition matrices

box). If both microphones work, the whole box is accepted. Otherwise, the whole box is rejected. If there are 45 excellent ones and 5 defective ones in a box, what is the probability that the box is accepted?

8. Five good light bulbs and two defective ones were mixed up unknowingly. To identify the defective ones for disposal, the bulbs are tested one by one. (a) If both defective ones are identified in the testing of the second one, what is the probability of this event? (b) If three bulbs had to be tested to get the two defective ones, what is the probability of such an event?

9. The life span of females in certain part of the world have been described using a continuous probability model. The probability that a selected female lives beyond the age x is

$$\exp\left(-\frac{x^2}{7688}\right)$$

What is the probability that (a) a randomly selected female does not live past the age of 75? (b) What is the probability that a female lives past 95 years? (c) Knowing that the people in a certain community live past 80 years, what is the probability that a female in that community will live past 95 years?

10. Life span of male elephants is modeled using a continuous probability model. The probability that a selected male lives past the age x is $\exp\left(-\frac{x}{46}\right)$

   (a) What is the probability that a randomly selected male will live past 55 years?
   (b) If it is known a certain male elephant is alive at 55, what is the probability that it might make it to 60?

11. Life span of birds of a certain type is described using a continuous probability model. The probability that they do not live past x years is expressed as

$$p(x) = \frac{1}{2}\left[1 - \exp\left(-\frac{x}{5}\right)\right] + \frac{1}{4}\left[1 - \exp\left(-\frac{x}{8}\right)\right] + \frac{1}{4}\left[1 - \exp\left(-\frac{x}{15}\right)\right]$$

   (a) Test whether the given expression represents a valid probability.
   (b) If it represents a valid probability, determine the probability that a parakeet lives past 10 years.

12. Two numbers are drawn, each in the range $\{0,1\}$. What is the probability that their sum is less than 1 while their product is greater than 1/5?

13. A dart is thrown at a dartboard of unit radius. Estimate the probability that the dart will fall outside the largest square that can be fitted within the circle.

14. Two signals reach the receiver independently, each in an interval of 0–2 minutes. Jamming occurs if the time difference between their arrival is less than 0.2 min. What is the probability that jamming will take place?

15. The probability that a person lives for at least 80 years is 0.7, while the probability that a person lives beyond 90 years is 0.6. What is the probability that a randomly selected person will survive to become 90 given that the chosen individual is surviving at 80?

16. At a computer repair facility, the probability that k computers are dropped for service is described in terms of the probability

$$\frac{3^k}{k!}e^{-3}$$

On any given day, what is the probability that (a) exactly five computers are received for service, (b) at most five computers are received, and (c) at least six are received for service?

17. An organization needs 100 computers. It buys 40 from manufacturer A, 35 from B, and 25 from C. The manufacturer A certifies that only 2% may be defective, while B and C certify that their products are only 1% defective. If a computer is chosen randomly at the organization, what is the probability that it is not defective?

18. Of the three manufacturers of motherboards, # 1 claims that 90% of its boards will last beyond 6 years, while # 2 claims that 85% of its boards will last beyond 6 years, and # 3 claims that 92% of it boards will last past 6 years. What is the probability that any chosen board fails within the first 6 years?

19. Two coins are available for toss, a fair coin and a biased coin with $p(H) = 0.6$. We now pick a coin at random and toss it twice. We notice that heads came up both times. What is the probability that (a) the biased coin was tossed and (b) the fair coin was tossed?

20. In certain parts of the world, tuberculosis (a very treatable and curable illness) is present in 25% of the population. A simple chest X-ray can be used as a diagnostic tool with an accuracy of 99% detection of TB if it is present. In only 0.5% of the cases, non-infected people receive a positive diagnosis. If a person is selected at random and the diagnosis is positive, what is the probability that the person is actually infected?

21. Professors often attempt to determine if the submissions by the students are genuine or copied off the Internet. The program that performs this task is only 95% accurate in correctly identifying a genuine submission and 80% accurate in correctly identifying copies. Based on the past statistics, 15% of the student turned in copied work. If a work is identified as a copy by the program, what is the probability that it is indeed a sample of copied work?

22. In a university, 90% of humanities/social sciences students, 70% of the science students, 60% of the engineering students, and 50% of the business students use library frequently. The university has a student body comprising of 30% humanities/social science, 25% science, 15% engineering, and the rest business majors.

   (a) Draw a transition chart describing the library use using probability concepts.
   (b) What % or what is the probability that a student from any major is using the library collection?
   (c) What is the probability that a student using the library collection is an engineering student?

23. In a ternary (three-level) communication system, 3 is transmitted 3 times more frequently than a 1, while a 2 is transmitted 2 times more frequently than 1. Probability of receiving a 2 when 1 is transmitted is $\alpha/2$ (same for receiving a 3). Probability of receiving a 1 when 2 is transmitted is $\beta/2$ (same for receiving a 3). The probability of receiving a 2 when 3 is transmitted is $\gamma/2$ (same for receiving a 1).

   (a) Draw the transition or the structure of the channel indicating the various probabilities.
   (b) At the receiver a "1" is observed. Using part (a) or directly, estimate the probability that the observed 1 was transmitted as a 1.
   (c) Using part (a), (b), or directly, what is the probability of receiving a "2" or a "3"?

24. "1"s and "−1"s are transmitted in a 0.6:0.4 ratio in a communication system. Because of the noise in the channel, "1" is received as a "1" 50% of the time, and "1" is received as a "−1" 40% of the time, and for the rest of the time, "1" is received as a "0." Similarly, "−1" is received as a "−1" 50% of the time, and "−1" is received as a "1" 40% of the time, and for the rest of the time, "−1" is

received as a "0." (i) Show the transition chart. (ii) What is the probability of receiving a "0"? (iii) If a "1" is observed, what is the probability that it was transmitted as a "1"?

25. Professor Shankar has made the following observation regarding students' performance in ECE 361. If a student reads the material ahead of the lecture, the student follows the lecture very well 95% of the time (student is ahead). If the student comes unprepared, the student is able to absorb the material only 10% of the time (student falls behind). There are three meetings (lectures or recitations) every week. If a student is well-prepared at the beginning of week # 1, what is the probability that the student is well-prepared by the start of week # 2?

26. The flu season starts in late fall. CDC estimates that it is about 60% successful. On the other hand, the chance of contracting flu among the unvaccinated population is 75%. If only 30% of the people get vaccinated (not acceptable), what is the probability that a person selected at random will be afflicted with flu this season? Show the transition chart and display the transition matrix.

27. The binary network circuit is characterized by the following transition matrix for the transmission of +1's and −1's. Note that while +1 is received correctly 80% of the time, −1 is received correctly only 75% of the time. If transmission requires four of these networks connected in series, what is the probability of receiving +1? If +1 is received, what is the probability that it was transmitted as a +1?

28. A student comes to the class late 40% of the time. If the class meets five times a week, (a) find the probability that the student is late for at least three classes in a week and (b) the probability that the student is not late at all in any given week.

29. Extremely sensitive messages are sent multiple times to ensure that they are received correctly before being carried out. This is necessary because the channel through which the message is sent has noise, and therefore, there is a 0.01% chance that it is received incorrectly. If a message is sent five times and a decision to carry out the command in the message is taken only with four or more correct decisions, what is the probability of a correct decision?

30. You are given the following data sets: X represents the data collected from an infrared (IR) receiver when there was no target in front of the transmitter, and Y represents the data collected from an IR receiver when a target was present in the field of view of the transmitter. With these data sets, obtain a priori probabilities.

Obtain the probabilities of false alarm, miss, and error rates of the threshold set at 3, 5, and 6. We expect the **presence of the target** to result in higher values of the measurand.

```
    X=[1.6978   7.7368   8.7591   1.9001   0.7486   0.2048   6.1228, 1.6926
2.2690  13.2931];
    Y=[11.4976,4.5624,17.0159,5.1402,8.9841,8.9526,2.0939,6.1407];
```

31–45 For each problem, two sets of data are given: one is obtained when NO TARGET is present in the field and the second one when a TARGET is present in the field of view. For each of these data sets, determine P_F, P_M, PPV, and error rate using the median of the pooled (combined sets) data.

46–105 For each problem, two sets of data are given: one is obtained when NO TARGET is present in the field and the second one when a TARGET is present in the field of view. For each of these data sets, determine P_F, P_M, PPV, and error rate using the median of the pooled (combined sets) data.

106–215 (110 data sets are provided)

A column of samples is given, with the first 40 entries corresponding to TARGET ABSENT and the remaining 30 entries corresponding to TARGET PRESENT. Obtain the histograms and appropriate ksdensity(.) fits and determine the peaks in each case. Use the midpoint of the peaks as the threshold (closest to one of the sample values), and obtain the following:

Display a scatter plot of the data.
Display the histograms, fits, peaks, and the midpoint. Additionally, obtain:
A priori probabilities
Probabilities of miss and false alarm
Positive predictive value and error rate
Confusion matrix and transition matrix
Display the partitioned data.

Problems 216–325 For the data sets in the previous exercise set, use ksdensity(.) to obtain the fit. Use the intersection of the densities (closest to one of the sample values) as the threshold, and obtain the following:

Display a scatter plot of the data.
Display the fits and the intersection. Additionally, obtain:
A priori probabilities
Probabilities of miss and false alarm
Positive predictive value and error rate
Confusion matrix and transition matrix
Display the partitioned data.

Problems 326–425 A column of samples is given, with the first 60 entries corresponding to TARGET ABSENT and the remaining 60 entries corresponding to TARGET PRESENT. Obtain the histograms and appropriate fits and determine the peaks in each case. Use the intersection of the densities (closest to one of the sample values) and obtain the following:

Display a scatter plot of the data
Display the histograms, fits, peaks, and the midpoint. Additionally, obtain:
A priori probabilities
Probablities of miss, false alarm
Positive predictive value, error rate
Confusion matrix and transition matrix
Display the partioned data

Problems 426–525 Use the data set from the previous problems (326–425). Obtain the histograms and appropriate fits and determine the peaks in each case. Use the midpoint of the peaks as the threshold (closest to one of the sample values) and obtain the following:

Display a scatter plot of the data
Display the histograms, fits, peaks, and the midpoint. Additionally, obtain:
A priori probabilities
Probablities of miss, false alarm
Positive predictive value, error rate
Confusion matrix and transition matrix
Display the partioned data

Problems 526–575 A column of samples is given, with the first 70 entries corresponding to TARGET ABSENT and the remaining 30 entries corresponding to TARGET PRESENT. Obtain the histograms and appropriate fits and determine the peaks in each case. Use the intersection of the densities (closest to one of the sample values) and obtain the following:

Display a scatter plot of the data
Display the histograms, fits, peaks, and the midpoint. Additionally, obtain:
A priori probabilities
Probablities of miss, false alarm
Positive predictive value, error rate
Confusion matrix and transition matrix
Display the partioned data

References

Eng J (2012) Teaching receiver operating characteristic analysis: an interactive laboratory exercise. Acad Radiol 19:1452–1456

Fawcett T (2006) An introduction to ROC analysis. Pattern Recogn Lett 27:861–874

Ghahramani S (2000) Fundamentals of Probability. Prentice-Hall, New Jersey

Greiner M, Pfeiffer D, Smith RD (2000) Principles and practical application of the receiver-operating characteristic analysis for diagnostic tests. Prev Vet Med 45(1–2):23–41

Kohavi R, Provost F (1998) Glossary of terms. Mach Learn 30:271–274

Leon-Garcia, A (1994) Probability and Random Processes for Electrical Engineering. Addison-Wesley, New York

Papoulis A, Pillai U (2002) Probability, random variables, and stochastic processes. McGraw-Hill, New York

Peebles, PZ (2001) Probability, Random Variables and Random Signal Principles. McGraw-Hill, New York

Ross S (2006) A first Course in Probability. Prentice-Hall, New Jersey

Shankar PM (2019) Pedagogy of Bayes' rule, confusion matrix, transition matrix, and receiver operating characteristics. Comput Appl Eng Educ 27(2):510–518. https://doi.org/10.1002/cae.22093

Stark H, Woods JW (1986) Probability, Random Process and Estimation Theory for Engineers. Prentice-Hall, New Jersey

Yates RD, Goodman DJ (2005) Probability and Stochastic Processes: A friendly Introduction For Electrical and Computer Engineers. Wiley, New Jersey

Ziemer RD (1997) Elements of Engineering Probability & Statistics. Prentice-Hall, New Jersey

Chapter 3
Concept of a Random Variable

3.1 Introduction

When measurements or observations are undertaken, the results often are random. For example, when a coin is tossed, the outcomes are heads or tails. If a fruit is picked (without looking) from a basket containing different fruits, it may be an apple, pear, plum, or any other ones in the basket. In the case of customers at a teller window in a bank, the number of customers being served per hour is random. When the received power in a wireless base station is measured, it may lie within a certain range of values. When the lifetimes of an electronic system are observed, they will be random. In all these cases, a simple way to model these outcomes is based on the concept of a random variable. A random variable is defined as a function that maps a set of outcomes in an experiment to a set of real values. For example, if one tosses a coin resulting in "heads" or "tails," a random variable can be created to map "heads" and "tails" into a set of numbers that will be discrete [1 for heads and 0 for tails or 1 for heads and -1 for tails]. Similarly, temperature measurements taken providing a continuous set of outcomes can be mapped into a continuous random variable (in this case, the temperature measured is numeric, and the mapping is one-to-one). By grouping sets of values, one can also create a discrete random variable as it is done in the case of analog-to-digital (A/D) converters. Because the outcomes are random, we can specify events as a random variable taking a specific value or taking values less than or greater than a specified value, etc. Thus the study of random variables is essential for the understanding and modeling of phenomena occurring around us. Random variables are characterized by their densities and distribution functions, some of which have unique names and forms.

Electronic supplementary material: The online version of this chapter (https://doi.org/10.1007/978-3-030-56259-5_3) contains supplementary material, which is available to authorized users.

3.2   Distribution and Density Functions

Since the observations or outcomes are random, we need an appropriate statistical description to explain their behavior so that we may be able to understand and then predict how the outcomes might change. We use X to represent the random variable and x the value that it may take. Using X and x, we can now define events such as $\{X = x\}$, $\{X < x\}$ or $\{X > x\}$, or $\{x_1 \leq X \leq x_2\}$. Note that x is a real number that can take discrete or continuous values lying in the range $-\infty$ to $+\infty$ depending on the outcomes and x_1 and x_2 constitute specific values. It is also possible that we have an experiment where the outcomes may take both continuous and discrete values (mixed). An experiment where the outcomes are mixed arises when a clipper circuit is used. Such experiments result in mixed random variables.

Consider the event, $\{X \leq x\}$. The probability associated with this event is defined as the cumulative distribution function (CDF) of the random variable. Thus the CDF of the random variable X, $F_X(x)$, is

$$F_X(x) = \text{Prob}\{X \leq x\} = P\{X \leq x\} \tag{3.1}$$

It is cumulative because the event includes all outcomes (cumulatively) from the lowest value permissible to x (including x). For example, if we are measuring noise that takes positive and negative values in the range $-\infty$ to ∞, Eq. (3.1) implies

$$F_X(x) = P\{-\infty \leq X \leq x\} \tag{3.2}$$

If one is measuring power (which only takes positive values), Eq. (3.1) implies

$$F_X(x) = P\{0 \leq X \leq x\} \tag{3.3}$$

There is no need to use Eq. (3.2) or Eq. (3.3), and Eq. (3.1) captures all that is necessary to explain the meaning of the CDF. CDF is a measure of the probability, and therefore, CDF will always be positive with an upper limit of unity,

$$0 \leq F_X(x) \leq 1 \tag{3.4}$$

The rate of change of the CDF w.r.t. to x or the derivative of the CDF is defined as the probability density function (pdf),

$$f_X(x) = \frac{d}{dx} F_X(x) \tag{3.5}$$

Note that often the words density and distribution are used to describe the probability density function. For example, an exponential density or exponential distribution refers to a case where the probability density follows the exponential model.

3.2.1   *Characteristics of the CDF and pdf*

From the definition of the CDF in Eq. (3.1), it can be seen that CDF implies monotonically increasing (cumulative) behavior. For two values of x, namely, x_1 and x_2, with $x_1 \leq x_2$, we can write

$$P\{x_1 \leq X \leq x_2\} = F_X(x_2) - F_X(x_1) \geq 0 \tag{3.6}$$

Since the CDF is monotonically increasing w.r.t. x, its derivative w.r.t. x, namely, the probability density function, will always be positive. This implies that

$$0 \leq f_X(x) \leq \infty \tag{3.7}$$

The presence of ∞ in Eq. (3.7) is acceptable because the pdf represents the rate of change of the CDF. Using Eq. (3.5), the CDF can be expressed in integral form as

$$F_X(x) = \int_{-\infty}^{x} f_X(y)dy \tag{3.8}$$

Equation (3.8) implies the normalization associated with the pdf because

$$F_X(-\infty) = 0 \tag{3.9}$$

$$F_X(\infty) = 1 \tag{3.10}$$

Equations (3.9) and (3.10) lead to

$$F_X(\infty) = \int_{-\infty}^{\infty} f_X(x)dx = 1 \tag{3.11}$$

This means that for any function $g(x)$ to be a valid probability density function,

$$\int_{-\infty}^{\infty} g(x)dx = 1, \qquad 0 \leq g(x) \leq \infty \tag{3.12}$$

Equation (3.6) can now be written as

$$P\{x_1 \leq X \leq x_2\} = F_X(x_2) - F_X(x_1) = \int_{x_1}^{x_2} f_X(x)dx \tag{3.13}$$

3.2.2   Discrete vs. Continuous Random Variables

Consider the roll of a die. In this case, the outcomes 1, 2, 3, 4, 5, and 6 will be discrete. If the die is a fair one,

$$P(X = k) = \frac{1}{6}, k = 1, 2, .., 6 \tag{3.14}$$

Since the outcomes are discrete, using the basic definition in Eq. (3.1), the CDF becomes

$$F_X(x) = \sum_{k=1}^{6} P(X = k)U(x - k) \tag{3.15}$$

In Eq. (3.15), $U(.)$ is the unit step function,

$$U(x) = \begin{cases} 0, & x < 0 \\ 1, & x \geq 0 \end{cases} \tag{3.16}$$

The derivative of the step function results in a δ-function and conversely,

$$U(x) = \int_{-\infty}^{x} \delta(y)dy \tag{3.17}$$

Details on the properties of unit step and δ-functions are given later in Sect. 3.9.6. This CDF associated with the random variable representing the roll of a die is shown in Fig. 3.1 illustrating the use of unit step functions. Differentiating Eq. (3.15), the pdf becomes

$$f_X(x) = \sum_{k=1}^{6} P(X = k)\delta(x - k) \tag{3.18}$$

The pdf is shown alongside the CDF in Fig. 3.1.

Thus, for a discrete random variable, the pdf will consist of delta functions, and CDF will be made up of unit step functions. Another function, defined specifically in connection with discrete random variables, is the probability mass function (PMF). The PMF is expressed as

$$PMF = P(X = k) \tag{3.19}$$

PMF has the property that

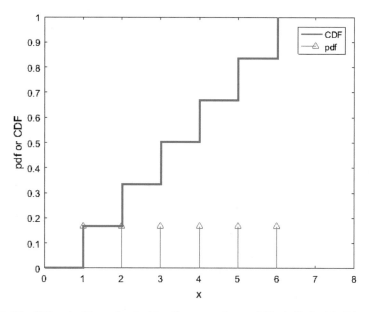

Fig. 3.1 The CDF and pdf associated with a discrete random variable (roll of a fair die)

$$\sum_k PMF = \sum_k P(X = k) = 1 \qquad (3.20)$$

In general, if there are n discrete outcomes $(x_k, k = 1,2,..n)$, the CDF and pdf of a discrete random variable become

$$F_X(x) = \sum_{k=1}^{n} P(X = x_k)U(x - x_k) \qquad (3.21)$$

$$f_X(x) = \sum_{k=1}^{n} P(X = x_k)\delta(x - x_k) \qquad (3.22)$$

The only condition imposed on x_k is that it must be real. Otherwise, x_k may take integer, non-integer, negative, or positive values.

Let us understand the differences between a discrete random variable and a continuous random variable by examining two events, $A = \{X < x\}$ and $B = \{X \leq x\}$. As an example of a continuous case, consider the simple example of an exponential density (we will examine this density later in this chapter)

$$f_X(x) = \frac{1}{b} \exp\left(-\frac{x}{b}\right) U(x) \Rightarrow \frac{1}{b} \exp\left(-\frac{x}{b}\right), x \geq 0, \ b > 0 \qquad (3.23)$$

The associated CDF is

$$F_X(x) = 1 - \exp\left(-\frac{x}{b}\right), x \geq 0 \qquad (3.24)$$

The pdf and CDF are shown in Fig. 3.2.

For the purposes of comparing a discrete and continuous random variable, let us choose a value of $x = 2$. As an example of a discrete random variable, let us choose the experiment of a roll of a fair die. In this case, the event $\{X < 2\}$ and the event $\{X = 1\}$ will be identical. The probability of the event A is $(1/6)$ because there is a single outcome, $x = 1$, associated with it. For the event $B = \{X \leq 2\}$, the outcomes will be $x = 1$ and $x = 2$, and probability of the event B will be 2/6.

$$P(A) = P[X < 2] = P[X = 1] = \frac{1}{6}$$
$$\qquad (3.25)$$
$$P(B) = P[X \leq 2] = P[X = 1] + P[X = 2] = \frac{2}{6}$$

From Eq. (3.25), we see that the probabilities of the events $\{X < 2\}$ and $\{X \leq 2\}$ are not equal.

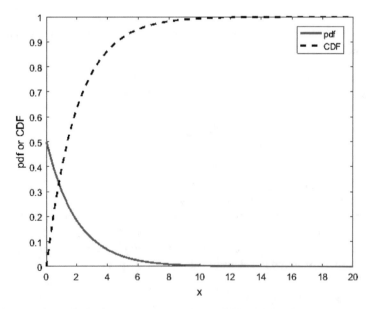

Fig. 3.2 The pdf and CDF of an exponential random variable with $b = 2$

Let us now look at the case of a continuous random variable described as having a pdf in Eq. (3.23). To help us delineate the two events A and B, let us go back to Eq. (3.6) and choose

$$x_1 = x$$
$$x_2 = x + \Delta x \tag{3.26}$$

Equation (3.6) becomes

$$P\{x_1 \leq X \leq x_2\} = F_X(x + \Delta x) - F_X(x) = \Delta F \tag{3.27}$$

Using the definition of the pdf in Eq. (3.5), we can write Eq. (3.27) as

$$P\{x + \Delta x \leq X \leq x\} = \Delta F = f(x)\Delta x \tag{3.28}$$

For the events A and B, we can write

$$P(X \leq 2) = P(X < 2) + P(X = 2) \tag{3.29}$$

For a continuous random variable,

$$P(X = 2) = P\{2 \leq X \leq 2 + \Delta x\} = \Delta F = f(x)\Delta x|_{\Delta x \to 0} = 0 \tag{3.30}$$

In other words, for a continuous random variable, *the probability that the random variable takes a specific value is always zero*. In simple terms, for a continuous random variable,

$$P(X = x) = 0 \tag{3.31}$$

Therefore,

$$P(X \leq 2) = P(X < 2) = F_X(2) \tag{3.32}$$

For the case of the exponential density in Eq. (3.23), this means that

$$P(X \leq 2) = P(X < 2) = F_X(2) = 1 - \exp\left(-\frac{2}{b}\right) \tag{3.33}$$

It is clear that for a continuous random variable,

$$P[X \geq u] = P[X > u] \tag{3.34}$$

For a discrete random variable,

$$P[X \geq u] = P[X = u] + P[X > u] \tag{3.35}$$

Simply stated, it is acceptable to ignore the equal sign appearing with the less than and greater than signs for a *continuous random variable*. For the case of a discrete random variable, the events $\{X < k\}$ and $\{X \leq k\}$ may have different probabilities, and the presence of equality may make a difference. Note that the CDF is still defined with $\leq$ sign regardless of whether the random variable is discrete or continuous.

We will examine mixed variables and the associated densities when we study the transformations of random variables.

Let us look at a few examples of densities and cumulative distribution functions.

Example 3.1 A function of x is given as

$$h(x) = \frac{x}{k}, 0 < x < 2, \quad k > 0$$

If $h(x)$ must be a valid CDF, what is the value of k? If it is a valid CDF, what is the pdf?

Solution The range of the random variable is $[0, 2]$. This means that $h(0) = 0$ and $h(2) = 1$ if $h(x)$ is a valid CDF. These limiting values of the function $h(x)$ lead to

$$\frac{2}{k} = 1 \quad \Rightarrow \quad k = 2$$

$$F_X(x) = \frac{x}{2}, 0 < x < 2$$

$$f_X(x) = \frac{dF_X(x)}{dx} = \frac{1}{2}, 0 < x < 2$$

Note that the density function obtained above is positive. Therefore, it is a valid pdf. This also proves the other important property of a CDF; its derivative must be positive.

Example 3.2 A function of x is given as

$$g(x) = x + 2x^3, \quad 0 < x < a$$

Is $g(x)$ a valid pdf? If so, what is the value of a and obtain an expression for the CDF?

Solution If $g(x)$ is a valid pdf,

$$\int_0^a g(x)dx = 1 \Rightarrow \int_0^a (x + 2x^3)dx = \frac{a^2}{2}(a^2 + 1) = 1$$

Solving for a, we have

$$a = 1$$

$$F_X(x) = \int_0^x (y + 2y^3)dy = \frac{x^2}{2}(x^2 + 1), \quad 0 < x < 1$$

Example 3.3   For the pdf $f(x)$ given below, what is the value of a? What is the CDF?

$$f_X(x) = \frac{a}{x^2}, \quad 0 < b < x < \infty, \quad a > 0$$

Solution

$$\int_b^\infty \frac{a}{x^2}dx = \frac{a}{b} \quad \Rightarrow a = b$$

$$F_X(x) = \int_a^x f_X(y)dy = \int_a^x \frac{a}{y^2}dy = 1 - \frac{a}{x}, x > a$$

Example 3.4   Obtain the CDF for the case of the pdf

$$f(x) = \frac{1}{\pi\sqrt{1 - x^2}}, \quad -1 < x < 1$$

Solution

$$F_X(x) = \begin{cases} \int_{-1}^x \frac{1}{\pi\sqrt{1 - y^2}}dy = \frac{1}{2} + \frac{1}{\pi}\sin^{-1}(x), \ -1 < x < 0 \\ \int_{-1}^0 \frac{1}{\pi\sqrt{1 - y^2}}dy + \int_0^x \frac{1}{\pi\sqrt{1 - y^2}}dy = \frac{1}{2} + \frac{1}{\pi}\sin^{-1}(x), 0 < x < 1 \\ 1, x > 1 \end{cases}$$

The expression pertaining to the interval $0 < x < 1$ on the right side of the equation above clearly demonstrates the cumulative aspect of the CDF as seen from the from the first integral of the sum going from -1 to 0.

Example 3.5 You are given the following function $h(x)$. Under what conditions can it be a valid CDF?

$$h(x) = \exp(bx), b > 0$$

Solution To test the validity as the CDF, we must examine the values of the random variable that will result in the limiting values of the CDF, 0 and 1. If x_{min} and x_{max} are the limiting values, with $x_{max} > x_{min}$,

$$h(x_{min}) = 0 \Rightarrow \exp(bx_{min}) = 0 \qquad \Rightarrow x_{min} = -\infty$$
$$h(x_{max}) = 1 \Rightarrow \exp(bx_{max}) = 1 \qquad \Rightarrow x_{max} = 0$$

These limiting values suggest that $h(x)$ is a valid CDF with $-\infty < x < 0$ as the range of validity.

$$F_X(x) = \exp(bx), \ -\infty < x < 0$$

The pdf is therefore

$$f_X(x) = \frac{dF_X(x)}{dx} = b\exp(bx), \ -\infty < x < 0$$

Once again it is seen that the expression obtained by taking the derivative of the CDF is positive. This property completes the requirement for $h(x)$ to be a valid CDF.

Example 3.6 The CDF of a random variable is given below. Obtain the range of the random variable $(x > 0)$ and the pdf.

$$F_X(x) = 1 - \frac{25}{x^2}$$

Solution If it is a valid CDF,

$$F(x_{min}) = 0 \quad \Rightarrow 1 - \frac{25}{x_{min}^2} = 0$$
$$x_{min}^2 = 25 \quad \Rightarrow x_{min} = 5$$
$$F(x_{max}) = 1 \quad \Rightarrow 1 - \frac{25}{x_{max}^2} = 1$$
$$\frac{25}{x_{max}^2} = 0$$
$$x_{max} = \infty$$

$$F_X(x) = 1 - \frac{25}{x^2}, \qquad x > 5$$

$$f_X(x) = \frac{50}{x^3}, \qquad x > 5$$

The pdf being always positive demonstrates the third property of the CDF that its derivative is always positive.

Example 3.7 It is given that the probability that $X > t$ is given as

$$P(X > t) = \frac{1}{3} \exp\left(-\frac{t}{3}\right) + \frac{2}{3} \exp\left(-\frac{t}{5}\right), t > 0$$

What is the pdf of the random variable X?

Solution We are given

$$P(X > t) = \frac{1}{3} \exp\left(-\frac{t}{3}\right) + \frac{2}{3} \exp\left(-\frac{t}{5}\right), t > 0$$

By the definition of the CDF, the left-hand side can be expressed in terms of the CDF, and the given expression is rewritten as

$$1 - F_X(t) = \frac{1}{3} \exp\left(-\frac{t}{3}\right) + \frac{2}{3} \exp\left(-\frac{t}{5}\right)$$

Differentiating,

$$-f_X(t) = -\frac{1}{9} \exp\left(-\frac{t}{3}\right) - \frac{2}{15} \exp\left(-\frac{t}{5}\right)$$

Rewriting and replacing t by x, we have

$$f_X(x) = \left[\frac{1}{9} \exp\left(-\frac{x}{3}\right) + \frac{2}{15} \exp\left(-\frac{x}{5}\right)\right] U(x)$$

Rearranging the values, the pdf becomes

$$f_X(x) = \left[\frac{1}{3}\left(\frac{1}{3}\right) \exp\left(-\frac{x}{3}\right) + \frac{2}{3}\left(\frac{1}{5}\right) \exp\left(-\frac{x}{5}\right)\right] U(x)$$

This density may be viewed as a *mixture density* as described later in Sect. 3.9.4.

Example 3.8 The pdf of X is

$$f(x) = x \exp(-x), x > 0$$

What are the following probabilities?

$$P(2 < X < 4)$$
$$P(X < 3)$$
$$P(X > 5)$$

Solution The CDF of X is

$$F_X(x) = 1 - (1 + x)\exp(-x)$$

Note that X is a continuous random variable and the presence of the absence of equal sign will not impact the results on probability calculations in this example.

$$P(2 < X < 4) = F_X(4) - F_X(2) = 3e^{-2} - 5e^{-4}$$
$$P(X < 3) = F_X(3) = 1 - 4e^{-3}$$
$$P(X > 5) = 1 - F_X(5) = 6e^{-5}$$

Example 3.9 The probability mass function (PMF) is expressed as

$$\text{PMF} = P(X = k) = \frac{C}{k(k+1)}, k = 1, 2, \ldots$$

What value of C will make this a valid PMF?

Solution For a valid PMF

$$\sum_{k=1}^{\infty} P(X = k) = 1 = \sum_{k=1}^{\infty} \frac{C}{k(k+1)} = C$$

Therefore, C must be equal to unity. The summation may be carried using the symbolic toolbox in Matlab as shown in Appendix B. (See the results below.)

```
>> syms k integer
>> syms C
>> symsum(C/(k*(k+1)),k,1,inf)
ans =  C
```

Example 3.10 In a basket containing fruits, there are 8 apples, 3 pears, 5 oranges, and 4 peaches. Create a random variable for the experiment of picking a fruit randomly from the basket and obtain its pdf and CDF.

Solution This experiment requires us to set up a mapping from fruits to numbers. Let us identify apples as -2, pears as -1, oranges as 1, and peaches as 2. This means that

$$P(X = -2) = \frac{8}{20}, \quad P(X = -1) = \frac{3}{20}, \quad P(X = 1) = \frac{5}{20}, \quad P(X = 2) = \frac{4}{20}$$

The pdf of the model to describe the picking a fruit from the basket is

$$f_X(x) = \frac{8}{20}\delta(x + 2) + \frac{3}{20}\delta(x + 1) + \frac{5}{20}\delta(x - 1) + \frac{4}{20}\delta(x - 2)$$

The corresponding CDF is

$$F_X(x) = \frac{8}{20}U(x + 2) + \frac{3}{20}U(x + 1) + \frac{5}{20}U(x - 1) + \frac{4}{20}U(x - 2)$$

Note that the choice of the mapped values is arbitrary.
The pdf and CDF are shown in Figure Ex. 3.10.

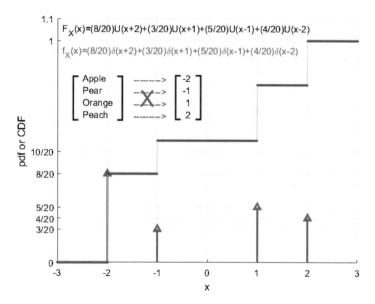

Figure Ex. 3.10

Example 3.11 A function $h(x)$ is given as

$$h(x) = Kx^2, \quad -2 < x < 2$$

For $h(x)$ to be a valid pdf, what should be the value of K? What is the CDF?

Solution For $h(x)$ to be a valid pdf,

$$\int\limits_{-2}^{2} Kx^2 dx = \frac{16}{3}K = 1$$

Therefore,

$$K = \frac{3}{16}$$

CDF is

$$F_X(x) = \begin{cases} 0, \; x < -2 \\ \int\limits_{-2}^{x} \frac{3}{16}y^2 dy = \frac{1}{2} + \frac{x^3}{16}, \; -2 < x < 0 \\ \int\limits_{-2}^{0} \frac{3}{16}y^2 dy + \int\limits_{0}^{x} \frac{3}{16}y^2 dy = \frac{1}{2} + \frac{x^3}{16}, 0 < x < 2 \\ 1, x > 2 \end{cases}$$

The CDF can also be written as

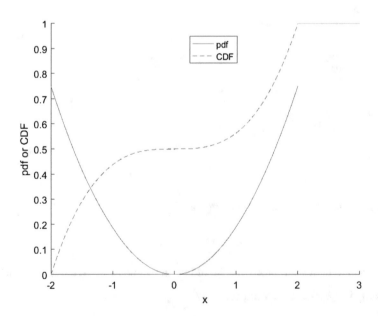

Figure Ex.3.11

$$F_X(x) = \begin{cases} 0, & x < -2 \\ \dfrac{1}{2} + \dfrac{x^3}{16}, & -2 < x < 2 \\ 1, x > 2 \end{cases}$$

The pdf and CDF are shown in Figure Ex. 3.11

Example 3.12 An example of a *mixed* random variable has the following CDF:

$$F_X(x) = \begin{cases} 0, & x < 0 \\ \dfrac{x}{8}, & 0 \le x < 4 \\ \dfrac{2}{3}, & 4 \le x < 6 \\ \dfrac{5}{6}, & 6 \le x < 8 \\ 1, & x \ge 8 \end{cases}$$

Plot the CDF. What is the probability that $X < 3$ and probability that $X = 4$?

Solution The CDF is shown in Figure Ex.3.12.1.

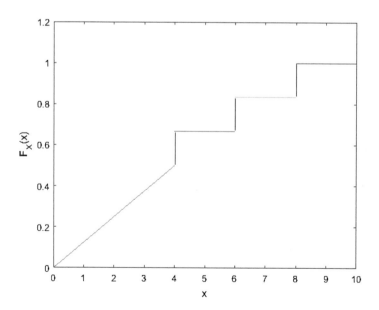

Figure Ex.3.12.1

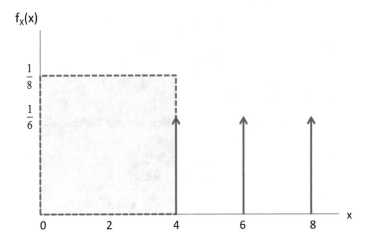

Figure Ex.3.12.2

$$P(X < 3) = P(X \le 3) = \frac{3}{8}$$

$$P(X = 4) = P(X \le 4) - P(X < 4) = \frac{2}{3} - \frac{1}{2} = \frac{1}{6}$$

It is clear that a mixed random variable leads to a CDF made up of continuous functions and step functions. The density of mixed variables will consist of continuous functions and delta functions. The corresponding pdf is shown in Figure Ex.3.12.2. It shows the existence of three delta functions each of height (1/6) at $x = 4$, 6, and 8.

Example 3.13 A discrete random variable has the following PMF. Find out the value of C.

$$P(X = k) = \frac{C}{k}, \quad k = 1, 3, 5, 7$$

What are the following probabilities?

$$P(X \le 3)$$
$$P(X < 5)$$
$$P(X = 5)$$

Solution Given the PMF, we must have

$$\sum_{k=1,3,5,7} P(X=k) = \sum_{k=1,3,5,7} \frac{C}{k} = 1$$

$$C\frac{176}{105} = 1 \quad \Rightarrow \quad C = \frac{105}{176}$$

$$P(X \le 3) = P(X=1) + P(X=3) = C + \frac{C}{3}$$

$$P(X < 5) = P(X=1) + P(X=3) = P(X \le 3) = C + \frac{C}{3}$$

$$P(X = 5) = \frac{C}{5}$$

Example 3.14 The pdf of a random variable is $(b > 0)$

$$f(x) = \frac{1}{10} \exp\left[-\frac{|x-3|}{b}\right], \quad -\infty \le x \le \infty$$

Obtain the value of b and the CDF.

Solution

$$1 = \int_{-\infty}^{\infty} \frac{1}{10} \exp\left[-\frac{|x-3|}{b}\right] dx = \int_{-\infty}^{3} \frac{1}{10} \exp\left[\frac{x-3}{b}\right] dx + \int_{3}^{\infty} \frac{1}{10} \exp\left[\frac{-(x-3)}{b}\right] dx$$

$$= \frac{b}{5}$$

This means that

$$b = 5$$

$$f(x) = \frac{1}{10} \exp\left[-\frac{|x-3|}{5}\right], \quad -\infty \le x \le \infty$$

The pdf above may be written as the general expression

$$f(x) = \frac{1}{2b} \exp\left[-\frac{|x-3|}{b}\right], \quad -\infty \le x \le \infty, \quad b > 0$$

Note that one must be careful in integrating to find the CDF since the pdf has a maximum at $x = 3$.

$$F_X(x) = \begin{cases} \displaystyle\int_{-\infty}^{x} \frac{1}{2b} \exp\left[\frac{(y-3)}{b}\right] dy = \frac{1}{2}\exp\left[\frac{(x-3)}{b}\right], & -\infty \le x \le 3 \\[2em] \displaystyle\int_{-\infty}^{3} \frac{1}{2b} \exp\left[\frac{(y-3)}{b}\right] dy + \int_{3}^{x} \frac{1}{2b} \exp\left[\frac{-(y-3)}{b}\right] dy = 1 - \frac{1}{2}\exp\left[\frac{-(x-3)}{b}\right], & x > 3 \end{cases}$$

The $f(x)$ in Example 3.14 is the Laplace density function.

Example 3.15 For any continuous random variable with constant β (with β lying in the range of values of the random variable), show that $h(x)$ given below is a valid pdf.

$$h(x) = \begin{cases} \dfrac{f_X(x)}{1 - F_X(\beta)}, & x > \beta \\ 0, & \text{otherwise} \end{cases}$$

Solution If $h(x)$ is a valid pdf,

$$\int h(x)dx = 1$$

This means that

$$\int_{\beta}^{\infty} \frac{f_X(x)}{1 - F_X(\beta)} dx = \frac{1}{1 - F_X(\beta)} \int_{\beta}^{\infty} f_X(x)dx = \frac{1}{1 - F_X(\beta)}[1 - F_X(\beta)] = 1$$

$$\int_{\beta}^{\infty} \frac{f_X(x)}{1 - F_X(\beta)} dx = 1$$

Indeed, $h(x)$ is a valid density. It will be shown later in Sect. 3.5 that $h(x)$ is an example of the pdf of X conditioned on a specific event.

3.3   Moments, Characteristic Functions (CHF), Moment Generating Functions (MGF), and Laplace Transforms

While every random variable has a probability density function and cumulative distribution function, random variables may also be characterized by several other properties. For example, results of every experimental observation are reported in terms of a mean and standard deviation. It is also known that the smaller the standard deviation, the better is the quality of the measurement. The mean and standard deviation are among a number of parameters that can be grouped into a single entity, the moments of the random variable.

The nth moment of a continuous random variable is defined as

$$M_n = \langle X^n \rangle = E[X^n] = \int_{-\infty}^{\infty} x^n f_X(x)dx \qquad (3.36)$$

The first-order moment ($n = 1$) is called the mean or the average of the random variable, and $E[.]$ or $E(.)$ represents the expectation operator with the mean expressed as

$$\text{mean} = m_1 = \langle X \rangle = E(X) = E[X] = \int_{-\infty}^{\infty} x f_X(x)dx \qquad (3.37)$$

Notice that $\langle . \rangle$ also represents the expectation operator. The variance of a random variable is defined as

$$\text{var}(X) = \langle X^2 \rangle - \langle X \rangle^2 \qquad (3.38)$$

Note that for any random variable X,

$$\text{var}(X) \geq 0 \qquad (3.39)$$

We can also define the central moments (about the mean) as

$$\langle (X - m_1)^n \rangle = E[(X - m_1)^n] = \int_{-\infty}^{\infty} (x - m_1)^n f(x)dx \qquad (3.40)$$

Comparing Eqs. (3.38) and (3.40), it is seen that the variance is the second central moment. Standard deviation (std. dev) is

$$\text{std.dev} = \sqrt{E(X^2) - [E(X)]^2} \qquad (3.41)$$

If the variance of a random variable is zero, the variable is deterministic. Note that

$$\text{var}(X) \geq 0 \quad \Rightarrow E(X^2) \geq [E(X)]^2 \qquad (3.42)$$

The moments of a discrete random variable are written in the form of a summation,

$$M_n = \langle X^n \rangle = E[X^n] = \sum_k [x_k]^n P[X = x_k] \tag{3.43}$$

Thus the mean of a discrete random variable is

$$E[X] = \sum_k x_k P[X = x_k] \tag{3.44}$$

The mode of a density is the value of x at which the pdf is has a maximum, while median is the value of $x = x_M$ where $P\{X < x_M\} = P\{X > x_M\} = 1/2$. The expression for the median x_M is

$$\frac{1}{2} = F(x_M) = 1 - F(x_M) \Rightarrow \quad x_M = F^{-1}\left(\frac{1}{2}\right) \tag{3.45}$$

The characteristic function (CHF) of a random variable is the Fourier transform of the probability density function,

$$\phi_X(\omega) = E[\exp(jX\omega)] = \int_{-\infty}^{\infty} f_X(x) \exp(jx\omega) dx \tag{3.46}$$

Notice that the kernel is positive in Eq. (3.46), while the kernel is negative when Fourier transforms are defined in communication systems (Papoulis and Pillai 2002; Shankar 2017).

Knowing the characteristic function allows one to invert (inverse Fourier transform) it to obtain the probability density as

$$f(x) = \frac{1}{2\pi} \int_{-\infty}^{\infty} \phi_X(\omega) \exp(-jx\omega) d\omega \tag{3.47}$$

The moment generating function (MGF) is related to the characteristic function, and MGF is defined as

$$M_X(t) = E[\exp(Xt)] = \int_{-\infty}^{\infty} f_X(x) \exp(xt) dx \tag{3.48}$$

If the random variable exists only for positive values, it is also possible to define the Laplace transform associated with the random variable as (Block and Savits 1980; Shankar 2017)

$$L_X(s) = \int_0^\infty f_X(x) \exp(-xs)dx \qquad (3.49)$$

All these properties of continuous and discrete random variables are summarized later in this chapter.

It can be seen that for random variables such as the exponential, Rayleigh, or gamma which exist in the range $\{0, \infty\}$, the MGF can be expressed in terms of the Laplace transforms as

$$M_X(t) = L_X(-t) \qquad (3.50)$$

Moment generating functions are useful in obtaining the moments of random variables. Rewriting Eq. (3.48) leads to

$$M_X(t) = E[\exp(tX)] = E\left[1 + Xt + \frac{(Xt)^2}{2!} + \ldots + \frac{(Xt)^k}{k!} \ldots\right]$$

$$= 1 + tE(X) + \frac{t^2}{2}E(X^2) + \ldots \qquad (3.51)$$

From Eq. (3.51), it is clear that if $M_x(t)$ is expanded as a power series in t, the coefficients can provide the moments of X. The coefficient of t is the first moment, the coefficient of t^2 is half of the second moment, etc. As an example, we can express the MGF of an exponential random variable with a mean a as

$$M_X(t) = \frac{1}{1 - at} \qquad (3.52)$$

Expanding in Taylor series (Abramowitz and Segun 1972; Gradshteyn and Ryzhik 2000), the MGF in Eq. (3.52) is

$$M_X(t) = \frac{1}{1 - at} = 1 + at + a^2t^2 + 3a^3t^3 + \ldots \qquad (3.53)$$

By collecting the coefficients, we get the moments as

$$E(X) = \text{coefficient of } t = a$$
$$E(X^2) = 2! * \text{coefficient of } t = 2a^2 \qquad (3.54)$$

The concept of expectation can be extended to obtain the expected value of a function of a random variable. If $g(X)$ is a function of the random variable X, its expected value is

$$E[g(X)] = \int\limits_{-\infty}^{\infty} g(x) f_X(x) dx \tag{3.55}$$

It can be easily seen that the nth moment of a random variable in Eq. (3.36) is obtained from Eq. (3.55) by putting

$$g(X) = X^n \tag{3.56}$$

Some additional results on moments (**a** and **b** are constants) can be obtained by noting that expectation is a linear operation and that variance and standard deviation are always positive.

$$E[aX + b] = E[aX] + E[b] = aE[X] + b \tag{3.57}$$

$$\mathrm{var}(aX + b) = a^2 \mathrm{var}(X) \tag{3.58}$$

$$\mathrm{std.dev}(aX + b) = |a| \mathrm{std.dev}(X) \tag{3.59}$$

A couple of other interesting and practical properties and definitions associated with random variables are the survival rate and hazard rate.

The survival function or survival rate $S(x)$ is the probability of the event $\{X > x\}$,

$$S_X(x) = P(X > x) = 1 - F_X(x) \tag{3.60}$$

If the lifetime of an system is modeled as a random variable, $P\{X \leq x\}$ is the probability that system fails within the time frame of $\{0,x\}$. The *survival rate* $S(x)$ provides the probability that the system is functioning in a time frame $\{X > x\}$.

The hazard rate, hazard function, or failure rate $h(x)$ is the ratio of the pdf to the survival function,

$$h(x) = \frac{f(x)}{S(x)} = \frac{f(x)}{1 - F_X(x)} \tag{3.61}$$

It will be shown later in Sect. 3.5 that the failure rate is the conditional density.

Example 3.16 The density function of a random variable is

$$f_X(x) = \frac{1}{\pi\sqrt{1 - x^2}}, \quad -1 < x < 1$$

Find the first two moments and comment on the kth moment.

Solution

$$E(X) = \int_{-1}^{1} xf(x)dx = \int_{-1}^{1} x\frac{1}{\pi\sqrt{1-x^2}}dx = 0$$

$$E(X^2) = \int_{-1}^{1} x^2 f(x)dx = \int_{-1}^{1} x^2 \frac{1}{\pi\sqrt{1-x^2}}dx = \frac{1}{2}$$

$$E(X^2) = 2\int_{0}^{1} x^2 \frac{1}{\pi\sqrt{1-x^2}}dx = \frac{1}{2}$$

It can be seen that the odd moments ($k = 1,3,5,..$) will be zero because the pdf is an even function. The pdf is symmetric about $x = 0$. The even moments are obtained as

$$E(X^k) = 2\int_{0}^{1} x^k \frac{1}{\pi\sqrt{1-x^2}}dx = \frac{1}{\pi}\beta\left(\frac{1}{2}, \frac{k+1}{2}\right), k = 2, 4, 6, \ldots$$

Note that $\Gamma(m)$ is the gamma function of m and $\beta(m,n)$ is the beta function (Gradshteyn and Ryzhik 2000) expressed as

$$\beta(m,n) = \int_{0}^{1} t^{m-1}(1-t)^{n-1}dt = \frac{\Gamma(m)\Gamma(n)}{\Gamma(m+n)}$$

Example 3.17 A discrete random variable has the pdf (Example 3.10)

$$f_X(x) = \frac{8}{20}\delta(x+2) + \frac{3}{20}\delta(x+1) + \frac{5}{20}\delta(x-1) + \frac{4}{20}\delta(x-2)$$

Obtain its mean and variance.

Solution

$$E(X) = \sum_{k} P(X = x_k)x_k = \frac{8}{20}(-2) + \frac{3}{20}(-1) + \frac{5}{20}(1) + \frac{4}{20}(2) = -\frac{3}{10}$$

$$E(X^2) = \sum_{k} P(X = x_k)x_k^2 = \frac{8}{20}(-2)^2 + \frac{3}{20}(-1)^2 + \frac{5}{20}(1)^2 + \frac{4}{20}(2)^2 = \frac{28}{10}$$

$$\text{var}(X) = \frac{28}{10} - \left(-\frac{3}{10}\right)^2 = 2.71$$

Example 3.18 A random variable has the pdf

$$f_X(x) = \frac{2}{x^3}, x > 1$$

Obtain its mean and second moment.

Solution

$$E(X) = \int\limits_1^\infty \frac{2}{x^3} x\,dx = 2$$

$$E(X^2) = \int\limits_1^\infty \frac{2}{x^3} x^2\,dx = \infty$$

Example 3.19 A random variable has the pdf

$$f_X(x) = \exp(-x)U(x)$$

Obtain the mean of

$$Y = e^{-2X}$$
$$W = 3X^2 + 2X + 3$$

Solution

$$E(Y) = \int\limits_0^\infty yf(x)dx = \int\limits_0^\infty e^{-2x}e^{-x}dx = \frac{1}{3}$$

$$E(W) = E(3X^2 + 2X + 3) = E(3X^2) + E(2X + 3) + E(3)$$

$$E(W) = 3\int\limits_0^\infty x^2 f(x)dx + 2\int\limits_0^\infty xf(x)dx + 3\int\limits_0^\infty f(x)dx = 3\int\limits_0^\infty x^2 e^{-x}dx + 2\int\limits_0^\infty xe^{-x}dx + 3\int\limits_0^\infty e^{-x}dx$$

$$E(W) = 3*2 + 2*1 + 3*1 = 11$$

Example 3.20 A random variable has the pdf

$$f_X(x) = \frac{1}{3}\exp\left(-\frac{x}{3}\right)U(x)$$

Obtain its first, second, third, and k^{th} ($k = 1,2,3,..$) moments. Also obtain its characteristic function, moment generating function, and Laplace transform. Establish any relationship that may exist among these functions and the moments.

Solution

$$E(X) = \int_0^\infty xf(x)dx = 3$$

$$E(X^2) = \int_0^\infty x^2 f(x)dx = 18 = 2 * 3^2 = 3^2 * 2! = 3^2\Gamma(3)$$

$$E(X^3) = \int_0^\infty x^3 f(x)dx = 162 = 6 * 3^3 = 3^3\Gamma(4)$$

$$E(X^k) = \int_0^\infty x^k f(x)dx = 3^k\Gamma(k+1)$$

The characteristic function is

$$\phi_X(\omega) = E\left(e^{i\omega X}\right) = \int_0^\infty e^{j\omega x}\frac{1}{3}e^{-\frac{x}{3}}dx = \frac{1}{1-3j\omega}$$

$$M_X(t) = E\left(e^{tX}\right) = \int_0^\infty e^{tx}\frac{1}{3}e^{-\frac{x}{3}}dx = \frac{1}{1-3t}$$

Note that we can expand the exponential term within $E(.)$ in series and write the MGF as

$$M_X(t) = E\left(e^{tX}\right) = E\left[1 + tX + \frac{t^2}{2!}X^2 + \ldots + \frac{t^k}{k!}X^k + \ldots\right]$$

$$= 1 + tE(X) + \frac{t^2}{2}E\left(X^2\right) + \frac{t^3}{3!}E\left(X^3\right) + \ldots + \frac{t^k}{k!}E\left(X^k\right) + \ldots$$

Noting that the power series expansion for $(1-x)^{-1}$ (Abramowitz and Segun 1972),

$$\frac{1}{1-x} = 1 + x + x^2 + x^3 + \ldots$$

we can write the MGF directly as

$$M_X(t) = \frac{1}{1-3t} = 1 + 3t + (3t)^2 + (3t)^3 + \ldots + (3t)^k + \ldots$$

This means that

$$1 + tE\left(X\right) + \frac{t^2}{2}E\left(X^2\right) + \frac{t^3}{3!}E\left(X^3\right) + \ldots + \frac{t^k}{k!}E\left(X^k\right) + \ldots$$

$$= 1 + 3t + \left(3t\right)^2 + \left(3t\right)^3 + \ldots + \left(3t\right)^k + \ldots$$

Equating the terms in powers of t in the general expression for the MGF and the MGF of the density given above, we have

$$E(X) = 3$$
$$E(X^2) = 2 * 3^2 = 3^2 2! = 3^2 \Gamma(3)$$
$$E(X^3) = 6 * 3^3 = 3^3 \Gamma(4)$$
$$E(X^k) = k! * 3^k = 3^k \Gamma(k+1)$$

As we can see, if the MGF or the characteristic function exists, MGF or the characteristic function provides all the moments.

The Laplace transform of the density is

$$L_X(s) = \int\limits_0^\infty e^{-xs} \frac{1}{3} e^{-\frac{x}{3}} dx = \frac{1}{1 + 3s}$$

By replacing s by $(-t)$, we can see that the Laplace transform becomes the moment generating function. Thus, we can observe the relationships among the characteristic function, MGF, and the Laplace transform of the densities. It should be noted that it is not essential that all these exist. This aspect will become clear when we examine specific types of random variables and their density functions.

3.4 Continuous and Discrete Random Variables: Some Common Densities

We will now look at a number of densities commonly used in engineering applications.

Uniform Density

The uniform random variable is extensively used to describe the statistics of the phase measured. If a number is picked randomly in an interval $[a,b]$, the best model to describe the outcome is a uniformly distributed random variable $[a,b]$. The density associated with the uniform random variable is identified as the uniform density because the probability of picking a number in any range lying between $[a,b]$ is the same regardless of where the midpoint of the range of interest exists. For example, if b is 10 and a is 0, the probability of picking a number between 4 and 6 is the same as the probability of picking a number between 5 and 7 or between 8 and 10, hence the name uniform. The expression for the density is

$$f_X(x) = \frac{1}{b-a}, \quad a < x < b \tag{3.62}$$

The CDF is

$$F_X(x) = \int_a^x f(x)dx = \begin{cases} 0, x < a \\ \dfrac{x-a}{b-a}, & 0 < x < b \\ 1, x > b \end{cases} \tag{3.63}$$

The mean and variance, respectively, are

$$E(X) = \frac{a+b}{2}$$
$$\mathrm{var}(X) = \frac{(b-a)^2}{12} \tag{3.64}$$

Note that the uniform density has symmetry around the point $(a+b)/2$, and therefore, its mean is $(a+b)/2$.

Triangular Density

Triangular densities occur in engineering applications when the sum of two uniform random variables is examined. It can be shown that if two uniformly distributed independent random variables, each uniform in $[-a,a]$, are added, the density of the sum is considered to be the triangular density. The pdf and CDF are given as

$$f_X(x) = \begin{cases} \left(\dfrac{1}{2a}\right)\left(1 + \dfrac{x}{2a}\right), & -2a < x < 0 \\ \left(\dfrac{1}{2a}\right)\left(1 - \dfrac{x}{2a}\right), & 0 < x < 2a \\ 0, \text{otherwise} \end{cases} \tag{3.65}$$

$$F_X(x) = \begin{cases} 0, & x < -2a \\ \dfrac{(x+2a)^2}{8a^2}, & -2a < x < 0 \\ \dfrac{1}{2} + \left(\dfrac{4ax - x^2}{8a^2}\right), & 0 < x < 2a \\ 1, & x > 2a \end{cases}$$

The first two moments are

$$E(X) = 0$$
$$E(X^2) = \text{var}(X) = \frac{2}{3}a^2 \tag{3.66}$$

Exponential Density
Exponential density is used to describe a wide variety of phenomena such as the decay of radiation, duration of telephone calls, waiting time in a restaurant, power received at a wireless receiver, elapsed time between successive occurrence of certain events, lifetime of components such as computers, light bulbs, etc. The pdf and CDF are

$$f_X(x) = \frac{1}{a} \exp\left(-\frac{x}{a}\right) U(x), \quad a > 0 \tag{3.67}$$

$$F_X(x) = \left[1 - \exp\left(-\frac{x}{a}\right)\right], x > 0 \tag{3.68}$$

The mean and the variance are

$$E(X) = a$$
$$\text{var}(X) = a^2 \tag{3.69}$$

The general moments $(k > 0)$ are

$$E(X^k) = a^k \Gamma(k+1) \tag{3.70}$$

The Laplace transform of the density is

$$L_X(s) = \frac{1}{(1+as)} \tag{3.71}$$

It should be noted that the exponential density has the unique property that the ratio of the mean to standard deviation is always unity. Depending on the nature of the model, the parameter a may be the average power (density of power), average lifetime (density of lifetime), average call duration (density of call duration), etc. The density is represented as E(a), and the parameter a is the mean when the density is expressed in a form given in Eq. (3.67). The other property of the exponential density, namely, the memoryless property, will be discussed after the presentation of the concept of conditional densities.

Sometimes the exponential density is also expressed as

$$f_X(x) = \lambda \exp(-\lambda x)U(x), \quad \lambda > 0 \tag{3.72}$$

For the density in Eq. (3.72), the mean will be $\left(\frac{1}{\lambda}\right)$.

Throughout this book, exponential density will be represented as $E(a)$ in Eq. (3.67).

Gamma Density
While exponential density is used to model lifetimes and power, the flexibility of modeling can be expanded using the gamma density. The gamma pdf is expressed as $G(a,b)$ given by

$$G(a,b) = \frac{x^{a-1}}{b^a \Gamma(a)} \exp\left(-\frac{x}{b}\right) U(x), \quad a,b > 0 \tag{3.73}$$

In Eq. (3.73), $\Gamma(a)$ is the gamma function (Gradshteyn and Ryzhik 2000). The CDF of the gamma random variable is

$$F_X(x) = 1 - \frac{\Gamma\left(a, \frac{x}{b}\right)}{\Gamma(a)} \tag{3.74}$$

In Eq. (3.74), $\Gamma\left(a, \frac{x}{b}\right)$ is the upper incomplete gamma function given by (Abramowitz and Segun 1972; Gradshteyn and Ryzhik 2000)

$$\Gamma\left(a, \frac{x}{b}\right) = \int_{\frac{x}{b}}^{\infty} e^{-y} y^{a-1} dy \tag{3.75}$$

The lower incomplete gamma function is $\gamma\left(a, \frac{x}{b}\right)$ given by

$$\gamma\left(a, \frac{x}{b}\right) = \int_{0}^{\frac{x}{b}} e^{-y} y^{a-1} dy \tag{3.76}$$

The Laplace transform of the pdf is (Shankar 2017)

$$L_X(s) = \frac{1}{(1+bs)^a} \tag{3.77}$$

It can be seen that when $a = 1$, the gamma density becomes an exponential density.

$$G(1,b) = E(b) \tag{3.78}$$

The gamma random variable associated with the density in Eq. (3.73) has an order of a. It is also identified as the shape parameter. The moments of the gamma density are given by

$$E(X^k) = b^k \frac{\Gamma(a+k)}{\Gamma(a)} \tag{3.79}$$

Gamma density is often represented in a slightly different form in wireless when Nakagami fading is present (Nakagami 1960; Papoulis and Pillai 2002; Shankar 2015, 2017). In this case,

$$G\left(m, \frac{\Omega}{m}\right) = \left(\frac{m}{\Omega}\right)^m \frac{x^{m-1}}{\Gamma(m)} e^{-\frac{m}{\Omega}x} U(x), \quad m \geq \frac{1}{2} \tag{3.80}$$

With the gamma density expressed in Eq. (3.80), the mean is

$$E(X) = \Omega \tag{3.81}$$

Even though the value of m $\geq$ ½ in Eq. (3.81), it is not necessary to have this limitation. The only limitation is that $m > 0$.

While gamma density offers a wider range of statistical modeling when used in place of the exponential density, there are a few additional special cases of the gamma density. These are the Erlang density and chi-squared density.

Erlang Density
When the order of the gamma density is an integer, the gamma density is identified as the Erlang density, commonly used in capacity calculations in computer and wireless networks (Papoulis and Pillai 2002). The Erlang pdf is

$$G(c, b) = \frac{x^{c-1}}{b^c(c-1)!} \exp\left(-\frac{x}{b}\right) U(x), \quad b > 0, c = 1, 2, 3, \ldots \tag{3.82}$$

Chi-Squared or Chi Square Density
Chi-squared density is used in statistical testing (goodness of fit). It is given by

$$G\left(\frac{\nu}{2}, 2\right) = \frac{x^{\frac{\nu}{2}-1}}{2^{\frac{\nu}{2}}\Gamma\left(\frac{\nu}{2}\right)} e^{-\frac{x}{2}} U(x), \quad \nu = 1, 2, 3, \ldots \tag{3.83}$$

The parameter ν represents the degrees of freedom of the chi-squared variate (Rohatgi and Saleh 2001; Papoulis and Pillai 2002). The mean of the chi-squared variate is 2, and the variance is 2.

Generalized Gamma Density

The modeling of lifetimes and power can be improved if the gamma density becomes the generalized gamma density. The pdf of the generalized gamma is given as

$$G(a, b, k) = f_X(x) = \frac{kx^{ka-1}}{b^{ka}\Gamma(a)} \exp\left(-\left(\frac{x}{b}\right)^k\right)U(x), \quad k, a, b > 0 \qquad (3.84)$$

Note that

$$G(a, b, 1) = G(a, b) \qquad (3.85)$$

The generalized density also is expressed in different forms, and it is often referred to as the Stacy density (Stacy 1962; Shankar 2017).

Another variation is

$$f_X(x) = \frac{x^{\frac{a}{k}-1}}{kb^a\Gamma(a)} \exp\left(-\frac{x^{\frac{1}{k}}}{b}\right)U(x), \quad k, a, b > 0 \qquad (3.86)$$

Weibull Density

Another density function used to describe the fading in wireless systems, lifetime of components, reliability of large-scale and small-scale systems, etc. is the Weibull density. The Weibull density is

$$f_X(x) = \frac{\beta}{\eta^\beta} x^{\beta-1} \exp\left(-\left(\frac{x}{\eta}\right)^\beta\right)U(x), \quad \beta, \eta > 0 \qquad (3.87)$$

The Weibull density becomes an exponential density when the shape parameter β is unity.

Rayleigh Density

While most of the densities described so far are used to model power or intensity (besides other observable parameters), Rayleigh density is generally used to model the amplitude (positive) in communication systems (wireless less, radar, sonar). The Rayleigh density is

$$R(b) = f_X(x) = \frac{x}{b^2} \exp\left(-\frac{x^2}{2b^2}\right)U(x), b > 0 \qquad (3.88)$$

The CDF of the Rayleigh random variable is

$$F_X(x) = 1 - \exp\left(-\frac{x^2}{2b^2}\right) \tag{3.89}$$

The mean and the second moment of the Rayleigh density are

$$E(X) = b\sqrt{\frac{\pi}{2}}$$
$$E(X^2) = 2b^2 \tag{3.90}$$

Just as in the case of exponential density characterized by the ratio of mean to standard deviation being equal to unity, the Rayleigh density is characterized by the ratio of mean to standard deviation being equal to 1.91.

Nakagami Density
A density related to the Rayleigh is the Nakagami density for the amplitude or envelope signals expressed as (Nakagami 1960)

$$f(x) = \frac{2}{\Gamma(m)}\left(\frac{m}{\Omega}\right)^m x^{2m-1} \exp\left(-\frac{m}{\Omega}x^2\right), \quad x \geq 0, \; m \geq \frac{1}{2}, \Omega > 0 \tag{3.91}$$

$$E(X^k) = \frac{\Gamma\left(m+\frac{k}{2}\right)}{\Gamma(m)}\left(\frac{\Omega}{m}\right)^{\frac{k}{2}} \tag{3.92}$$

$$E(X^2) = \Omega \tag{3.93}$$

$$m = \frac{E^2(X^2)}{\text{var}(X^2)} \tag{3.94}$$

The Nakagami density in Eq. (3.91) offers a more general way of modeling outcomes in place of the Rayleigh density. It can be easily seen that when $m = 1$, the Nakagami density becomes the Rayleigh density. Eq. (3.91) is the Nakagami density associated with the amplitude, and the power described using this model will follow the density given in Eq. (3.80). If X is the amplitude, X^2 represents the power. This implies that if the amplitude is Rayleigh distributed, power will be exponentially distributed.

Rician Density
Another density used to model the amplitude is the Rician density with a pdf given by

$$f(x) = \frac{x}{\sigma^2} \exp\left(-\frac{s^2 + x^2}{2\sigma^2}\right) I_0\left(\frac{s}{\sigma^2}x\right), \quad x \geq 0, s \geq 0 \tag{3.95}$$

In Eq. (3.95), $I_0(.)$ is the modified Bessel function of the first kind of order 0 (Abramowitz and Segun 1972).

$$E\left(X^2\right) = s^2 + 2\sigma^2 \qquad E\left(X^4\right) = 8s^2\sigma^2 + s^4 + 8\sigma^2 \tag{3.96}$$

When $s = 0$, Eq. (3.95) becomes the Rayleigh density. A version of the Rician density is also used in describing the statistics of the power or intensity (Shankar 2017). Rician density is described in Chap. 4.

Lognormal Density

A density function that is used to describe the power fluctuations in wireless systems or the lifetimes of components is the lognormal density. The lognormal density has the pdf (Limpert et al. 2001; Shankar 2015, 2017)

$$f_X(x) = \frac{K}{x\sqrt{2\pi\sigma^2}} \exp\left(-\frac{\left(10\log_{10}(x) - \mu\right)^2}{2\sigma^2}\right) U(x) \tag{3.97}$$

In Eq. (3.97),

$$K = \frac{10}{\log_e(10)} \tag{3.98}$$

Furthermore, both σ and μ are in decibel units.

While all the densities described so far with the exception of the uniform density exist only for positive values of the observables, some of the measurements also result in values that lie in the range of $-\infty$ to ∞. We will now look at a few examples of such densities.

Gaussian Density

The normal or the Gaussian density is used to model noise in electronic systems. The pdf of the Gaussian random variable is

$$N\left(\mu, \sigma^2\right) = \frac{1}{\sqrt{2\pi\sigma^2}} \exp\left(-\frac{(x-\mu)^2}{2\sigma^2}\right), \quad -\infty < x < \infty \tag{3.99}$$

The mean of the Gaussian variable is μ and the variance is σ^2. The CDF is given by

$$F_X(x) = \frac{1}{2}\left[1 - \mathrm{erf}\left(\frac{\mu - x}{\sigma\sqrt{2}}\right)\right] = \frac{1}{2}\left[1 + \mathrm{erf}\left(\frac{x - \mu}{\sigma\sqrt{2}}\right)\right] \tag{3.100}$$

In Eq. (3.100), erf(.) is the error function (Gradshteyn and Ryzhik 2000). The characteristic function of the Gaussian variable is

$$\phi_X(\omega) = \exp\left(-\frac{\omega^2\sigma^2}{2} + jm\omega\right) \tag{3.101}$$

Laplacian Density

Laplacian density is also used to model noise (spikier than Gaussian noise) under certain conditions. The probability density function is

$$f_X(x) = \frac{1}{2b} \exp\left(-\frac{|x-a|}{b}\right) \tag{3.102}$$

The cumulative distribution function is

$$F_X(x) = \begin{cases} \frac{1}{2} \exp\left(-\frac{(a-x)}{b}\right), x < a \\ 1 - \frac{1}{2} \exp\left(-\frac{(x-a)}{b}\right), x > a \end{cases} \tag{3.103}$$

The mean and variance of the Laplacian are

$$E(X) = a \tag{3.104}$$

$$\text{var}(X) = 2b^2 \tag{3.105}$$

The characteristic function is

$$\phi_X(\omega) = \frac{\exp(ia\omega)}{1 + b\omega^2} \tag{3.106}$$

Cauchy Density

Cauchy density is unique. Its moments are not defined. The pdf is

$$f_X(x) = \frac{1}{\pi b\left[1 + \frac{(x-a)^2}{b^2}\right]} \tag{3.107}$$

The CDF is

$$F_X(x) = \frac{1}{2} + \frac{1}{\pi} \tan^{-1}\left(\frac{x-a}{b}\right) \tag{3.108}$$

The characteristic function is

$$\phi_X(\omega) = \exp(ia\omega - |\omega|b) \tag{3.109}$$

We looked at several continuous random variables. A number of discrete random variables also find application in engineering.

Binomial Distribution

The discrete random variable with the binomial distribution arises when we study the outcomes of Bernoulli trails. The PMF is

$$P(X = k) = C_k^n p^k (1 - p)^{n-k}, k = 0, 1, 2, \ldots, n \tag{3.110}$$

The CDF is

$$F_X(x) = \sum_{k=0} C_k^n p^k (1 - p)^{n-k} U(X - k) \tag{3.111}$$

In Eq. (3.110), n is the number of trials with binary outcomes of success or failure, and p is the probability of success. The mean and variance are

$$\begin{aligned} E(X) &= np \\ \mathrm{var}(X) &= np(1 - p) \end{aligned} \tag{3.112}$$

Geometric Distribution

Geometric distribution arises when we conduct an experiment until we see success. If success is seen at the nth trial, the random variable that describes the outcome is the geometric random variable. If p is the probability of success, the PMF of a geometric random variable is

$$P(X = n) = p(1 - p)^{n-1}, \quad n = 1, 2, 3, \ldots \tag{3.113}$$

The CDF is

$$F_X(x) = \sum_n p(1 - p)^{n-1} U(X - n) \tag{3.114}$$

An interesting result is the one obtained when the random variable takes the values from $n = 1$ to $n = k$. The probability that the variables takes the values from $n = 1$ to $n = k$ will be summation of the quantity in Eq. (3.113) from $n = 1$ to $n = k$ as

$$F_X(k) = \sum_{n=1}^{k} p(1 - p)^n = 1 - (1 - p)^k \tag{3.115}$$

The mean and variance of the geometric random variable is

$$\begin{aligned} E(X) &= \frac{1}{p} \\ \mathrm{var}(X) &= \frac{1 - p}{p^2} \end{aligned} \tag{3.116}$$

Negative Binomial Distribution

The notion of the geometric random variable can be generalized to a case when we conduct an experiment until we have success r times (instead of the first success for the geometric variable). The PMF is

$$P(X = n) = C_{r-1}^{n-1} p^r (1 - p)^{n-r}, \quad n = r, r + 1, \ldots \tag{3.117}$$

The mean and variance are

$$E(X) = \frac{r}{p}$$
$$\text{var}(X) = \frac{r(1 - p)}{p^2} \tag{3.118}$$

Poisson Distribution

The Poisson random variable arises when we are modeling the outcomes such as the emission of electrons, photons, or other particles. It is also used to model the number of calls received by a call center for assistance. The PMF of a Poisson random variable is

$$P(X = k) = \frac{\lambda^k}{k!} \exp(-\lambda), \quad \lambda > 0, \quad k = 0, 1, 2, 3, \ldots \tag{3.119}$$

In Eq. (3.119), λ is the average number of events in the observation interval. The Poisson random variable has the unique property that its mean and variance are equal.

$$E(X) = \text{var}(X) = \lambda \tag{3.120}$$

An overview of the densities and their properties is given in Sect. 3.10.

3.5 Conditional Densities and Conditional Distribution Functions

Often in statistical analysis, one has prior information on the outcomes. For example, someone observes that the signal measured always appears to be larger than (or smaller than, as the case may be) a certain value even though the random variable can take any value (from $-\infty$ to ∞). Such prior knowledge of the trends allows us to introduce the concept of conditional distribution functions and conditional densities. Let X be a random variable with a pdf $f_X(x)$ and CDF $F_X(x)$. Let us define A as the event

$$A = \{X > a\} \tag{3.121}$$

Let us treat this as prior knowledge regarding the experiment. If so, we are interested in finding out the conditional CDF expressed as

$$F_{X|A}(x|A) = P(X \le x | X > a) \tag{3.122}$$

Using the concept of conditional probabilities introduced in Chap. 2, we can write Eq. (3.122) as

$$F_{X|A}(x|A) = P(X \le x | X > a) = \frac{P(X \le x, X > a)}{P(X > a)} = \frac{P(X \le x, X > a)}{1 - F_x(a)} \tag{3.123}$$

The numerator of Eq. (3.123) is the joint probability, and the denominator is the probability associated with the known event (marginal probability or a priori probability). It is clear that if the random variable takes values less than a (i.e., $X < x$, with $x = a$), the joint probability in the numerator will be zero since the events $\{X < a\}$ and $\{X > a\}$ are mutually exclusive. If $X < x$ with $x > a$, there will be common events, and the numerator will be non-zero. This is shown by the patterned area in Fig. 3.3 between $X = x$ and $X = a$.

$$F_{X|A}(x|A) = \frac{P(X \le x, X > a)}{1 - F_x(a)} = \begin{cases} 0, & x < a \\ \dfrac{F_X(x) - F_X(a)}{1 - F_x(a)}, & x \ge a \end{cases} \tag{3.124}$$

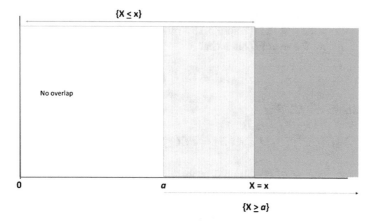

Fig. 3.3 Concept of the conditional density and regions of interest $\{X > a\}$. Regions are shown for demonstration only (lower limit taken to be 0). The shape of the pdf is chosen to be a rectangle even though the analysis is valid regardless of the shape of the pdf so long as it exists from $-\infty$ to ∞. Conditioning event is $A = \{X > a\}$

The conditional density is obtained by differentiating the CDF in Eq. (3.124) as

$$f_{X|A}(x|A) = \frac{d}{dx}\left[F_{X|A}(x|A)\right] = \begin{cases} 0, & x < a \\ \dfrac{f_X(x)}{1 - F_x(a)}, & x \geq a \end{cases} \tag{3.125}$$

Note that the conditional density exists only when the random variable takes values larger than a. We can verify that the conditional density is a valid density by integration as

$$\int_{-\infty}^{\infty} f_{X|A}(x|A)dx = \int_{a}^{\infty} \frac{f_X(x)}{1 - F_x(a)}dx = \frac{1}{1 - F_x(a)}\left[1 - \int_{-\infty}^{a} f_X(x)dx\right] \tag{3.126}$$

Note that the quantity in the square bracket in Eq. (3.126) is

$$\left[1 - \int_{-\infty}^{a} f_X(x)dx\right] = 1 - F_X(a) \tag{3.127}$$

Using Eq. (3.127), Eq. (3.126) becomes

$$\int_{-\infty}^{\infty} f_{X|A}(x|A)dx = 1 \tag{3.128}$$

In other words, the conditional density is no different from any other pdf. Notice the similarity of the pdf in Example 3.15 to the density in Eq. (3.125)

Now, consider the event B such that

$$B = \{X < b\} \tag{3.129}$$

The conditional CDF is

$$F_{X|B}(x|B) = P(X \leq x|X < b) \tag{3.130}$$

Using the concept of conditional probabilities introduced earlier, we can write Eq. (3.130)as

$$F_{X|B}(x|B) = P(X \leq x|X < b) = \frac{P(X \leq x, X < b)}{P(X < b)} = \frac{P(X \leq x, x < b)}{F_x(b)} \tag{3.131}$$

The case is illustrated in Fig. 3.4.

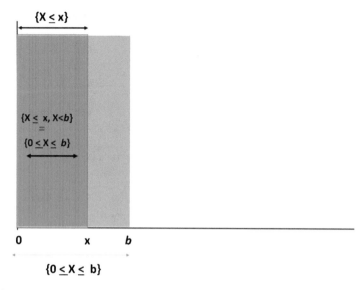

Fig. 3.4 Concept of conditional density and regions of interest $\{X < b\}$. Regions are shown for demonstration only. The shape of the pdf is chosen to be a rectangle even though the analysis is valid regardless of the shape of the pdf so long as it exists from $-\infty$ to ∞. Conditioning event is $B = \{X < b\}$

The conditional CDF becomes

$$F_{X|B}(x|B) = \frac{P(X \le x, X < b)}{F_x(b)} = \begin{cases} \dfrac{F_X(x)}{F_x(b)}, & x < b \\ \dfrac{F_X(b)}{F_x(b)}, & x \ge b \end{cases}$$

$$= \begin{cases} \dfrac{F_X(x)}{F_x(b)}, & x < b \\ 1, & x \ge b \end{cases} \tag{3.132}$$

The conditional density becomes

$$f_{X|B}(x|B) = \frac{d}{dx}\left[F_{X|B}(x|B)\right] = \begin{cases} \dfrac{f_X(x)}{F_x(b)}, & x < b \\ 0, & x \ge b \end{cases} \tag{3.133}$$

It can be easily observed that the conditional density integrates to unity,

$$\int_{-\infty}^{\infty} f_{X|B}(x|B)dx = \int_{-\infty}^{b} \frac{f_X(x)}{F_x(b)}dx = \frac{1}{F(b)}\int_{-\infty}^{b} f_X(x)dx = \frac{F(b)}{F(b)} = 1 \tag{3.134}$$

One can now examine the most general case where the known event (conditioning event) C is (a and b are arbitrary values with $b > a$)

$$C = \{a < X < b\} \tag{3.135}$$

The probability of this event is

$$P(C) = F_X(b) - F_X(a) \tag{3.136}$$

The conditional CDF now becomes

$$F_{X|C}(x \mid C) = \frac{P(X \le x, a < X < b)}{P(a < X < b)} = \begin{cases} 0, & x < a \\ \dfrac{F_X(x) - F_X(a)}{F_X(b) - F_X(a)}, & a < x < b \\ \dfrac{F_X(b) - F_X(a)}{F_X(b) - F_X(a)} = 1, & x > b \end{cases} \tag{3.137}$$

The conditional density becomes

$$f_{X|C}(x|C) = \frac{d}{d}\left[F_{X|C}(x|C)\right] = \begin{cases} 0, x < a \\ \dfrac{f_X(x)}{F_X(b) - F_X(a)}, & a < x < b \\ 0, x > b \end{cases} \tag{3.138}$$

All three conditional densities are illustrated in Fig. 3.8 for the case of a marginal density of

$$f_X(x) = \frac{x}{9}\exp\left(-\frac{x^2}{18}\right)U(x) \tag{3.139}$$

The value of $a = 3$ and $b = 5$ is chosen for this illustration. It is easily seen from the equations and Fig. 3.5 that the conditional densities have larger maxima than the marginal density.

It is worthwhile to note that conditional densities can also be expressed as marginal densities. For the conditional density (conditioning event, $X > 3$), we can write

$$f_Y(y) = \begin{cases} \dfrac{1}{\exp\left(-\dfrac{1}{2}\right)}\dfrac{y}{3^2}\exp\left(-\dfrac{y^2}{2(3)^2}\right), & y > 3 \\ 0, & \text{elsewhere} \end{cases} \tag{3.140}$$

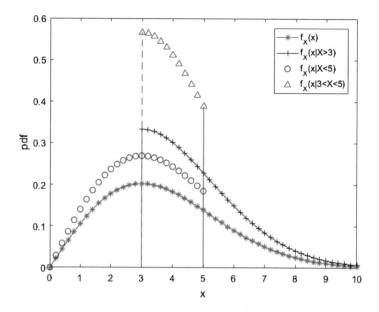

Fig. 3.5 Marginal and conditional densities for a Rayleigh density $f_X(x) = \dfrac{x}{3^2} e^{-\frac{x^2}{2(3^2)}} U(x)$

Equation (3.140) is equivalent to the conditional density in Eq. (3.125) with Y as the new random variable, defined as

$$Y = (X|X > 3) \tag{3.141}$$

Similarly, we can write

$$f_Z(z) = \begin{cases} \dfrac{1}{\left[1 - \exp\left(-\dfrac{5^2}{2(3)^2}\right)\right]} \dfrac{z}{3^2} \exp\left(-\dfrac{z^2}{2(3)^2}\right), & z < 5 \\ 0, & \text{elsewhere} \end{cases} \tag{3.142}$$

Equation (3.142) is equivalent to the conditional density in Eq. (3.133) with Z as a new random variable,

$$Z = (X|X \leq 5) \tag{3.143}$$

Additionally, we can write

$$f_W(w) = \begin{cases} \dfrac{1}{\left[\exp\left(-\dfrac{3^2}{2(3)^2}\right) - \exp\left(-\dfrac{5^2}{2(3)^2}\right)\right]} \dfrac{w}{3^2} \exp\left(-\dfrac{w^2}{2(3)^2}\right), & 3 \le w \le 5 \\ 0, & \text{elsewhere} \end{cases}$$

(3.144)

It is easily seen that Eq. (3.144) is equivalent to the conditional density in Eq. (3.138) with W as the new variable,

$$W = (X|3 \le X \le 5)$$

(3.145)

3.5.1   Total Probability and Bayes' Rule

We now revisit total probability calculations using Bayes' theorem. The total probability is

$$P(A) = \sum_k P(A|B_k)P(B_k)$$

(3.146)

Since A is an event, we can define A within the domain of the random variable X as

$$A = \{X \le x\}$$

(3.147)

Equation (3.147) helps us to rewrite Eq. (3.146) in terms of CDF as

$$P(A) = P(X \le x) = F_X(x) = \sum_k F_X(x|B_k)P(B_k)$$

(3.148)

Differentiating Eq. (3.148) with respect to x, we get an expression for the pdf as

$$f_X(x) = \sum_k f_X(x|B_k)P(B_k)$$

(3.149)

Note that $B_1, B_2, \ldots$ form mutually exclusive partitions of the universal set S. We can now write Bayes' rule for any two events, event C and the event $\{X \le x\}$, as

$$P(C|X \le x) = \frac{P(X \le x|C)P(C)}{P(X \le x)} = \frac{F(x|C)}{F(x)}P(C)$$

(3.150)

Expanding further by defining the conditioning event to be $x_1 \le X \le x_2$, Eq. (3.150) becomes

$$P(C|x_1 \le X \le x_2) = \frac{P(x_1 \le X \le x_2|C)P(C)}{P(x_1 \le X \le x_2)}$$

$$= \frac{F(x_2|C) - F(x_1|C)}{F(x_2) - F(x_1)}P(C) \qquad (3.151)$$

While $P\{X = x\} = 0$ when X is a continuous random variable, one can still define the conditional density. To accomplish this, we define using Eq. (3.28)

$$P[X = x] = P[x \le X \le x + \Delta x] = f(x)\Delta x|_{\Delta x \to 0} \qquad (3.152)$$

The conditional probability in Eq. (3.151) for the case of the conditioning event $\{X = x\}$ becomes

$$P(C|x \le X \le x + \Delta x) = \frac{F(x + \Delta x|C) - F(x|C)}{F(x + \Delta x) - F(x)}P(C)$$

$$= \frac{f(x|C)\Delta x}{f(x)\Delta x}P(C) \qquad (3.153)$$

Simplifying, Eq. (3.153) becomes

$$P(C|X = x) = P(C|x \le X \le x + \Delta x) = \frac{f(x|C)}{f(x)}P(C) \qquad (3.154)$$

It should be noted that the conditional CDF has all the properties of CDF and conditional density has all the properties of the probability density. This means that

$$\int_{-\infty}^{\infty} f(x|C)dx = 1 \qquad (3.155)$$

Rearranging Eq. (3.154), we have

$$P(C|X = x)f(x) = f(x|C)P(C) \qquad (3.156)$$

Integrating both sides of Eq. (3.156), we have

$$\int P(C|X = x)f(x)dx = \int f(x|C)P(C)dx = P(C)\int f(x|C)dx = P(C) \qquad (3.157)$$

We made use of Eq. (3.155) to simplify the right-hand side of Eq. (3.157). From Eq. (3.157), we can obtain an expression for $P(C)$ as

$$P(C) = \int P(C|X = x)f(x)dx \tag{3.158}$$

Using Eqs. (3.154) and (3.158), we have

$$f(x|C) = \frac{P(C|X = x)f(x)}{P(C)} = \frac{P(C|X = x)f(x)}{\int P(C|X = x)f(x)dx} \tag{3.159}$$

Equation (3.159) represents *Bayes' theorem for continuous random variables*. We will see applications of Bayes' theorem when we examine modeling of engineering systems.

3.5.2   Memoryless Property

The concept of conditional densities can be invoked to explore the memoryless property exhibited by exponential and geometric random variables.

Consider a computer chip with an average lifetime of T units modeled in terms of an exponential random variable. This means that the CDF of the lifetime X is

$$F_X(x) = 1 - \exp\left(-\frac{x}{T}\right) \tag{3.160}$$

If it is known that the computer chip is functioning beyond a time of T_1 units, we can find out the probability that it will function for another period of at least T_2 units. This probability is

$$P(X > T_1 + T_2 | X > T_1) = \frac{P(X > T_1 + T_2, X > T_1)}{P(X > T_1)}$$

$$= \frac{P(X > T_1 + T_2)}{P(X > T_1)} \tag{3.161}$$

Using the CDF, we have

$$P(X > T_1 + T_2 | X > T_1) = \frac{1 - F_X(T_1 + T_2)}{1 - F_X(T_1)} = \frac{\exp\left(-\frac{T_1 + T_2}{T}\right)}{\exp\left(-\frac{T_1}{T}\right)}$$

$$= \exp\left(-\frac{T_2}{T}\right) = P(X > T_2) \tag{3.162}$$

In other words, the probability that the system will function for at least another period of T_2 units does not depend on its past history. This is the memoryless property associated with the exponential random variable.

Let us consider the case of an experiment being done until we see success. If we have k successes so far, we can find the probability of observing at least another **r** successes. This is the case of a geometric variable and the probability of seeing k or less successes is given in Eq. (3.115). This value is

$$P(X \leq k) = 1 - (1 - p)^k \tag{3.163}$$

We are seeking the following probability:

$$P(X > r + k | X > k) = \frac{P(X > r + k, X > k)}{P(X > k)} = \frac{P(X > r + k)}{P(X > k)} \tag{3.164}$$

Using Eq. (3.163), the probability of having at least another r successes is

$$P(X > r + k | X > k) = \frac{P(X > r + k)}{P(X > k)} = \frac{(1 - p)^{r+k}}{(1 - p)^k} = (1 - p)^r$$

$$= P(X > r) \tag{3.165}$$

Once again, we see that having had *k* successes does not affect having at least another *r* successes. This is the memoryless property associated with the geometric random variable.

3.6   Uniform Density Revisited

Before we look at a few examples based on a number of densities described above, it is worthwhile to revisit the uniform density because it may be described as both a *continuous* density and a *discrete* density.

Among the continuous densities that have specific names, a broad means to classify the densities is based on the range of validity of the variable. Either the random variable exists only for positive values $(0 \rightarrow \infty)$, or it exists everywhere $(-\infty \rightarrow \infty)$. For example, it is understood that Rayleigh, exponential, and gamma densities imply a range of validity of $(0 \rightarrow \infty)$ and Gaussian, Laplacian, and Cauchy imply a range of validity of $(-\infty \rightarrow \infty)$. The exception to this broad demarcation is the case of a uniform random variable.

Uniform density implies that the probability density is a constant for the operating range of the random variable. It finds primary use in modeling the phase. If the phase is uniformly distributed in the range $\{0, 2\pi\}$, we have a random variable that only takes positive values, truncated at 2π. It is also possible to model the phase being uniform in the range $\{-\pi, \pi\}$ in which case the random variable takes both positive and negative truncated values. Another application of uniform density is the use of a uniform random number in the range $\{0,1\}$ as the generator for simulating other random numbers such as Gaussian, exponential, Rayleigh, etc.

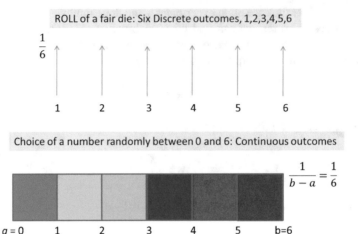

Fig. 3.6 Association between continuous uniform and discrete uniform

A uniform random variable can also be considered as a *discrete* random variable. This is explained within the context of discrete random variables. An example of a discrete uniform random variable is the case of a roll of a fair die. Since the probability of observing any one of the six numbers equal, the PMF of a discrete uniform random variable is

$$P(X = k) = \frac{1}{6}, k = 1, 2, 3, 4, 5, 6 \tag{3.166}$$

A pictorial description of the continuous uniform and discrete uniform is shown in Fig. 3.6. Two experiments are undertaken. First is the roll of a fair die. Second is the process of choosing a number randomly lying between $a = 0$ and $b = 6$.

Roll of die: $P(1) = P(2) = P(3) = P(4) = P(5) = P(6) = 1/6$.

Choice of a number: The numbers are continuous and they span a length of 6 as seen from Fig. 3.6. We are dealing with probability, and therefore, using the concept of areas, the experiment is modeled in terms of a rectangle of length 6 and height $(1/6)$ so that the area is 1. If X represents the number chosen, $P(0 < X < 6) = P$ $(0 \leq X < 6) = P(0 < X \leq 6) = P(0 \leq X \leq 6) =$ area of the rectangle $= 1$. This means that probability of picking a number between 1.5 and 2.5 is P $(1.5 < X < 2.5) = (2.5-1.5)/6 = 1/6$. Similarly, the probability of picking a number between 3 and 5 is $P(3 < X < 5) = (5-3)/6 = 2/6$.

In other words, $P(x < X < x + \Delta x) = \Delta x/6$. This means that P(number picked equals 5) $= P(5 < X < 5 + \Delta x) = \Delta x/6$ with $\Delta x \rightarrow 0$. Therefore, ***P(number picked = 5) = P(number picked = 4) = P(number picked = 0) = 0.***

Analog-to-digital conversion We conduct a third experiment from the second one, namely the continuous random variable. We take all the numbers between 0 and

1 and designate these as **A**, numbers between 1 and 2 as **B**, numbers between 2 and 3 as **C**, numbers between 3 and 4 as **D**, numbers between 4 and 5 as **E**, and numbers between 5 and 6 as **F**. (The random variable in this case is continuous, and therefore, equality is ***unimportant***.)

We have.

$P(A) = P(0 < X < 1) = 1/6$	$P(B) = P(1 < X < 2) = 1/6$	$P(C) = P(2 < X < 3) = 1/6$
$P(D) = P(3 < X < 4) = 1/6$	$P(E) = P(4 < X < 5) = 1/6$	$P(F) = P(5 < X < 6) = 1/6$

It is now possible to designate the set $[A, B, C, D, E, F]$ as outcomes $[1, 2, 3, 4, 5, 6]$ and we get the experiment of the roll of a fair die. We can thus establish the connection between rectangular discrete outcomes and rectangular continuous outcomes.

There is also another class of random variable referred to as the mixed variable (see Example 3.12 done earlier). It will be discussed again following the presentation of the transformation of random variables.

Example 3.21 Radar returns (power) are modeled in terms of the following (gamma) density:

$$f(x) = x \exp(-x) U(x)$$

(a) What is the probability that the received power exceeds 2 units?
(b) What is the probability that the received power is less than 6 units?
(c) What is the probability that the received power lies between 2 and 6 units?
(d) What is the probability that the received power is exactly 5 units?
(e) If past observations have shown that the received power is always more than 3 units, what is the probability that the received power will be less than 6 units?

Solution The CDF of the random variable is

$$F(x) = 1 - (1 + x) \exp(-x)$$

The probability that the received power exceeds 2 units $= 1 - F(2) = 3\exp(-2)$.
Probability that the power is less than 6 units $= F(6) = 1 - 7\exp(-6)$.
Probability that the power lies between 2 and 6 units $= F(6) - F(2) = 3\exp(-2) - 7\exp(-6)$.
Received power is exactly 5 units $= 0$; X is a continuous random variable. The probability that it takes a specific value is zero.
Probability that the received power is less than 6 units|power received is more than 3 units.

$$P(X < 6|X > 3) = \frac{P(X < 6, X > 3)}{P(X > 3)} = \frac{P(3 < X < 6)}{P(X > 3)} = \frac{F(6) - F(3)}{1 - F(3)}$$

$$P(X < 6|X > 3) = \frac{F(6) - F(3)}{1 - F(3)} = \frac{4\exp(-3) - 7\exp(-6)}{4\exp(-3)}$$

Example 3.22 A number is picked randomly in the range [0,7]. What is the probability that the number picked is between 3 and 5? What is the probability that the number picked is less than 4 knowing that the number usually picked appears to be in in the range [3,5]?

Solution X is a uniform variable with a pdf and CDF of

$$f(x) = \frac{1}{7}, 0 < x < 7$$
$$F(x) = \frac{x}{7}, 0 < x < 7$$

Probability that the number picked is between 3 and 5 $= F(5) - F(3) = 5/7 - 3/7 = 2/7$.

Probability that the number picked is less than 4|number is always between 3 and 5.

$$P(X < 4|3 < X < 5) = \frac{P(X < 4, 3 < X < 5)}{P(3 < X < 5)} = \frac{P(3 < X < 4)}{P(3 < X < 5)} = \frac{F(4) - F(3)}{F(5) - F(3)}$$
$$= \frac{1/7}{2/7} = \frac{1}{2}$$

Example 3.23 Speed on a highway is modeled as a Rayleigh random variable with an average speed of 60 MPH. (a) If speeds exceeding 75 are considered as excessive, obtain the pdf of the excessive speed of the drivers. What is the expected value of the excessive speed? (b) If there is a requirement that one should not drive below 40 MPH, obtain the pdf of these slow speed drivers who will be pulled over. What is the expected value of these slow speed drivers? (c) What is the probability density of the speed for the vehicles that obey the rules of the road (i.e., 40 MPH to 75 MPH)? What is the expected value of the speed of drivers who stay within 40 and 75 MPH?

Solution The speed is Rayleigh distributed with a density of

$$f(x) = \frac{x}{b^2} \exp\left(-\frac{x^2}{2b^2}\right) U(x)$$
$$F(x) = 1 - \exp\left(-\frac{x^2}{2b^2}\right)$$

We are given that

$$60 = b\sqrt{\frac{\pi}{2}} \qquad \Rightarrow b = 48MPH$$

(a) The conditional CDF of the speed given that the speed exceeds 75 MPH is

$$F(x|X > 75) = \frac{P(X < x, X > 75)}{P(X > 75)} = \begin{cases} 0, x < 75 \\ \dfrac{F_X(x) - F_X(75)}{1 - F_X(75)}, x > 75 \end{cases}$$

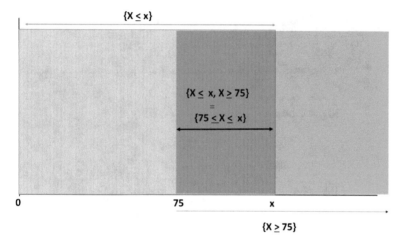

Differentiating, we have the pdf of the excessive speed as

$$f(x|X > 75) = \begin{cases} 0, x < 75 \\ \dfrac{f_X(x)}{1 - F_X(75)}, x > 75 \end{cases}$$

Using the CDF of the Rayleigh variable,

$$f(x|X > 75) = \frac{\frac{x}{b^2} \exp\left(-\frac{x^2}{2b^2}\right)}{\exp\left(-\frac{75^2}{2b^2}\right)} = \frac{x}{b^2} \exp\left(-\frac{(x^2 - 75^2)}{2b^2}\right), \qquad x > 75$$

$$E(X|X > 75) = \int_0^\infty xf(x|X > 75)dx = \int_{75}^\infty x\frac{x}{b^2} \exp\left(-\frac{(x^2 - 75^2)}{2b^2}\right)dx = 99.1MPH$$

(b) We are now seeking the speed with the condition that the speed is less than 40 MPH.

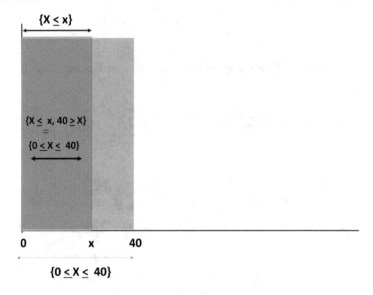

$$F(x|X < 40) = \frac{P(X < x, X < 40)}{P(X < 40)} = \begin{cases} \dfrac{F_X(x)}{F_X(40)}, x < 40 \\ \dfrac{F_X(40)}{F_X(40)}, x > 40 \end{cases}$$

The conditional pdf is

$$f(x|X < 40) = \begin{cases} \dfrac{f_X(x)}{F_X(40)}, x < 40 \\ 0, x > 40 \end{cases} = \begin{cases} \dfrac{\dfrac{x}{b^2} \exp\left(-\dfrac{x^2}{2b^2}\right)}{1 - \exp\left(-\dfrac{40^2}{2b^2}\right)}, & x < 40 \\ 0, & x > 40 \end{cases}$$

$$E(X|X < 40) = \int\limits_0^\infty xf(x|X < 40)dx = \int\limits_0^{40} x\frac{\dfrac{x}{b^2}\exp\left(-\dfrac{x^2}{2b^2}\right)}{\left(1 - \exp\left(-\dfrac{40^2}{2b^2}\right)\right)}dx = 25.73 \text{ MPH}$$

(c) We are now seeking the conditional density of the speed given that the speed is limited to the range 40–75 MPH.

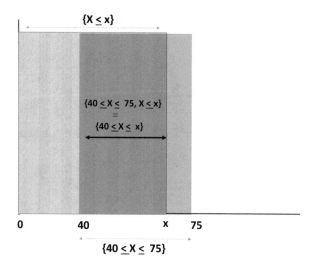

$$F(x|40 \leq X \leq 75) = \frac{P(X < x, 40 \leq X \leq 75)}{P(40 \leq X \leq 75)}$$

$$= \begin{cases} 0, & x < 40 \\ \dfrac{F_X(x) - F_X(40)}{F_X(75) - F_X(40)}, & 40 \leq x \leq 75 \\ 1, & x \geq 75 \end{cases}$$

Differentiating, we have

$$f(x|40 \leq X \leq 75) = \begin{cases} \dfrac{f_X(x)}{F_X(75) - F_X(40)}, & 40 \leq x \leq 75 \\ 0, & \text{elsewhere} \end{cases}$$

$$E(X|40 < X < 75) = \int\limits_0^\infty xf(x|40 < X < 75)dx$$

$$= \int\limits_{40}^{75} x \frac{x}{b^2} \frac{\exp\left(-\frac{x^2}{2b^2}\right)}{\exp\left(-\frac{40^2}{2b^2}\right) - \exp\left(-\frac{75^2}{2b^2}\right)} dx = 56.78 \text{ MPH}$$

This example is also solved using random number generation in Appendix B.

Example 3.24 The random variable X has the probability mass function (PMF) as

$$P[X = k] = \frac{c}{k}, \quad k = 2, 4, 8$$

What is the value of the constant c?
What is $E[X]$?

Solution If PMF is valid, $\displaystyle\sum_{k=2,4,8} \frac{c}{k} = 1 \Rightarrow c\left[\frac{1}{2} + \frac{1}{4} + \frac{1}{8}\right] \Rightarrow c\frac{7}{8} = 1 \Rightarrow c = \frac{8}{7}$

$$E(X) = \sum_k P(X = x_k)x_k = \sum_{k=2,4,8} \frac{c}{k}k = 3c = \frac{24}{7}$$

Example 3.25 The probability density function of (lifetime of an electronic component in years) X is

$$f_X(x) = Ax\exp\left(-\frac{x^2}{32}\right)U(x)$$

(a) What value of A will make this a valid pdf?
(b) What is the expression for CDF of X?
(c) What is the probability that the electronic component lasts beyond 5 years?
(d) What is the probability that the component lasts beyond 4 years, knowing it that it will last at least 2 years?
(e) What is the probability that it will fail within 6 years, given that it has always lasted beyond 3 years?
(f) What is the probability that it will fail within 6 years, given that normally these units tend to fail within 4–7 years?

Solution
(a) From simple observation, it is seen that X is Rayleigh distributed. Therefore, $A = (1/16)$.

$$F(x) = 1 - \exp\left(-\frac{x^2}{32}\right)$$

$$1 - F(5) = \exp\left(-\frac{25}{32}\right) = 0.4578$$

$$P[X > 4|X > 2] = \frac{P[X > 4, X > 2]}{P[X > 2]} = \frac{P[X > 4]}{P[X > 2]} = \frac{1 - F(4)}{1 - F(2)}$$

$$= \frac{\exp\left(-\frac{16}{32}\right)}{\exp\left(-\frac{4}{32}\right)} = \exp\left(-\frac{3}{8}\right) = 0.6873$$

$$P[X < 6 | X > 3] = \frac{P[X < 6, X > 3]}{P[X > 3]} = \frac{F(6) - F(3)}{1 - F(3)} = \frac{\exp\left(-\frac{9}{32}\right) - \exp\left(-\frac{36}{32}\right)}{\exp\left(-\frac{9}{32}\right)}$$

$$= 0.57$$

$$P[X < 6 | 4 < X < 7] = \frac{P[X < 6, 4 < X < 7]}{P[4 < X < 7]} = \frac{F(6) - F(4)}{F(7) - F(4)}$$

$$= \frac{\exp\left(-\frac{16}{32}\right) - \exp\left(-\frac{36}{32}\right)}{\exp\left(-\frac{16}{32}\right) - \exp\left(-\frac{49}{32}\right)} = 0.72$$

Example 3.26 X is a random variable used in modeling the received power in a wireless system (X is exponential). The average received power is 9 units. If amplitude is used as a means to make a decision on the received signal, what is the probability that the received signal amplitude will be less than 3 units?

Solution

$$f_X(x) = \frac{1}{9} \exp\left(-\frac{x}{9}\right)$$

Since amplitude is the square root of the power, we are seeking

$$P\left(\sqrt{X} < 3\right) = P(X < 9) = F(9) = 1 - \exp\left(-\frac{9}{9}\right) = 1 - e^{-1}$$

Example 3.27 X is $N(5,4)$. Obtain the following probabilities using the normal probability table in Sect. 3.9.7. In each case, express the Z_{score} to be used.

(a) $|X| > 5$

(b) $|X - 4| < 3$

(c) $-4 < X < 8$

(d) $X > 9$

(e) $3X + 5 > 10$

Solution   Since the CDF of a Gaussian is expressed in terms of error functions, one needs either access to Matlab, Maple, or Mathematica or the normal probability table. The Gaussian probability table is given in Sect. 3.9.7. The table contains the CDF values of a standard Gaussian, $N(0,1)$. Use of the Gaussian probability table necessitates the creation of the z-value or the z-score (see Sect. 3.9.7). The values start at $x = 0$. For $N(0, \sigma^2)$, the CDF can be expressed in terms of $F(z_{\text{score}})$ with

$$z_{score} = z = \frac{x - \mu}{\sigma}$$

$$F(z) = \frac{1}{\sqrt{2\pi}} \int\limits_{-\infty}^{z} \exp\left(-\frac{x^2}{2}\right) dx, \quad z \geq 0$$

$$F(z) = \frac{1}{2} + \frac{1}{\sqrt{2\pi}} \int\limits_{0}^{z} e^{-\frac{x^2}{2}} dx, \quad z \geq 0$$

When z is negative (see Fig. 3.40 in the Summary section),

$$F(-z) = \frac{1}{\sqrt{2\pi}} \int\limits_{-\infty}^{-z} \exp\left(-\frac{x^2}{2}\right) dx = 1 - F(z), \, z > 0$$

(a) $P(|X| > 5) = P(X < -5) + P(X > 5) = F_X(-5) + 1 - F_X(5) =$

$F_z\left(\dfrac{-5-5}{2}\right) + 1 - F_z\left(\dfrac{-5+5}{2}\right) = F_z(-5) + 1 - F_z(0) = 1 - F_z(5) + 1 - F_z(0)$

$= 1 - 1 + 1 - .5 = 0.5$

(b) $P(|X - 4| < 3) = P(-3 < X - 4 < 3) = P(1 < X < 7)$ add 4 to all terms

$P(|X - 4| < 3) = F_X(7) - F_X(1) = F_z\left(\dfrac{7-5}{2}\right) - F_z\left(\dfrac{1-5}{2}\right) = F_z(1) - F_z(-2)$

$P(|X - 4| < 3) = F_z(1) - \left[1 - F_z(2)\right] = 0.8413 - \left[1 - 0.9772\right]$

$= 0.8413 - 0.1228 = 0.8185$

(c) $P(-4 < X < 8) = F_X(8) - F_X(-4) = F_Z\left(\dfrac{3}{2}\right) - F_z(-4.5)$

$P(-4 < X < 8) = F_z(1.5) - \left[1 - F_z(4.5)\right] = 0.9332 - 0 = 0.9332$

(d) $P(X > 9) = 1 - F_X(9) = 1 - F_z(2) = 1 - 0.9772 = 0.0228$

(e) $P(3X + 5 > 10) = P(3X > 5) = 1 - F_X\left(\dfrac{5}{3}\right) = 1 - F_z\left(\dfrac{\dfrac{5}{3} - 5}{2}\right)$

$= 1 - F_z(-1.66) = F_z(1.66) = 0.9515$

Example 3.28 The number of computers being serviced is a Poisson random variable with an average of 8/day. What is the probability that number of computer serviced is exactly equal to 8, at least 8 or at most 8?

Solution The mean, $\lambda = 8$

Probability that exactly 8 computers are serviced in a day: $P(X = 8) = \frac{8^k}{8!}e^{-8}$

Probability that the number of computers serviced is at most 8: $p_1 = \sum\limits_{k=0}^{8} \frac{8^k}{k!}e^{-8}$.

Probability that the number of computers serviced is at least 8: $p_2 = 1 - \sum\limits_{k=0}^{7} \frac{8^k}{k!}e^{-8}$.

These summations may be evaluated using symbolic toolbox in Matlab as shown in Appendix B.

Example 3.29 X is a uniform variable in the range $[-2,3]$. What is the probability that $X^2 \leq 4$?

Solution

$$f_X(x) = \frac{1}{5}, \quad -2 < x < 3$$

$$F_X(x) = \begin{cases} 0, x < -2 \\ \dfrac{x+2}{5}, \quad -2 < x < 3 \\ 1, x > 3 \end{cases}$$

$$P(X^2 < 4) = P(-2 < X < 2) = F_X(2) - F_X(-2) = F_X(2) - 0 = F_X(2) = \frac{4}{5}$$

Note that equality sign is not important because X is a continuous random variable.

Example 3.30 X is an exponential random variable with a mean a. What is $E(X|X > w), \quad w > 0$?

Solution Note that $X|X > w$ is a conditioned variable. Therefore, the conditioned density is

$$f(x|X > w) = \frac{f_X(x)}{1 - F_X(w)} = \frac{1}{a}\frac{\exp\left(-\frac{x}{a}\right)}{\exp\left(-\frac{w}{a}\right)}, \quad x > w$$

$$E(X|X > w) = \int\limits_{w}^{\infty} xf(x|X > w)dx = \int\limits_{w}^{\infty} x\frac{1}{a}\frac{\exp\left(-\frac{x}{a}\right)}{\exp\left(-\frac{w}{a}\right)}dx = a + w$$

```
%% verification using simulation: choose a=3 and w=4;
xx=exprnd(3,1,1e6);
xw=xx(xx>4);% conditioned set of random numbers
mean(xw)
ans: 6.9955
```

3.7 Transformation of Random Variables

In engineering problems, often the measured quantity needs to be further processed. For example, if signal strength is measured as voltage or current, there is an interest in determining the statistics of the intensity or power. Given the real and imaginary parts of a signal, it may be of interest to determine the statistics of the phase. Another example would be the determination of the statistics of the power in decibel units from the statistics of the power measured in W or mW. Other examples of interest include are half-wave rectifiers, full-wave rectifiers, clippers, limiters, A-D converters, aggregators (discrete samples measured in several levels are aggregated or compressed to fewer levels), etc.

Transformation may be classified in different ways and terms. If X is the input and Y is the output, we can write the transformation in Fig. 3.7 as

$$Y = g(X) \tag{3.167}$$

If Eq. (3.167) only has a single solution, the transformation is monotonic. Examples will be

$$Y = aX + b$$
$$Y = \frac{c}{X} \tag{3.168}$$

In Eq. (3.168), a, b, and c are constants (real). On the other hand, if X is a random variable taking both positive and negative values (Gaussian, Laplacian, Cauchy, etc.), one can get an example of a non-monotonic transformation,

$$Y = aX^2 \tag{3.169}$$

It should be noted that the transformation in Eq. (3.169) will be monotonic if X only takes positive values (exponential, Rayleigh, gamma, etc.). Therefore, the monotonic or non-monotonic nature depends on the transforming function $g(.)$ and the range of the original or input random variable. The non-monotonic transformation is also referred to as one-to-many transformation, while monotonic transformation may be classified as one-to-one. Another example of the non-monotonic transformation (X is a uniform random variable) will be

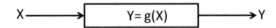

Fig. 3.7 Concept of the transformation of a random variable

$$Y = \tan(X),$$
$$f_X(x) = \frac{1}{2\pi}, 0 < x < 2\pi \tag{3.170}$$

It is also possible to create mixed random variables where the transformation leads to a new random variable which takes continuous and discrete values. As an example, consider the case of a Gaussian random variable X and the transformation (a and b are positive numbers) Y as

$$Y = \begin{cases} -a, & X \leq -a \\ bX, & -a < X < a \\ a, & X > a \end{cases} \tag{3.171}$$

In Eq. (3.171), the transformation leads to discrete outputs when

$$|X| > a \tag{3.172}$$

The continuous nature of Y exists when

$$|X| < a \tag{3.173}$$

While Eq. (3.171) is an example of continuous to *mixed*, one can also have examples of continuous-to-*discrete* transformation. If X is Gaussian, we can define the following transformation:

$$Y = \begin{cases} -a, & X \leq -a \\ b, & -a < X < a \\ a, & X > a \end{cases} \tag{3.174}$$

It is also possible to have discrete-to-discrete transformation as well. An example of a discrete-to-discrete transformation may be the case of (data) compression used in communication systems where the range of discrete data from -8 to 8 is compressed to -4 to 4. In terms of complexity, discrete-to-discrete transformations are easier than the rest in obtaining the density following the transformation. The continuous-to-continuous transformation may be one-to-one (Fig. 3.8a) or one-to-many (Fig. 3.8b). In the case shown in Fig. 3.8b, for every value of Y, there are two values of X.

Consider the case of a monotonic transformation given in Eq. (3.167). A display of monotonic transformation is shown in Fig. 3.8a. For each value of X, there exists only a single value of Y. This means that when X takes values between x and $x + \Delta x$, there is a corresponding set for Y, namely, y and $y + \Delta y$. In terms of probability, this translates to

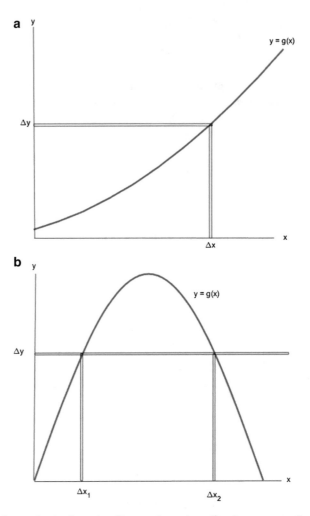

Fig. 3.8 (**a**) Monotonic transformation. For any change in x, there is a corresponding change in y. (**b**) Non-monotonic transformation. Any change in y is coming from two corresponding components in x (level of non-monotonicity of 2)

$$P(x \leq X < x + \Delta x) = P(y < Y < y + \Delta y) \qquad (3.175)$$

Using the definition of the CDF, Eq. (3.175) becomes

$$f_X(x)\Delta x = f_Y(y)\Delta y \qquad (3.176)$$

Rewriting, Eq. (3.176) leads to

$$f_Y(y) = f_X(x)\frac{\Delta x}{\Delta y} = \frac{f_X(x)}{\frac{\Delta y}{\Delta x}} \tag{3.177}$$

When $\Delta x \to 0$, Eq. (3.177) becomes

$$f_Y(y) = \frac{f_X(x)}{\left|\frac{dy}{dx}\right|}\Bigg|_{x=g^{-1}(y)} \tag{3.178}$$

The absolute sign in the denominator takes care of the case when dy/dx is negative (monotonically decreasing transformation) and ensures that the pdf is always positive as it must be. The final step is to express the right-hand side in terms of y.

It is also possible to get the pdf of Y using the fundamental definition of the CDF. Using the definition, the CDF of Y is

$$F_Y(y) = P(Y \le y) = P(g(X) \le y) = P\left[X \le g^{-1}(y)\right] = F_X\left(g^{-1}(y)\right) \tag{3.179}$$

If the transformation is monotonically decreasing, the CDF becomes

$$F_Y(y) = P(Y \le y) = P\left[X > g^{-1}(y)\right] = 1 - F_X\left(g^{-1}(y)\right) \tag{3.180}$$

The density of Y is obtained by differentiating the CDF w.r.t. y.

Consider a case of a non-monotonic transformation. A non-monotonic transformation implies that for every value of Y, there are at least two values of X which lead to the same value of Y. For simplicity, let us start with the case when the degree of non-monotonicity is two depicted in Fig. 3.8b. This means that Eq. (3.175) becomes

$$P(y < Y < y + \Delta y) = P(x_1 < X < x_1 + \Delta x_1) + P(x_2 < X < x_2 + \Delta x_2) \tag{3.181}$$

In Eq. (3.181), x_1 and x_2 are the two values of x that lead to the same value of y. Equation (3.178) now becomes

$$f_Y(y) = \frac{f_X(x)}{\left|\frac{dy}{dx}\right|}\Bigg|_{x_1=g^{-1}(y)} + \frac{f_X(x)}{\left|\frac{dy}{dx}\right|}\Bigg|_{x_2=g^{-1}(y)} \tag{3.182}$$

The concept can now be extended to the case of N value of x giving rise to the same value of y. In this case the density of Y becomes

$$f_Y(y) = \sum_{k=1}^{N} \frac{f_X(x)}{\left|\frac{dy}{dx}\right|}\Bigg|_{x_k=g^{-1}(y)} \tag{3.183}$$

It can be seen that Eq. (3.178) is special case of Eq. (3.183) with $N = 1$.

Example 3.31 X is an exponential random variable of parameter a. Obtain the pdf of $Y = X + 2$.

Solution

$$f_X(x) = \frac{1}{a} \exp\left(-\frac{x}{a}\right) U(x)$$

$$Y = X + 2 \implies \frac{dy}{dx} = 1$$

The transformation is a monotonically increasing one. Therefore,

$$f_Y(y) = \frac{f_X(x)}{\left|\frac{dy}{dx}\right|}\Bigg|_{x=y-2} = \frac{1}{a} \exp\left(-\frac{(y-2)}{a}\right) U(y-2)$$

Notice that the minimum value of the random variable Y is now 2. The solution to the transformation will be incomplete without specifying the range of the output variable. Using the concept of the CDF, we can write an expression for the CDF of Y as

$$F_Y(y) = P(Y \le y) = P(X + 2 \le y) = P(X \le y - 2) = F_X(y - 2)$$

Differentiating the CDF, we get

$$f_Y(y) = \frac{d}{dy} F_X(y - 2) = f_X(y - 2) = \frac{1}{a} \exp\left(-\frac{(y-2)}{a}\right) U(y-2)$$

Example 3.32 For the case of a Rayleigh random variable X with parameter b, obtain the pdf of

$$Y = X^2$$

Solution This is a monotonic transformation since X only takes positive values. We have

$$f_X(x) = \frac{x}{b^2} \exp\left(-\frac{x^2}{2b^2}\right) U(x)$$

$$Y = X^2 \Rightarrow \frac{dy}{dx} = 2x = 2\sqrt{y}$$

$$f_Y(y) = \left. \frac{f_X(x)}{\left| \frac{dy}{dx} \right|} \right|_{x=\sqrt{y}} = \frac{1}{2b^2} \exp\left(-\frac{y}{2b^2} \right) U(y)$$

Using the concept of the CDF, we have

$$F_Y(y) = P(Y \le y) = P(X^2 \le y) = P(X \le \sqrt{y}) = F_X(\sqrt{y})$$

Differentiating, we have

$$f_Y(y) = \frac{d}{dy} F_X(\sqrt{y}) = \frac{1}{2\sqrt{y}} f_X(\sqrt{y}) = \frac{1}{2b^2} \exp\left(-\frac{y}{2b^2} \right) U(y)$$

An important observation is that *the square of the Rayleigh variable will be an exponential variable*. In other words, if X represents Rayleigh distributed amplitude, Y represents the pdf of the power which is exponential. The converse is also true. The density of the square root of an exponential random variable will be a Rayleigh density.

Example 3.33 X is a Laplacian random variable with a pdf

$$f_X(x) = \frac{1}{2} e^{-|x|}$$

Obtain the pdf of $Y = X^2$.

Solution Since X takes both positive and negative values, the transformation is non-monotonic with two roots, $x = \pm\sqrt{y}$. This means that the pdf of Y will be the sum of two terms,

$$f_Y(y) = \left. \frac{f_X(x)}{\left| \frac{dy}{dx} \right|} \right|_{x=+\sqrt{y}} + \left. \frac{f_X(x)}{\left| \frac{dy}{dx} \right|} \right|_{x=-\sqrt{y}}$$

$$\frac{dy}{dx} = 2x = 2\sqrt{y}$$

$$f_Y(y) = \frac{1}{2\sqrt{y}} e^{-\sqrt{y}} U(y)$$

Using the concept of the CDF, the CDF of Y becomes

$$F_Y(y) = P(Y < y) = P(X^2 < y) = P(-\sqrt{y} < X < \sqrt{y}) = F_X(\sqrt{y}) - F_X(-\sqrt{y})$$

Since X is a symmetric (symmetry around the origin) random variable,

$$F_X(-\sqrt{y}) = 1 - F_X(\sqrt{y})$$

Making note of the symmetry, the CDF of Y becomes

$$F_Y(y) = F_X(\sqrt{y}) - F_X(-\sqrt{y}) = 2F_X(\sqrt{y}) - 1$$

Differentiating, we get the pdf as

$$f_Y(y) = \frac{1}{2\sqrt{y}} e^{-\sqrt{y}} U(y)$$

Example 3.34 X is $N(\mu, \sigma^2)$. Obtain the pdf of $Y = |X|$.

Solution The transformation certainly is non-monotonic with two roots,

$$x = \pm y$$

$$\frac{dy}{dx} = 1$$

$$f_Y(y) = \frac{f_X(x)}{\left|\frac{dy}{dx}\right|}\bigg|_{x=+y} + \frac{f_X(x)}{\left|\frac{dy}{dx}\right|}\bigg|_{x=-y} = \frac{1}{\sqrt{2\pi\sigma^2}} \left[e^{-\frac{(y-\mu)^2}{2\sigma^2}} + e^{-\frac{(y+\mu)^2}{2\sigma^2}} \right] U(y)$$

Using concept of the CDF, the CDF of Y will be

$$F_Y(y) = P(Y < y) = P(|X| < y) = P(-y < X < y) = F_X(y) - F_X(-y)$$

Differentiating the CDF, we get the pdf given by

$$f_Y(y) = f_X(y) + f_X(-y) = \frac{1}{\sqrt{2\pi\sigma^2}} \left[e^{-\frac{(y-\mu)^2}{2\sigma^2}} + e^{-\frac{(y+\mu)^2}{2\sigma^2}} \right] U(y)$$

Example 3.35 X is uniform in $[-1,1]$. Obtain the pdf of $Y = 1/X$.

Solution For this uniform variable,

$$f_X(x) = \frac{1}{2}, \ -1 < x < 1$$

$$F_X(x) = \begin{cases} 0, & x < -1 \\ \dfrac{x+1}{2}, & -1 < x < 0 \\ \dfrac{1}{2} + \dfrac{x}{2}, & 0 < x < 1 \\ 1, & x > 1 \end{cases}$$

The transformation $Y = 1/X$ is monotonically decreasing as seen by its derivative (being negative),

$$\frac{dy}{dx} = -\frac{1}{x^2}$$

$$f_Y(y) = \frac{f_X(x)}{\left|\frac{dy}{dx}\right|}\Bigg|_{x=\frac{1}{y}} = \frac{1}{2y^2}, \quad |y| > 1$$

Notice that the range of the variable Y is $-\infty < y < 1$ and $1 > y > \infty$ or written collectively as $|y| > 1$. The range is obtained by examining the limits of X and finding out the limiting values of y.
Using the concept of the CDF,

$$F_Y(y) = P(Y < y) = P\left(\frac{1}{X} < y\right) = P\left(X > \frac{1}{y}\right) = 1 - F_X\left(\frac{1}{y}\right)$$

Differentiating,

$$f_Y(y) = \frac{1}{y^2} f_X\left(\frac{1}{y}\right) = \frac{1}{2y^2}, \quad |y| > 1$$

Notice that $f_X(1/y) = 1/2$.

Example 3.36 Consider a uniformly distributed random variable in $[-1,3]$. Obtain the pdfs of

$$Y = X^2$$

$$W = \frac{1}{X^2}$$

Solution The density function of X is

$$f_X(x) = \frac{1}{4}, \quad -1 \le x \ge 3$$

These transformations are unique because each of them contains both monotonic and non-monotonic parts. The transformation $Y = X^2$ is monotonic in $1 < x < 3$, while it is non-monotonic in $-1 < x < 1$.

$$\frac{dy}{dx} = 2x = 2\sqrt{y}$$

Let us do the transformation in pieces.

$$-1 \leq x \leq 1 \quad \Leftrightarrow \quad 0 \leq y \leq 1 \qquad \text{Non-monotonic}$$

$$f_Y(y) = \frac{f_X(x)}{\left|\frac{dy}{dx}\right|}\Bigg|_{x=+\sqrt{y}} + \frac{f_X(x)}{\left|\frac{dy}{dx}\right|}\Bigg|_{x=-\sqrt{y}} = \frac{1}{8\sqrt{y}} + \frac{1}{8\sqrt{y}} = \frac{1}{4\sqrt{y}}, 0 \leq y \leq 1$$

$$1 \leq x \leq 3 \quad \Leftrightarrow \quad 1 \leq y \leq 9 \qquad \text{Monotonic}$$

$$f_Y(y) = \frac{f_X(x)}{\left|\frac{dy}{dx}\right|}\Bigg|_{x=+\sqrt{y}} = \frac{1}{8\sqrt{y}}, 1 \leq y \leq 9$$

Combining, we can write the pdf for Y as

$$f_Y(y) = \begin{cases} \dfrac{1}{4\sqrt{y}}, & 0 \leq y \leq 1 \\[2mm] \dfrac{1}{8\sqrt{y}}, & 1 < y \leq 9 \end{cases}$$

The CDF of Y is

$$F_Y(y) = \begin{cases} 0, & y < 0 \\ \int_0^y \dfrac{1}{4\sqrt{z}}dz, & 0 \leq y \leq 1 \\ \int_0^1 \dfrac{1}{4\sqrt{z}}dz + \int_1^y \dfrac{1}{8\sqrt{z}}dz, & 1 \leq y \leq 9 \\ 1, & y > 9 \end{cases} = \begin{cases} 0, & y < 0 \\ \dfrac{\sqrt{y}}{2}, & 0 \leq y \leq 1 \\ \dfrac{1}{4} + \dfrac{\sqrt{y}}{4}, & 1 \leq y \leq 9 \\ 1, & y > 9 \end{cases}$$

The transformation $W = \frac{1}{x^2}$ is monotonic in $1 < x < 3$, while it is non-monotonic in $-1 < x < 1$.

$$\frac{dw}{dx} = -\frac{2}{x^3} = -2w^{\frac{3}{2}}$$

$$-1 \leq x \leq 1 \quad \Leftrightarrow \quad w \geq 1 \qquad \text{Non-monotonic}$$

$$f_W(w) = \frac{f_X(x)}{\left|\frac{dw}{dx}\right|}\Bigg|_{x=+\frac{1}{\sqrt{w}}} + \frac{f_X(x)}{\left|\frac{dw}{dx}\right|}\Bigg|_{x=-\frac{1}{\sqrt{w}}} = \frac{1}{8w^{\frac{3}{2}}} + \frac{1}{8w^{\frac{3}{2}}} = \frac{1}{4w^{\frac{3}{2}}}, w > 1$$

$$1 \leq x \leq 3 \quad \Leftrightarrow \quad \frac{1}{9} \leq w \leq 1 \qquad \text{Monotonic}$$

$$f_W(w) = \frac{f_X(x)}{\left|\frac{dw}{dx}\right|}\Bigg|_{x=+\frac{1}{\sqrt{w}}} = \frac{1}{8w^{\frac{3}{2}}}, \frac{1}{9} \leq w \leq 1$$

Combining,

$$
f_W(w) = \begin{cases} \dfrac{1}{8w^{\frac{3}{2}}}, & \dfrac{1}{9} \le w \le 1 \\[2ex] \dfrac{1}{4w^{\frac{3}{2}}}, & w > 1 \end{cases}
$$

The CDF is

$$
F_W(w) = \begin{cases} 0, & w < \dfrac{1}{9} \\[2ex] \dfrac{3}{4} - \dfrac{1}{4\sqrt{w}}, & \dfrac{1}{9} \le w \le 1 \\[2ex] 1 - \dfrac{1}{2\sqrt{w}}, & w > 1 \end{cases}
$$

In this example, the transformation contained monotonic and non-monotonic parts.

Example 3.37 X uniform in $[0, \pi]$. Obtain the density of

$$
Y = \sin(X)
$$
$$
Z = \cos(X)
$$

Solution The transformation from X to Y is non-monotonic because there are two roots in the range $[0, \pi]$. Therefore,

$$
f_Y(y) = \frac{f_X(x)}{\left|\frac{dy}{dx}\right|}\Bigg|_{x=x_1=\sin^{-1}(y)} + \frac{f_X(x)}{\left|\frac{dy}{dx}\right|}\Bigg|_{x=x_2=\sin^{-1}(\pi-y)}
$$

$$
\frac{dy}{dx} = \cos(x) = \sqrt{1-y^2}
$$

$$
f_Y(y) = \frac{1}{\pi\sqrt{1-y^2}} + \frac{1}{\pi\sqrt{1-y^2}} = \frac{2}{\pi\sqrt{1-y^2}}, \quad 0 \le y \le 1
$$

The transformation from X to Z is *monotonic* because there is only one root in $[0, \pi]$.

$$
f_Z(z) = \frac{f_X(x)}{\left|\frac{dz}{dx}\right|}\Bigg|_{x=\cos^{-1}(z)}
$$

$$
\frac{dz}{dx} = -\sin(x) = -\sqrt{1-z^2}
$$

$$
f_Z(z) = \frac{1}{\pi\sqrt{1-z^2}}, \quad -1 \le z \le 1
$$

Notice that the ranges of Y and Z are different.

Example 3.38 X is a random variable taking value of -2, -1, 0, 1, 2, and 3 with probabilities, respectively, of $(1/8)$, $(1/4)$, $(1/8)$, $1/8$, $1/8$, and $1/4$. If $Y = X^2$, obtain the density of Y.

Solution We have

$$f_X(x) = \frac{1}{8}\delta(x+2) + \frac{1}{4}\delta(x+1) + \frac{1}{8}\delta(x) + \frac{1}{8}\delta(x-1) + \frac{1}{8}\delta(x-2) + \frac{1}{4}\delta(x-3)$$

$$Y = [0, 1, 4, 9]$$

$$P(Y = 0) = P(X = 0) = \frac{1}{8}$$

$$P(Y = 1) = P(X = -1) + P(X = 1) = \frac{1}{4} + \frac{1}{8} = \frac{3}{8}$$

$$P(Y = 4) = P(X = -2) + P(X = 2) = \frac{1}{8} + \frac{1}{8} = \frac{2}{8}$$

$$P(Y = 9) = P(X = 3) = \frac{1}{4} = \frac{2}{8}$$

The pdf of Y is

$$f_Y(y) = \frac{1}{8}\delta(y) + \frac{3}{8}\delta(y-1) + \frac{2}{8}\delta(y-4) + \frac{2}{8}\delta(y-9)$$

3.8   Mixed Random Variables and Mixed Transformations

Consider a system which generates numbers randomly between 0 and 10. However, a decision is made to "clip" the output such that if the generated numbers exceed 5, those numbers will be assigned a fixed value of 7. *We will now define a random variable to model this arrangement.*

It is clear that the original generator is such that the probability of seeing a number between 1 and 2 is $(1/10)$, probability of seeing a number between 5 and 6 is also $(1/10)$, etc. But the new arrangement is that the numbers in the range $[0,5]$ appear as they are and the rest will appear as 7. Thus, the arrangement has two events, one has numbers in the range $[0,5]$ and the other has a fixed value of 7.

The event of the 7: The probability of this event, $P[7] = P[\text{original numbers lie between 5 and 10}] = (1/2)$.

The event: numbers between 0 and 5: The probability of this event is $(1/2)$. This means that the probability of seeing a number between 0 and 1 now is still $(1/10)$, probability of seeing a number between 4 and 5 is still $(1/10)$, and so on. The random variable created is referred to as a mixed variable since it has outcomes that are continuous and as well as a discrete one. The probability density function is shown in Fig. 3.9.

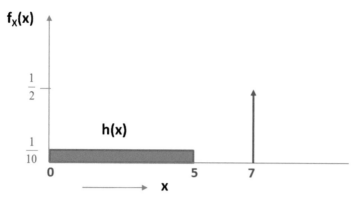

Fig. 3.9 Probability density function of a mixed random variable

The probability density function $f(x)$ becomes

$$f_X(x) = h(x) + P[X = 7]\delta(x - 7) = h(x) + \frac{1}{2}\delta(x - 7), \quad 0 \leq x \leq 7 \quad (3.184)$$

The function $h(x)$ is

$$h(x) = \frac{1}{10}, \quad 0 \leq x \leq 5 \quad (3.185)$$

The function $h(x)$ may also be written as

$$h(x) = \frac{1}{10}[U(x) - U(x - 5)] \quad (3.186)$$

The cumulative distribution function is shown in Fig. 3.10.
The cumulative distribution function becomes

$$F_X(x) = \begin{cases} \dfrac{x}{10}, & 0 \leq x \leq 5 \\ \dfrac{1}{2}, & 5 \leq x < 7 \\ 1, & x \geq 7 \end{cases} \quad (3.187)$$

Example 3.39 Consider the case of a Rayleigh distributed random variable X with the pdf,

$$f_X(x) = \frac{x}{4} \exp\left(-\frac{x^2}{8}\right) U(x)$$

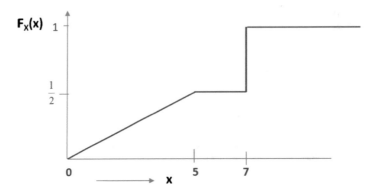

Fig. 3.10 Cumulative distribution function of a mixed random variable with a pdf shown in Fig. 3.9

Obtain the density of

$$Y = \begin{cases} 1, X \le 2 \\ X, X > 2 \end{cases}$$

Solution It is seen that the output contains a discrete outcome of $Y = 1$ with a probability

$$P(Y = 1) = P(X \le 2) = 1 - F_X(2) = 1 - e^{-\frac{4}{8}} = 1 - e^{-\frac{1}{2}}$$

For $Y > 2$, the transformation is monotonic and leads to

$$f_{Y_c}(y) = \frac{y}{4} \exp\left(-\frac{y^2}{8}\right), y > 2$$

The subscript C of Y merely reflects the fact that the density above pertains to the continuous part of Y. Combining the discrete and continuous outcomes, the density of Y becomes

$$f_Y(y) = \left(1 - e^{-\frac{1}{2}}\right)\delta(y - 1) + \frac{y}{4}\exp\left(-\frac{y^2}{8}\right)U(y - 2)$$

The CDF is

$$F_Y(y) = \left(1 - e^{-\frac{1}{2}}\right)U(y - 1) + \left[\int_2^y \frac{z}{4}\exp\left(-\frac{z^2}{8}\right)dz\right]U(y - 2)$$

Simplifying, CDF becomes

$$F_Y(y) = \left(1 - \exp\left(-\frac{1}{2}\right)\right)U(y - 1) + \left[\exp\left(-\frac{1}{2}\right) - \exp\left(-\frac{y^2}{8}\right)\right]U(y - 2)$$

The pdf and CDF are shown in Figures Ex.3.39a and Ex.3.39b.

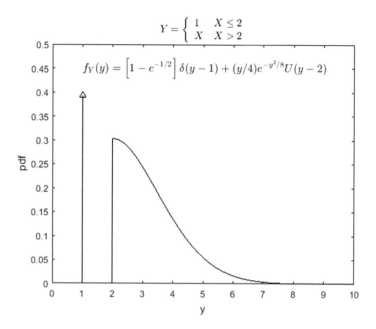

Figure Ex.3.39a

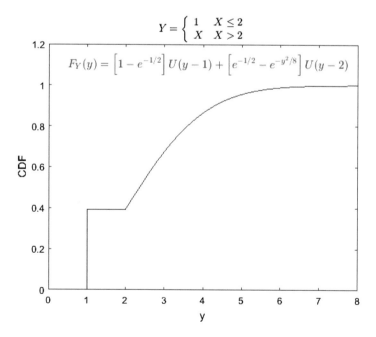

Figure Ex.3.39b

Example 3.40 We will now look at a case of the transformation of continuous variable to a discrete random variable (analog-to-digital conversion). If X is $N(0,\sigma^2)$, obtain the density of

$$Y = \begin{cases} -1, X < 0 \\ 1, X > 0 \end{cases}$$

Solution We have two outcomes with probabilities equal to ½ given as

$$P(Y = -1) = P(X < 0) = \frac{1}{2}$$
$$P(Y = 1) = P(X > 0) = \frac{1}{2}$$

The pdf of Y is

$$f_Y(y) = \frac{1}{2}\delta(y + 1) + \frac{1}{2}\delta(y - 1)$$

Example 3.41 Obtain the pdf of a half-wave rectifier when the input is Gaussian, $N(0,\sigma^2)$.

Solution The system of a half-wave rectifier is characterized by the input (X) output (Y) relationship

$$Y = \begin{cases} 0, X < 0 \\ X, X > 0 \end{cases}$$

The relationship above is also represented as

$$Y = \frac{1}{2}(X + |X|)$$

Based on the examples discussed above, we expect a delta function for the output pdf. If X is $N(0,\sigma^2)$,

$$P(Y = 0) = \frac{1}{2}$$

The density of the output now becomes

$$f_Y(y) = \frac{1}{2}\delta(y) + \frac{1}{\sqrt{2\pi\sigma^2}} \exp\left(-\frac{y^2}{2\sigma^2}\right) U(y)$$

Example 3.42 If X is a continuous random variable with a pdf $f_X(x)$ and CDF of $F_X(x)$, obtain the density of

$$Y = F(X)$$

Solution This transformation is monotonic, and the density of Y can be written as

$$f_Y(y) = \frac{f_X(x)}{\left|\frac{dy}{dx}\right|} = \frac{f_X(x)}{f_X(x)} = 1, \quad 0 < y < 1$$

The range of Y is obvious from the fact that Y is the CDF of a continuous random variable. The pdf $f_Y(y)$ shows that Y is a uniform random variable in the range $[0,1]$, regardless of the form of $f_X(x)$. This also means that if CDF is available, it is possible to generate samples of the random variable X from samples of uniform random numbers in the range $[0,1]$.

3.9 Miscellaneous Topics

3.9.1 Hazard Rates and Reliability

As explained in previous sections, some density functions such as the Rayleigh, exponential, gamma, and Weibull that exist only for positive values are used in the modeling of lifetime of engineering systems (electronic components, computer chips, bridges, structures, etc.). The hazard rate was defined in connection with the definition of the CDF, and it was suggested that it is related to the condition density defined in Eq. (3.125). To understand this relationship, consider the following scenario in the modeling of the lifetime of an engineering system. If we know that the system has survived and it is still functioning at time x, the probability of this event, i.e., it is still functioning at $X = x$, is the reliability of the system which was defined in Eq. (3.60) as the survival rate $S(x)$. The reliability $R(x)$ is

$$R(x) = S(x) = P(X > x) = 1 - F_X(x) \tag{3.188}$$

Equation (3.188) can be interpreted as a measure or fraction of the components or system that will *fail after* time x or that the system is still functioning. Differentiating Eq. (3.188), we have

$$R'(x) = -f_X(x) \tag{3.189}$$

Consider now the case of $\{X > \alpha\}$, i.e., the system is surviving beyond time instant α. The conditional probability that it lasts beyond α was given in Eq. (3.123) as

$$F_{X|X>a}(x|X > a) = \frac{P(X \leq x, X > a)}{1 - F_x(a)} \tag{3.190}$$

The conditional density was given in Eq. (3.125) as

$$f(x|X > a) = \frac{f_X(x)}{1 - F_x(a)}, \quad x \geq a \tag{3.191}$$

If we replace a by x, we get the hazard rate or failure rate $h(x)$ or $\lambda(x)$ which is obtained by replacing the left-hand side of Eq. (3.191) by $\lambda(x)$ as

$$\lambda(x) = \frac{f_X(x)}{1 - F_x(x)}, \quad x \geq 0 \tag{3.192}$$

We can appreciate the implications of the failure rate by exploring Eq. (3.192) further. Rewriting Eq. (3.192) using the relationship between the pdf and CDF, we have

$$\lambda(x) = \frac{\left[\frac{dF(x)}{dx}\right]}{1 - F_x(x)} \tag{3.193}$$

Equation (3.193) expands to

$$\lambda(x)dx = \frac{1}{1 - F(x)} dF_x(x) \tag{3.194}$$

Integrating both sides,

$$\int_0^x \lambda(x)dx + k = -\log_e(1 - F(x)) \tag{3.195}$$

In Eq. (3.195), k is an unknown constant. Rewriting Eq. (3.195), we have

$$\log_e(1 - F(x)) = -\int_0^x \lambda(x)dx + k_0 \tag{3.196}$$

Note that the constant now is k_0. Solving Eq. (3.196), we have

$$1 - F(x) = e^{k_0} \exp\left(-\int_0^x \lambda(x)dx\right) \tag{3.197}$$

To find the constant k_0, put $x = 0$ and $F(0) = 0$. This leads to

$$1 - F(x) = e^{k_0} \exp\left(-\int_0^x \lambda(x)dx\right) \underset{x=0}{\Rightarrow} 1 = e^{k_0} \Rightarrow k_0 = 1 \qquad (3.198)$$

This means that the CDF is

$$F(x) = 1 - \exp\left(-\int_0^x \lambda(x)dx\right) \qquad (3.199)$$

In other words, if the failure rate is known, it is possible to obtain the CDF and therefore the pdf.

One interesting corollary associated with reliability is its relationship to the mean of a random variable that only takes positive values. The mean time to failure (MTTF) is the mean of the random variable (note: x is always positive). Consider the mean of a random variable X defined earlier (for $x > 0$)

$$E[X] = \int_0^\infty xf(x)dx \qquad (3.200)$$

Let us express x as a simple integral

$$x = \int_0^x dy \qquad (3.201)$$

Expectation in Eq. (3.200) now becomes

$$E[X] = \int_0^\infty xf(x)dx = \int_0^\infty \left(\int_0^x dy\right)f(x)dx \qquad (3.202)$$

Equation (3.202) implies that $y < x$, and making use of this, we change the order of integration. Rewriting, Eq. (3.202) becomes

$$E[X] = \int_0^\infty \left(\int_y^\infty f(x)dx\right)dy = \int_0^\infty ([1 - F(y)])dy = \int_0^\infty R(y)dy = \int_0^\infty R(x)dx \qquad (3.203)$$

Equation (3.203) establishes the relationship between the reliability of a random variable $R(x)$ and its mean.

Example 3.43 It has been observed that the failure rate increases with time. Find the reliability, density, and distribution functions of the lifetime of the system.

Solution We have the failure rate as

$$\lambda(t) = a + ct$$

Note that a and c are constants.

$$\lambda(t) = a + ct = \frac{f(t)}{1 - F_T(t)} = \frac{f(t)}{R(t)} = -\frac{R'(t)}{R(t)}$$

$$-\frac{R'(t)}{R(t)} = a + ct$$

Rewriting,

$$-\frac{\frac{d}{dt}R(t)}{R(t)} = a + ct$$

$$-\frac{dR}{R} = [a + ct]\,dt$$

Integrating, we have

$$-\log_e(R) = at + \frac{c}{2}t^2$$

The reliability is

$$R = \exp\left(-at - \frac{c}{2}t^2\right)$$

Noting that the reliability and CDF are related, we have

$$R(t) = 1 - F_T(t) = \exp\left(-at - \frac{c}{2}t^2\right)$$

The CDF becomes

$$F_T(t) = 1 - \exp\left(-at - \frac{c}{2}t^2\right)$$

We may take $a = 0$ (this step does not violate the increase in failure rate with time); the CDF becomes

$$F_T(t) = 1 - \exp\left(-\frac{c}{2}t^2\right)$$

The corresponding density is

$$f_T(t) = ct \exp\left(-\frac{c}{2}t^2\right)U(t)$$

Note that the lifetime is Rayleigh distributed.

3.9.2 Error Rates

In Chap. 2, the concept of error was introduced in connection with the analysis of a binary channel using Bayes' rule. The transmission of 1's and 0's (or any two-level system) through a noisy channel resulted in two forms of errors, the probability of false alarm ("0" detected as a "1") and the probability of miss ("1" detected as a "0"). Treating transmission rates of 0's and 1's as the a priori probabilities, we have

$$P(H_0) = p, \ \ 0 < p < 1$$
$$P(H_1) = q = 1 - p \tag{3.204}$$

In decision theory (to be discussed in Chap. 5), this modeling is represented in terms of two hypotheses H_1 and H_0, with H_1 representing the transmission of a "1" and H_0 representing the transmission of a "0." The received signals will be

$$r(t_0) = s(t_0) + n(t_0) \tag{3.205}$$

In Eq. (3.205), t_0 is the observation time instant, $s(.)$ is the transmitted signal, and $n(.)$ is the noise.

In Chap. 2, the channel was described in terms of transition probabilities, namely, $P(R_1|T_1)$, $P(R_1|T_0)$, $P(R_0|T_0)$, and $P(R_0|T_1)$. We now treat the channel as having Gaussian noise of zero mean and variance σ^2 suggesting that the noise is $N(0, \sigma^2)$. We can describe the probability densities of the received voltage associated with the transmission channel as

$$f(r|H_0) = \frac{1}{\sqrt{2\pi\sigma^2}} \exp\left[-\frac{(r+\mu)^2}{2\sigma^2}\right]$$
$$f(r|H_1) = \frac{1}{\sqrt{2\pi\sigma^2}} \exp\left[-\frac{(r-\mu)^2}{2\sigma^2}\right] \tag{3.206}$$

We made use of the transformation of Gaussian variables to obtain the result in Eq. (3.206) because the received signals will be Gaussian with identical variances and mean values dictated by the voltages corresponding to the two levels, +1 and −1. The bit "1" is transmitted with a voltage μ, and "0" is transmitted with a voltage$-\mu$. Using simple logic, it would seem that we can decide on whether the received signal belongs to hypothesis H_0 or H_1 as (for details, see Chap. 5)

$$f(H_1|r) \underset{H_0}{\overset{H_1}{\underset{<}{>}}} f(H_0|r) \tag{3.207}$$

Invoking Bayes' rule, Eq. (3.207) becomes

$$f(r|H_1)P(H_1) \underset{H_0}{\overset{H_1}{\underset{<}{>}}} f(r|H_0)P(H_0) \tag{3.208}$$

Equation (3.208) leads to

$$\left[\frac{f(r|H_1)}{f(r|H_0)}\right] \underset{H_0}{\overset{H_1}{\underset{\leq}{>}}} \left[\frac{P(H_0)}{P(H_1)}\right] \tag{3.209}$$

The left-hand side of Eq. (3.209) is known as the likelihood ratio, and the entire equation is identified as the likelihood ratio test (LRT). If both hypotheses are equally likely, Eq. (3.209) becomes

$$\left[\frac{f(r|H_1)}{f(r|H_0)}\right]_{H_0}^{H_1} \underset{<}{\overset{>}{}} 1 \tag{3.210}$$

Taking logarithms, the left-hand side becomes the log likelihood ratio, and it is (Van Trees 1968; Helstrom 1968; Papoulis and Pillai 2002; Smits et al. 2007; Lawton 2009; Rota and Antolini 2014)

$$\log \left[\frac{f(r|H_1)}{f(r|H_0)}\right]_{H_0}^{H_1} \underset{<}{\overset{>}{}} 0 \tag{3.211}$$

Substituting the appropriate densities from Eq. (3.206) in Eq. (3.211), we see that the threshold is

$$r = \text{thr} = \frac{\mu + (-\mu)}{2} = 0 \tag{3.212}$$

It can be seen that the threshold (thr) is the midpoint (in this case at voltage = 0) because of the symmetry of the two densities. In general, if the levels of the two signals are a_1 and a_0, the threshold will be

$$\text{thr} = \frac{(a_1 + a_0)}{2} \tag{3.213}$$

Using the threshold, we may obtain the probabilities of false alarm and miss. The probability of false alarm (P_F) is

$$P_F = P(R_1|T_0) = \int_{\text{thr}}^{\infty} f(r|H_0)dr = \int_0^{\infty} f(r|H_0)dr = \frac{1}{2}\left[1 - \text{erf}\left(\frac{\mu}{\sigma\sqrt{2}}\right)\right] \tag{3.214}$$

The probability of miss (P_M) is

$$P_M = P(R_0|T_1) = \int_{-\infty}^{\text{thr}} f(r|H_1)dr = \int_{-\infty}^{0} f(r|H_1)dr = \frac{1}{2}\left[1 - \text{erf}\left(\frac{\mu}{\sigma\sqrt{2}}\right)\right] \tag{3.215}$$

The two densities and the two probabilities are shown in Fig. 3.11. It can be easily seen that

$$P_F = P_M \tag{3.216}$$

The probability of error is

$$P(\text{error}) = P(\text{error}|H_0)P(H_0) + P(\text{error}|H_1)P(H_1) \tag{3.217}$$

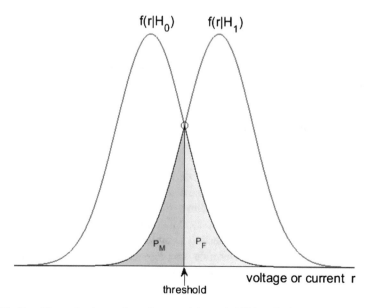

Fig. 3.11 Densities under the two hypotheses and the probabilities of miss and false alarm

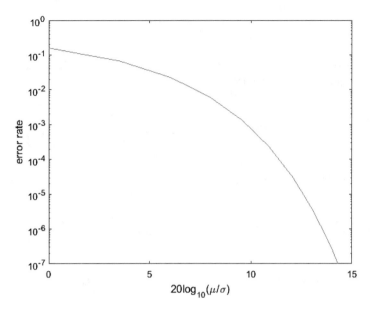

Fig. 3.12 Error rates in a Gaussian channel

If 0's and 1's are transmitted with equal probability, the probability of error is

$$P(\text{error}) = P_F P(H_0) + P_M P(H_1) = P_F = P_M \tag{3.218}$$

The error rate is plotted in Fig. 3.12. It can be seen that as the value of the mean (for a fixed value of σ) increases, the error goes down.

Note that in a very general case, such as in the case of medical diagnostics, the a priori probabilities are not equal. In fact they are very different and often at $P(H_1)$ may only be 1e-5 or such small fractions. Under these conditions, the probabilities of false alarm and miss hold different meanings. Probability of miss represents the probability that the presence of the disease is missed in a diagnostic test, while probability of false alarm implies that the presence of the disease is indicated when no disease is actually present. Consequently, error rate is not an effective measure of the performance of the receiver (in this case, the diagnostic test). In a binary communication system, the two hypotheses simply represent two levels, and from a system theory viewpoint, any error made in the detection of one of the levels has the same meaning as the error made in the detection of the other level, and because of this fact, error rate is a reasonable tool to estimate the performance. From the brief discussion here, it can easily be understood that the error rate holds less significance in diagnostic testing, whether the testing is undertaken in medicine or as a part of machine vision or robotics to determine the presence or absence of a target (Zweig and Campbell 1993; Fawcett 2006; Brown and Davis 2006; Eng 2012).

In addition to having unequal a priori probabilities and the limited role of the error rates, it is also possible to have the density functions of the received signal under the two mutually exclusive hypotheses depart from Gaussian, and they might be different under the two hypotheses. These factors necessitate the need for a different strategy to obtain an optimum value of the threshold. In Chap. 2, the threshold was chosen to be the point of intersection of the two densities.

Before we look at the strategy for finding the optimal threshold, let us first consider the case of hypothesis testing when the two densities are different and non-Gaussian (Helstrom 1968).

The density functions seen in radar, sonar, and lidar receivers are not Gaussian. Consider (as an example) the case where the density functions are (Rayleigh for the hypothesis H_0 and Nakagami for the hypothesis H_1: we may choose Nakagami for H_0 as well with $m_0 = 1$)

$$f(r|H_0) = \frac{r}{b^2} \exp\left[-\frac{r^2}{2b^2}\right] = 2\left(\frac{m_0}{c_0}\right)^{m_0} \frac{r^{2m_0-1}}{\Gamma(m_0)} \exp\left[-\frac{m_0}{c_0}r^2\right]U(r)$$

$$f(r|H_1) = 2\left(\frac{m_1}{c_1}\right)^{m_1} \frac{r^{2m_1-1}}{\Gamma(m_1)} \exp\left[-\frac{m_1}{c_1}r^2\right]U(r)$$

$$(3.219)$$

The density functions along with the probabilities of false alarm and miss (evaluated at the threshold where the densities intersect) are shown in Fig. 3.13 with an arbitrarily chosen threshold.

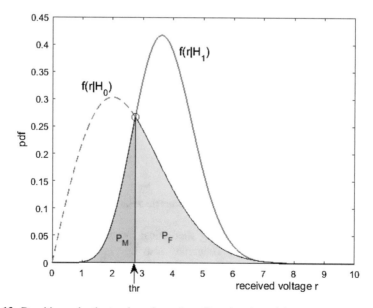

Fig. 3.13 Densities under the two hypotheses (non-Gaussian channels)

In this case, the probability of false alarm is

$$P_F = \exp\left(-\frac{thr^2}{2b^2}\right) = 1 - \frac{\gamma\left(m_0, \frac{m_0}{c_0} thr^2\right)}{\Gamma(m_0)} \qquad (3.220)$$

$$P_M = \frac{\gamma\left(m_1, \frac{m_1}{c_1} thr^2\right)}{\Gamma(m_1)} \qquad (3.221)$$

Compared to the case of a Gaussian channel with P_F and P_M in Eqs. (3.214) and (3.215), the values of P_F and P_M in Eqs. (3.220) and (3.221) are not equal. This means that the effects of having false alarm and miss are different in non-Gaussian channels and error rates are not the best indicator of the performance.

Thus, for two reasons, one being unequal a priori probabilities and non-Gaussian nature of the densities, we need a different way of characterizing the performance of the channel as well as a method for obtaining the optimum threshold. The methodology based on the receiver operating characteristics (ROC) curves offers a means to accomplish this task (Helstrom 1968; Van Trees 1968; Hanley and McNeil 1982; Metz 2006).

3.9.3   Receiver Operating Characteristics Curves

One way to characterize the channel and estimate its performance is through the use of the receiver operating characteristics (ROC) curve. The ROC curve is the plot of the probability of false alarm versus the probability of detection as the threshold is varied. The probability of detection, P_D (Chap. 2), is

$$P_D = 1 - P_M \qquad (3.222)$$

It can be seen that for the case of the Gaussian channel presented earlier as seen in Eq. (3.218),

$$P_D = 1 - P_F \qquad (3.223)$$

For the non-Gaussian channel, this symmetric relationship between P_F and P_M does not exist.

ROC plots may be obtained using theoretical approaches which rely on the availability of density functions. They may also be created from data collected from experiments (machine vision, radar detection, robotics, etc.) or clinical trials (Hanley and McNeil 1982, 1983; DeLong et al. 1988; Bradley 1997).

3.9.3.1 ROC Curves Using Probability Density Functions

The ROC plot is obtained by choosing a probability of false alarm, determining the threshold (thr) and estimating the probability of detection at that threshold, and then plotting the probability of false alarm vs. the probability of detection as the probability of false alarm values span from 0 to 1 (van Trees 1968; Helstrom 1968; Shankar 2017). For the Gaussian channel, this means that

$$P_F = \frac{1}{2}\left[1 - \mathrm{erf}\left(\frac{\mu + \mathrm{thr}}{\sigma\sqrt{2}}\right)\right] \tag{3.224}$$

$$P_D = \frac{1}{2}\left[1 + \mathrm{erf}\left(\frac{\mu - \mathrm{thr}}{\sigma\sqrt{2}}\right)\right] \tag{3.225}$$

For a range of values starting from 1e-7 through 1, P_F may be inverted to obtain the threshold (thr) from Eq. (3.224), and the corresponding probability of detection can be obtained from Eq. (3.225).

The probability of detection is also referred to the *sensitivity*, while the *specificity* is equal to $1-P_F$. The ROC curve for a Gaussian channel is shown in Fig. 3.14 for two cases.

A similar procedure may be carried out to get the ROC curve for the non-Gaussian case described through Eqs. (3.220) and (3.221). Figure 3.15 shows the ROC plot for a non-Gaussian channel.

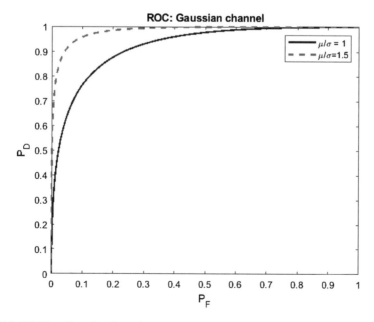

Fig. 3.14 ROC in a Gaussian channel

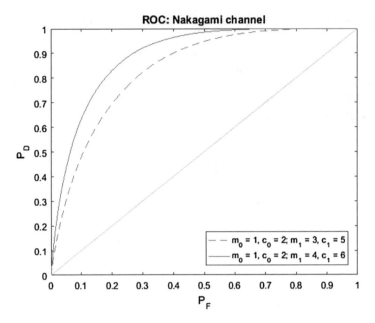

Fig. 3.15 ROC curve in a non-Gaussian channel

ROC plots allow the measure called the area under the ROC curve (AUC) normally expressed as A_z (Metz 2006). The formula for the area is

$$A_Z = \int_0^1 P_D(P_F)dP_F \qquad (3.226)$$

The area can be found graphically. The A_Z values for the Gaussian channel (Fig. 3.14) are

$$A_Z = \begin{cases} 0.9213, \frac{\mu}{\sigma} = 1 \\ 0.9830, \frac{\mu}{\sigma} = 1.5 \end{cases} \qquad (3.227)$$

In Eq. (3.227), the values of the mean and standard deviation correspond to Eqs. (3.224) and (3.225) with the threshold at the midpoint, i.e., thr = 0.

The A_Z values for the Nakagami channel (Fig. 3.15) are

$$A_Z = \begin{cases} 0.8377, m_0 = 1, c_0 = 2; m_1 = 3, c_1 = 5 \\ 0.8934, m_0 = 1, c_0 = 2; m_1 = 4, c_1 = 6 \end{cases} \qquad (3.228)$$

While the ROC curves described above relies on the availability of the densities of the observed parameters, the ROC curves may also be obtained from the data collected during testing and experimentation.

The importance and usefulness of the ROC plot and the area under the ROC curve can be surmised from the fact that AUC will be unity in the ideal case. This means that the curve actually traces the triangle with coordinates [0,0], [0,1], and [1,1] allowing us a standardized measure of performance normalized to unity. The ROC curve also allows the user to choose an operating point (P_F, P_D), obtain a threshold corresponding to the operating point, and conduct the interpretation of the data. The choice of an optimum operation point will be discussed next.

3.9.3.2   ROC Curves from Measured Data

An analysis of data sets was described in Chap. 2 (Example 2.22 and Example 2.23) where the data set collected was studied to obtain the probabilities of false alarm and miss as well as the positive predictive value. While the area under the ROC curve provides a normalized measure of the quality of the receiver, there is still a need to obtain an optimal threshold for practical applications so that one may decide whether the target is present or absent (or disease is present or absent following a diagnostic test).

An example to illustrate the use of ROC curve and a method for determination of the threshold is given next. The data collected from a simple machine vision test is given in Table 3.1. It gives 10 (N_0) values of the backscattered data when no target is present in the field of view and 5 (N_1) values of the data when a target is present. Smaller size data sets are chosen for illustrative purposes. The absence of the target is indicated by "0"s, while the presence of the target is indicated by "1"s. These values of 0's and 1's constitute the "gold standard" in medical statistics, while in machine vision and machine learning, we may simply refer to them as "labels." The set of labels is known because the experiment is conducted under controlled conditions with the operator knowledge of whether the target is present or absent. The analogous case in medical diagnostics is biopsy data which is considered the "gold standard." The term "gold standard" implies that any decision made must be compared to the "absolute truth" manifested in the "gold standard" for that particular receiver or diagnostic test.

Figure 3.16 summarizes the various steps involved in the calculation of the probabilities of false alarm and detection (Shankar 2019, Shankar 2020). The first two columns are the input data sets. The third column now consists of the "gold standard" or "binary labels," and the fourth column represents the corresponding

Table 3.1 Short data set used in the example consisting of a total of 15 values

[1.1428,0.3511,0.9526,0.8165,0.8395,0.9911,1.3645,1.3647,0.8563,1.4065]; target absent, $N_0 = 10$
[1.4308,1.4577,0.8309,1.1524,1.5498]; target present, $N_1 = 5$

Target Absent	Target Present	Binary Label	All Data	Sorted Label	Sorted Data	Threshold T	N_C	N_F	$1-P_M$	P_F	distance $\sqrt{P_F^2 + P_M^2}$
1.1428	1.4308	0	1.1428	1	1.5498	1.5498	0	0	0	0	1
0.3511	1.4577	0	0.3511	1	1.4577	1.4577	1	0	0.2	0	0.8
0.9526	0.8309	0	0.9526	1	1.4308	1.4308	2	0	0.4	0	0.6
0.8165	1.1524	0	0.8165	0	1.4065	1.4065	3	0	0.6	0	0.4
0.8395	1.5498	0	0.8395	0	1.3647	1.3647	3	1	0.6	0.1	0.412
0.9911		0	0.9911	0	1.3645	1.3645	3	2	0.6	0.2	0.447
1.3645		0	1.3645	1	1.1524	1.1524	3	3	0.6	0.3	0.5
1.3647		0	1.3647	0	1.1428	1.1428	4	3	0.8	0.3	0.361 ⇐
0.8563		0	0.8563	0	0.9911	0.9911	4	4	0.8	0.4	0.447
1.4065		0	1.4065	0	0.9526	0.9526	4	5	0.8	0.5	0.539
		1	1.4308	0	0.8563	0.8563	4	6	0.8	0.6	0.632
		1	1.4577	0	0.8395	0.8395	4	7	0.8	0.7	0.728
$N_0 = 10$	$N_1 = 5$	1	0.8309	1	0.8309	0.8309	4	8	0.8	0.8	0.825
		1	1.1524	0	0.8165	0.8165	5	8	1	0.8	0.8
		1	1.5498	0	0.3511	0.3511	5	9	1	0.9	0.9
						0.0000	5	10	1	1	1
$N = N_0 + N_1 = 15$			Descending order (data)			$N+1$ Values	1's above T	0's	$\frac{N_C}{N_1}$	$\frac{N_F}{N_0}$	$[P_F, 1-P_M] \rightarrow [0,1]$

Fig. 3.16 The various steps in the ROC analysis. The shortest distance to the top left corner of the ROC plot, [0.1], from the ROC curve shown by an arrow ($\Leftarrow$)

measured values (pooled values) rewritten from the first two columns. Columns 3 and 4 are now sorted in descending order of the values, and columns 5 and 6 represent the sorted results.

To estimate the probability of false alarm and probability of detection, it is necessary to choose a threshold. The highest value of the observed quantity is chosen as the first threshold, and the numbers of 1's *above* the threshold are counted as the number of correct decisions (N_C), and the numbers of 0's are counted as the "false" decisions (N_F). Since the values are positive, the lowest value of the threshold is chosen as "0," and this value becomes the $(N + 1)^{th}$ threshold with $N = N_0 + N_1$. It is clear that there will be $N + 1$ value of N_C as well as N_F. The probability of detection is obtained as N_C/N_1, and the probability of false alarm is N_F/N_0.

A plot of the ROC curve appears in Fig. 3.17. An important question pertains to the basis for the appropriate interpretation of the ROC curve. To answer this question, we need a strategy. The area under the ROC curve is 0.78.

A simple strategy is based on the realization that in our experiments, we need to have the maximum value of the probability of detection and the minimum value of the probability of false alarm. This is known as the *Neyman-Pearson* strategy (Helstrom 1968). This strategy is implemented by determining the shortest distance to the top left hand corner of the ROC plot $[P_F = 0; P_D = 1]$ as

$$d = \sqrt{P_F^2 + (1 - P_D)^2} \tag{3.229}$$

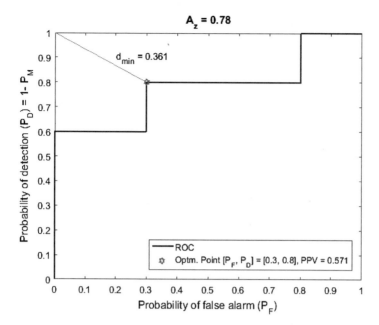

Fig. 3.17 ROC curve associated with the data in Table 3.1

$$d_{\min} = \sqrt{P_{F_{\min}}^2 + (1 - P_{D_{\max}})^2} \qquad (3.230)$$

It should be noted that $P_{F_{\min}}$ and $P_{D_{\max}}$ are based on the shortest distance to the top left corner.

The distances are given in the last column of Fig. 3.17, and the optimal operating point $[P_{F_{\min}}, P_{D_{\max}}]$ is [0.3, 0.8]. The threshold is chosen from the set of values of the data corresponding to the optimal operating point, and from Table 3.1, the optimum threshold is 1.1428.

Once a specific threshold is chosen to perform data analysis to determine whether a target is present or not, the next issue of interest is the percentage of truth that is associated with the decision. In other words, if a decision is made to conclude that a target exists on the basis of the threshold, how much trust can be placed on that decision? In other words, what are the probabilities of false alarm and miss at the optimum threshold? Issues similar to this were addressed in Chap. 2 where a posteriori probabilities were calculated. These issues can now be revisited now with the aid of the concept of densities and distribution functions of random variables. Let us identify the events D_p as target detected and D_n as target not detected. The transition matrix seen in Chap. 2 now becomes

$$T_X = \begin{bmatrix} P(D_n|H_0) & P(D_n|H_1) \\ P(D_p|H_0) & P(D_p|H_1) \end{bmatrix} \qquad (3.231)$$

In Eq. (3.231), the two hypotheses H_0 and H_1 are

$$P(\text{TargetAbsent}) = P(H_0)$$
$$P(\text{TargetPresent}) = P(H_1) \tag{3.232}$$

The probabilities of target detected and target not detected become

$$\begin{bmatrix} P(D_n) \\ P(D_p) \end{bmatrix} = \begin{bmatrix} P(D_n|H_0) & P(D_n|H_1) \\ P(D_p|H_0) & P(D_p|H_1) \end{bmatrix} \begin{bmatrix} P(H_0) \\ P(H_1) \end{bmatrix} = T_X \begin{bmatrix} P(H_0) \\ P(H_1) \end{bmatrix} \tag{3.233}$$

Let us now introduce the notion of random variable in the context of the data collected. The data set in any' experiment consists of random outcomes, and therefore, we may consider the values to belong to a random variable. We express the properties of the data collected in terms of two conditional density functions as

$$\begin{aligned} H_0 : \quad & f_V(v|H_0) \\ H_1 : \quad & f_V(v|H_1) \end{aligned} \tag{3.234}$$

In Eq. (3.234), $f_V(v)$ is the pdf of the random variable V. Using the definition of the probabilities of false alarm and miss defined earlier, we have

$$P_D = 1 - P_M = P(v|H_1 > v_T) = P(D_p|H_1) = 1 - F_V(v_T|H_1) = S_V(v_T|H_1)$$
$$P_F = P(v|H_0 > v_T) = P(D_p|H_0) = 1 - F_V(v_T|H_0) = S_V(v_T|H_0)$$
$$\tag{3.235}$$

In Eq. (3.235), v_T is the threshold. $F_V(.)$ is the cumulative distribution function and $S_V(.)$ is the survival function. The transition matrix in Eq. (3.231) now becomes

$$T_X = \begin{bmatrix} 1 - P_F & P_M \\ P_F & 1 - P_M \end{bmatrix} = \begin{bmatrix} F_V(v_T|H_0) & F_V(v_T|H_1) \\ S_V(v_T|H_0) & S_V(v_T|H_1) \end{bmatrix} \tag{3.236}$$

The association between the transition matrix defined in terms of simple probability concepts in Chap. 2 and the notion of random variables has now been established. This connection also establishes the link between data analytics and analysis of data using the concept of random variables. By virtue of the association between the transition matrix and the confusion matrix, one can also see the association between the confusion matrix and the distribution and survival functions of random variables.

One issue discussed in Chap. 2 in connection with the counts to obtain the probabilities of miss and false alarm pertained to the use of the phrase, "samples exceeding the value of threshold." Equation (3.236) shows that the probability of false alarm is the survival rate and probability of miss is the cumulative distribution function. This means that probability of false alarm should only include samples

Target Absent (H_0)					Target Present (H_1)		
0.114	0.827	1.619	2.834		0.203	3.656	6.324
0.159	0.844	1.637	3.364		1.364	3.89	6.55
0.485	0.873	1.669	3.92		1.619	4.158	7.085
0.518	1.118	1.757	4.054		1.789	4.484	7.947
0.597	1.133	1.87	4.332		2.866	5.102	8.135
0.609	1.195	2.092	4.809		2.974	5.481	8.334
0.683	1.21	2.203	5.413		3.292	5.613	9.358
0.775	1.573	2.357	6.319		3.478	5.726	10.753
0.799	1.596	2.4	12.761		3.556	6.254	13.175
0.818	1.609	2.544	13.17		3.617	6.269	13.778

Fig. 3.18 Data set #1

(values) *exceeding* the threshold. Probability of detection should only include samples (values) *exceeding* the threshold. Probability of miss on the other hand will include samples with values less than or equal to the threshold. This was discussed in connection with Fig. 3.16.

ROC analyses of three sets are shown as examples. The first one is based on a machine vision data consisting of 40 samples of target absent and 30 samples of target present (Fig. 3.18).

The confusion matrix and transition matrix based on the midpoint (see Chap. 2) as the threshold are shown below (Fig. 3.19).

The confusion matrix and transition matrix obtained using the intersection as the threshold are shown below (Fig. 3.20).

The ROC and the distances to the top left corner are shown including the distance (minimum one) for the optimum threshold (Fig. 3.21).

The confusion and transition matrices obtained using the optimal threshold are shown in Fig. 3.22.

The density fits with the optimum threshold are shown (P_F and P_M are shaded) in Fig. 3.23.

It is seen that the distance to the top left corner is lowest for the optimum threshold. The optimum threshold also leads to the lowest error rate (also matches the midpoint). This means that with this data set, the midpoint and optimal thresholds match.

The next example consists of the analysis of the data of size 60×60. The results are similar (Figs. 3.24, 3.25, 3.26, 3.27, 3.28, and 3.29).

The next example shows the results of a 70×30 set (Figs. 3.30, 3.31, 3.32, 3.33, 3.34, and 3.35).

Confusion Matrix

Threshold (v_T) = 2.834 (midpoint)

	Target Not Detected (D_n)	Target Detected (D_p)	Total Samples
Target Absent	31	False Alarm →9 ← N_F	40 ← N_0
Target Present	4 ← Miss	26 ← N_C	30 ← N_1
Total Decisions	35	35	70 ← N

$$\text{ERROR RATE} = \frac{N_F + (N_1 - N_C)}{N} = \frac{13}{70} \qquad \text{PPV} = \frac{N_C}{N_F + N_C} = \frac{26}{35}$$

Transition Matrix

$$T_X = \begin{bmatrix} P(D_n|H_0) & P(D_n|H_1) \\ P(D_p|H_0) & P(D_p|H_1) \end{bmatrix}$$

$$T_X = \begin{bmatrix} 1 - P_F & P_M \\ P_F & 1 - P_M \end{bmatrix} = \begin{bmatrix} 31 & 4 \\ 9 & 26 \end{bmatrix} \begin{bmatrix} \frac{1}{40} & 0 \\ 0 & \frac{1}{30} \end{bmatrix} = \begin{bmatrix} \frac{31}{40} & \frac{4}{30} \\ \frac{9}{40} & \frac{26}{30} \end{bmatrix}$$

a priori prob. →
$$\begin{bmatrix} P(H_0) \\ P(H_1) \end{bmatrix} = \frac{1}{N} \begin{bmatrix} N_0 \\ N_1 \end{bmatrix} = \frac{1}{70} \begin{bmatrix} 40 \\ 30 \end{bmatrix}$$

$$\begin{bmatrix} P(D_n) \\ P(D_p) \end{bmatrix} = T_X \begin{bmatrix} P(H_0) \\ P(H_1) \end{bmatrix} = \frac{1}{70} \begin{bmatrix} 35 \\ 35 \end{bmatrix}$$

Fig. 3.19 Transition and confusion matrices (data set #1, Fig. 3.18) associated with the midpoint

---

Confusion Matrix

Threshold (v_T) = 2.866 (intersection)

	Target Not Detected (D_n)	Target Detected (D_p)	Total Samples
Target Absent	31	False Alarm →9 ← N_F	40 ← N_0
Target Present	5 ← Miss	25 ← N_C	30 ← N_1
Total Decisions	36	34	70 ← N

$$\text{ERROR RATE} = \frac{N_F + (N_1 - N_C)}{N} = \frac{14}{70} \qquad \text{PPV} = \frac{N_C}{N_F + N_C} = \frac{25}{34}$$

Transition Matrix

$$T_X = \begin{bmatrix} P(D_n|H_0) & P(D_n|H_1) \\ P(D_p|H_0) & P(D_p|H_1) \end{bmatrix}$$

$$T_X = \begin{bmatrix} 1 - P_F & P_M \\ P_F & 1 - P_M \end{bmatrix} = \begin{bmatrix} 31 & 5 \\ 9 & 25 \end{bmatrix} \begin{bmatrix} \frac{1}{40} & 0 \\ 0 & \frac{1}{30} \end{bmatrix} = \begin{bmatrix} \frac{31}{40} & \frac{5}{30} \\ \frac{9}{40} & \frac{25}{30} \end{bmatrix}$$

a priori prob. →
$$\begin{bmatrix} P(H_0) \\ P(H_1) \end{bmatrix} = \frac{1}{N} \begin{bmatrix} N_0 \\ N_1 \end{bmatrix} = \frac{1}{70} \begin{bmatrix} 40 \\ 30 \end{bmatrix}$$

$$\begin{bmatrix} P(D_n) \\ P(D_p) \end{bmatrix} = T_X \begin{bmatrix} P(H_0) \\ P(H_1) \end{bmatrix} = \frac{1}{70} \begin{bmatrix} 36 \\ 34 \end{bmatrix}$$

Fig. 3.20 Transition and confusion matrices (data set #1, Fig. 3.18) associated with the intersection

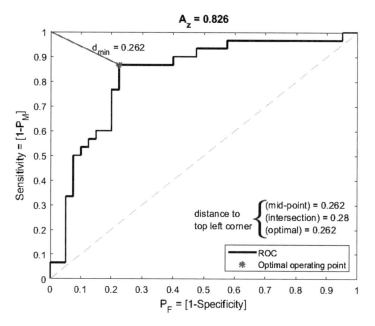

Fig. 3.21 ROC (data set #1 in Fig. 3.18)

Threshold (v$_T$) = 2.834 (optimal)

	Target Not Detected (D$_n$)	Target Detected (D$_p$)	Total Samples
Target Absent	31	**False Alarm** →9 ← N$_F$	40 ← N$_0$
Target Present	4 ← **Miss**	26 ← N$_C$	30 ← N$_1$
Total Decisions	35	35	70 ← N

$$\text{ERROR RATE} = \frac{N_F + (N_1 - N_C)}{N} = \frac{13}{70} \qquad \text{PPV} = \frac{N_C}{N_F + N_C} = \frac{26}{35}$$

$$T_X = \begin{bmatrix} P(D_n|H_0) & P(D_n|H_1) \\ P(D_p|H_0) & P(D_p|H_1) \end{bmatrix} = \begin{bmatrix} F_V(v_T|H_0) & F_V(v_T|H_1) \\ S_V(v_T|H_0) & S_V(v_T|H_1) \end{bmatrix}$$

$$T_X = \begin{bmatrix} 1 - P_F & P_M \\ P_F & 1 - P_M \end{bmatrix} = \begin{bmatrix} 31 & 4 \\ 9 & 26 \end{bmatrix} \begin{bmatrix} \frac{1}{40} & 0 \\ 0 & \frac{1}{30} \end{bmatrix} = \begin{bmatrix} \frac{31}{40} & \frac{4}{30} \\ \frac{9}{40} & \frac{26}{30} \end{bmatrix}$$

$$\text{a priori prob.} \rightarrow \begin{bmatrix} P(H_0) \\ P(H_1) \end{bmatrix} = \frac{1}{N} \begin{bmatrix} N_0 \\ N_1 \end{bmatrix} = \frac{1}{70} \begin{bmatrix} 40 \\ 30 \end{bmatrix}$$

$$\begin{bmatrix} P(D_n) \\ P(D_p) \end{bmatrix} = T_X \begin{bmatrix} P(H_0) \\ P(H_1) \end{bmatrix} = \frac{1}{70} \begin{bmatrix} 35 \\ 35 \end{bmatrix}$$

Transition Matrix   Confusion Matrix

Fig. 3.22 Confusion and transition matrices (data set #1 in Fig. 3.18) for the optimum threshold. The transition matrix is also expressed in terms of the CDF and survival function

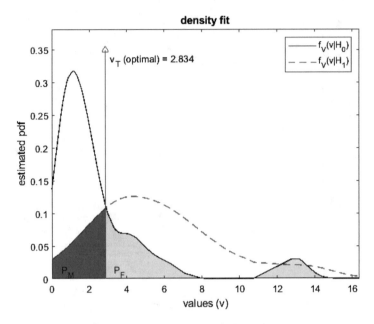

Fig. 3.23 P_F and P_M at optimum threshold (data set #1 in Fig. 3.18)

Target Absent (H_0)						Target Present (H_1)					
0.059	0.48	1.125	1.621	2.467	3.631	0.278	1.709	2.841	4.283	5.701	8.558
0.138	0.522	1.146	1.673	2.482	3.773	0.865	2.195	3.005	4.569	5.971	8.606
0.165	0.523	1.186	1.774	2.595	3.821	0.967	2.229	3.029	4.647	6.604	8.787
0.171	0.635	1.235	2.088	2.596	3.848	1.191	2.418	3.047	4.78	6.819	9.404
0.213	0.746	1.235	2.12	2.804	3.852	1.421	2.459	3.179	4.823	6.96	12.195
0.218	0.789	1.29	2.164	2.805	4.174	1.527	2.494	3.653	5.075	6.965	14.152
0.242	0.875	1.305	2.267	3.119	5.448	1.579	2.518	3.938	5.102	7.016	14.539
0.276	0.88	1.307	2.274	3.182	6.911	1.614	2.539	4.116	5.288	7.988	14.751
0.339	0.911	1.451	2.295	3.203	7.065	1.688	2.577	4.176	5.355	8.06	14.829
0.454	0.986	1.595	2.346	3.402	7.532	1.7	2.802	4.204	5.653	8.546	18.466

Fig. 3.24 Data set # 2

It is seen that the ROC curves provide the single measure of the performance in terms of the area under the curve (AUC or Az), while any direct relationship to the sample size is absent.

The optimal threshold is calculated without any indication on the sample size or more specifically the disease prevalence in the case of medical diagnostics. The lack of any association to the sample size and optimal threshold may lead to incorrect interpretation of the positive predictive values. The reasons for this can be traced to

Transition Matrix **Confusion Matrix**

Threshold (v_T) = 2.088 (midpoint)

	Target Not Detected (D_n)	Target Detected (D_p)	Total Samples
Target Absent	34	False Alarm →26 ← N_F	60 ← N_0
Target Present	11 ← Miss	49 ← N_C	60 ← N_1
Total Decisions	45	75	120 ← N

$$\text{ERROR RATE} = \frac{N_F + (N_1 - N_C)}{N} = \frac{37}{120} \quad \text{PPV} = \frac{N_C}{N_F + N_C} = \frac{49}{75}$$

$$T_X = \begin{bmatrix} P(D_n|H_0) & P(D_n|H_1) \\ P(D_p|H_0) & P(D_p|H_1) \end{bmatrix}$$

$$T_X = \begin{bmatrix} 1 - P_F & P_M \\ P_F & 1 - P_M \end{bmatrix} = \begin{bmatrix} 34 & 11 \\ 26 & 49 \end{bmatrix} \begin{bmatrix} \frac{1}{60} & 0 \\ 0 & \frac{1}{60} \end{bmatrix} = \begin{bmatrix} \frac{34}{60} & \frac{11}{60} \\ \frac{26}{60} & \frac{49}{60} \end{bmatrix}$$

$$\textit{a priori prob.} \rightarrow \begin{bmatrix} P(H_0) \\ P(H_1) \end{bmatrix} = \frac{1}{N} \begin{bmatrix} N_0 \\ N_1 \end{bmatrix} = \frac{1}{120} \begin{bmatrix} 60 \\ 60 \end{bmatrix}$$

$$\begin{bmatrix} P(D_n) \\ P(D_p) \end{bmatrix} = T_X \begin{bmatrix} P(H_0) \\ P(H_1) \end{bmatrix} = \frac{1}{120} \begin{bmatrix} 45 \\ 75 \end{bmatrix}$$

Fig. 3.25 Confusion and transition matrices (data set # 2, Fig. 3.24): midpoint as the threshold

Transition Matrix **Confusion Matrix**

Threshold (v_T) = 3.402 (intersection)

	Target Not Detected (D_n)	Target Detected (D_p)	Total Samples
Target Absent	50	False Alarm →10 ← N_F	60 ← N_0
Target Present	25 ← Miss	35 ← N_C	60 ← N_1
Total Decisions	75	45	120 ← N

$$\text{ERROR RATE} = \frac{N_F + (N_1 - N_C)}{N} = \frac{35}{120} \quad \text{PPV} = \frac{N_C}{N_F + N_C} = \frac{35}{45}$$

$$T_X = \begin{bmatrix} P(D_n|H_0) & P(D_n|H_1) \\ P(D_p|H_0) & P(D_p|H_1) \end{bmatrix}$$

$$T_X = \begin{bmatrix} 1 - P_F & P_M \\ P_F & 1 - P_M \end{bmatrix} = \begin{bmatrix} 50 & 25 \\ 10 & 35 \end{bmatrix} \begin{bmatrix} \frac{1}{60} & 0 \\ 0 & \frac{1}{60} \end{bmatrix} = \begin{bmatrix} \frac{50}{60} & \frac{25}{60} \\ \frac{10}{60} & \frac{35}{60} \end{bmatrix}$$

$$\textit{a priori prob.} \rightarrow \begin{bmatrix} P(H_0) \\ P(H_1) \end{bmatrix} = \frac{1}{N} \begin{bmatrix} N_0 \\ N_1 \end{bmatrix} = \frac{1}{120} \begin{bmatrix} 60 \\ 60 \end{bmatrix}$$

$$\begin{bmatrix} P(D_n) \\ P(D_p) \end{bmatrix} = T_X \begin{bmatrix} P(H_0) \\ P(H_1) \end{bmatrix} = \frac{1}{120} \begin{bmatrix} 75 \\ 45 \end{bmatrix}$$

Fig. 3.26 Confusion and transition matrices (data set # 2, Fig. 3.24): intersection as the threshold

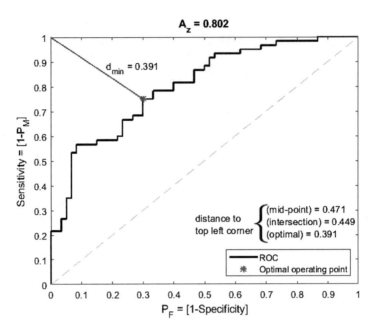

Fig. 3.27 ROC (data set # 2, Fig. 3.24)

Transition Matrix   Confusion Matrix

Threshold (v_T) = 2.482 (optimal)

	Target Not Detected (D_n)	Target Detected (D_p)	Total Samples
Target Absent	42	False Alarm → 18 ← N_F	60 ← N_0
Target Present	15 ← Miss	45 ← N_C	60 ← N_1
Total Decisions	57	63	120 ← N

$$\text{ERROR RATE} = \frac{N_F + (N_1 - N_C)}{N} = \frac{33}{120} \qquad \text{PPV} = \frac{N_C}{N_F + N_C} = \frac{45}{63}$$

$$T_X = \begin{bmatrix} P(D_n|H_0) & P(D_n|H_1) \\ P(D_p|H_0) & P(D_p|H_1) \end{bmatrix} = \begin{bmatrix} F_V(v_T|H_0) & F_V(v_T|H_1) \\ S_V(v_T|H_0) & S_V(v_T|H_1) \end{bmatrix}$$

$$T_X = \begin{bmatrix} 1 - P_F & P_M \\ P_F & 1 - P_M \end{bmatrix} = \begin{bmatrix} 42 & 15 \\ 18 & 45 \end{bmatrix} \begin{bmatrix} \frac{1}{60} & 0 \\ 0 & \frac{1}{60} \end{bmatrix} = \begin{bmatrix} \frac{42}{60} & \frac{15}{60} \\ \frac{18}{60} & \frac{45}{60} \end{bmatrix}$$

$$\text{a priori prob.} \rightarrow \begin{bmatrix} P(H_0) \\ P(H_1) \end{bmatrix} = \frac{1}{N} \begin{bmatrix} N_0 \\ N_1 \end{bmatrix} = \frac{1}{120} \begin{bmatrix} 60 \\ 60 \end{bmatrix}$$

$$\begin{bmatrix} P(D_n) \\ P(D_p) \end{bmatrix} = T_X \begin{bmatrix} P(H_0) \\ P(H_1) \end{bmatrix} = \frac{1}{120} \begin{bmatrix} 57 \\ 63 \end{bmatrix}$$

Fig. 3.28 Transition and confusion matrices (data set # 2, Fig. 3.24): optimal operating point

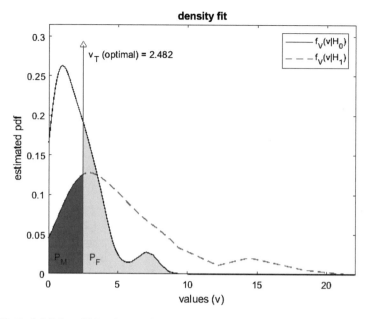

Fig. 3.29 Probabilities of false alarm and miss at optimal operating point (data set # 2, Fig. 3.24)

Target Absent (H_0)							Target Present (H_1)		
0.022	0.245	0.544	0.864	1.1	1.837	2.465	0.234	3.026	6.142
0.046	0.25	0.562	0.883	1.106	1.984	2.601	1.608	3.027	7.081
0.07	0.296	0.591	0.912	1.149	2.003	3.109	1.639	4.21	7.891
0.071	0.298	0.638	0.912	1.269	2.047	3.529	1.641	4.304	8.131
0.074	0.309	0.693	0.959	1.436	2.122	3.627	2.155	4.334	8.834
0.122	0.311	0.71	0.997	1.499	2.213	3.86	2.733	4.412	11.274
0.128	0.319	0.735	1.014	1.526	2.242	4.053	2.861	4.463	11.759
0.147	0.322	0.775	1.017	1.618	2.243	4.853	2.862	4.818	14.525
0.189	0.429	0.824	1.036	1.739	2.411	6.947	2.914	5.779	16.212
0.199	0.483	0.859	1.093	1.761	2.438	9.039	2.996	6.044	16.788

Fig. 3.30 Data set #3

the discussion of examples of binary channels involving cancer detection presented in Chap. 2 in connection with Bayes' rule. In those examples, PPV was calculated on the basis of a priori probabilities that were estimated from the sample sizes. This must be rectified for the proper estimation of PPV, and the recalculation of PPV is discussed in the next section.

Threshold (v_T) = 1.837 (midpoint)

Transition Matrix · Confusion Matrix *(left margin)*

	Target Not Detected (D_n)	Target Detected (D_p)	Total Samples
Target Absent	51	**False Alarm** →19 ← N_F	70 ← N_0
Target Present	4 ← **Miss**	26 ← N_C	30 ← N_1
Total Decisions	55	45	100 ← N

$$\text{ERROR RATE} = \frac{N_F + (N_1 - N_C)}{N} = \frac{23}{100} \qquad \text{PPV} = \frac{N_C}{N_F + N_C} = \frac{26}{45}$$

$$T_X = \begin{bmatrix} P(D_n|H_0) & P(D_n|H_1) \\ P(D_p|H_0) & P(D_p|H_1) \end{bmatrix}$$

$$T_X = \begin{bmatrix} 1 - P_F & P_M \\ P_F & 1 - P_M \end{bmatrix} = \begin{bmatrix} 51 & 4 \\ 19 & 26 \end{bmatrix} \begin{bmatrix} \frac{1}{70} & 0 \\ 0 & \frac{1}{30} \end{bmatrix} = \begin{bmatrix} \frac{51}{70} & \frac{4}{30} \\ \frac{19}{70} & \frac{26}{30} \end{bmatrix}$$

$$\text{a priori prob.} \rightarrow \begin{bmatrix} P(H_0) \\ P(H_1) \end{bmatrix} = \frac{1}{N} \begin{bmatrix} N_0 \\ N_1 \end{bmatrix} = \frac{1}{100} \begin{bmatrix} 70 \\ 30 \end{bmatrix}$$

$$\begin{bmatrix} P(D_n) \\ P(D_p) \end{bmatrix} = T_X \begin{bmatrix} P(H_0) \\ P(H_1) \end{bmatrix} = \frac{1}{100} \begin{bmatrix} 55 \\ 45 \end{bmatrix}$$

Fig. 3.31 Confusion and transition matrices (data set #3 in Fig. 3.30): midpoint

Threshold (v_T) = 2.601 (intersection)

Transition Matrix · Confusion Matrix *(left margin)*

	Target Not Detected (D_n)	Target Detected (D_p)	Total Samples
Target Absent	62	**False Alarm** →8 ← N_F	70 ← N_0
Target Present	5 ← **Miss**	25 ← N_C	30 ← N_1
Total Decisions	67	33	100 ← N

$$\text{ERROR RATE} = \frac{N_F + (N_1 - N_C)}{N} = \frac{13}{100} \qquad \text{PPV} = \frac{N_C}{N_F + N_C} = \frac{25}{33}$$

$$T_X = \begin{bmatrix} P(D_n|H_0) & P(D_n|H_1) \\ P(D_p|H_0) & P(D_p|H_1) \end{bmatrix}$$

$$T_X = \begin{bmatrix} 1 - P_F & P_M \\ P_F & 1 - P_M \end{bmatrix} = \begin{bmatrix} 62 & 5 \\ 8 & 25 \end{bmatrix} \begin{bmatrix} \frac{1}{70} & 0 \\ 0 & \frac{1}{30} \end{bmatrix} = \begin{bmatrix} \frac{62}{70} & \frac{5}{30} \\ \frac{8}{70} & \frac{25}{30} \end{bmatrix}$$

$$\text{a priori prob.} \rightarrow \begin{bmatrix} P(H_0) \\ P(H_1) \end{bmatrix} = \frac{1}{N} \begin{bmatrix} N_0 \\ N_1 \end{bmatrix} = \frac{1}{100} \begin{bmatrix} 70 \\ 30 \end{bmatrix}$$

$$\begin{bmatrix} P(D_n) \\ P(D_p) \end{bmatrix} = T_X \begin{bmatrix} P(H_0) \\ P(H_1) \end{bmatrix} = \frac{1}{100} \begin{bmatrix} 67 \\ 33 \end{bmatrix}$$

Fig. 3.32 Confusion and transition matrices (data set #3 in Fig. 3.30): intersection

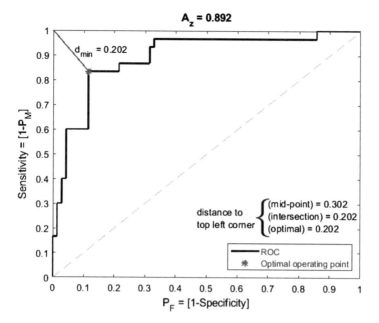

Fig. 3.33 ROC (data set #3 in Fig. 3.30)

Transition Matrix   Confusion Matrix

Threshold (v_T) = 2.601 (optimal)

	Target Not Detected (D_n)	Target Detected (D_p)	Total Samples
Target Absent	62 **False Alarm** →8 ← N_F		70 ← N_0
Target Present	5 ← **Miss**	25 ← N_C	30 ← N_1
Total Decisions	67	33	100 ← N

$$\text{ERROR RATE} = \frac{N_F + (N_1 - N_C)}{N} = \frac{13}{100} \qquad \text{PPV} = \frac{N_C}{N_F + N_C} = \frac{25}{33}$$

$$T_X = \begin{bmatrix} P(D_n|H_0) & P(D_n|H_1) \\ P(D_p|H_0) & P(D_p|H_1) \end{bmatrix} = \begin{bmatrix} F_V(v_T|H_0) & F_V(v_T|H_1) \\ S_V(v_T|H_0) & S_V(v_T|H_1) \end{bmatrix}$$

$$T_X = \begin{bmatrix} 1 - P_F & P_M \\ P_F & 1 - P_M \end{bmatrix} = \begin{bmatrix} 62 & 5 \\ 8 & 25 \end{bmatrix} \begin{bmatrix} \frac{1}{70} & 0 \\ 0 & \frac{1}{30} \end{bmatrix} = \begin{bmatrix} \frac{62}{70} & \frac{5}{30} \\ \frac{8}{70} & \frac{25}{30} \end{bmatrix}$$

$$\textit{a priori prob.} \rightarrow \begin{bmatrix} P(H_0) \\ P(H_1) \end{bmatrix} = \frac{1}{N} \begin{bmatrix} N_0 \\ N_1 \end{bmatrix} = \frac{1}{100} \begin{bmatrix} 70 \\ 30 \end{bmatrix}$$

$$\begin{bmatrix} P(D_n) \\ P(D_p) \end{bmatrix} = T_X \begin{bmatrix} P(H_0) \\ P(H_1) \end{bmatrix} = \frac{1}{100} \begin{bmatrix} 67 \\ 33 \end{bmatrix}$$

Fig. 3.34 Confusion and transition matrices (data set #3 in Fig. 3.30): optimal operating point

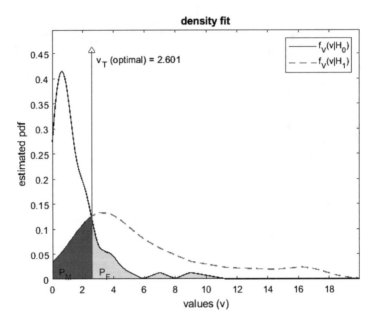

Fig. 3.35 Probabilities of false alarm and miss at the optimal operating point (data set #3 in Fig. 3.30)

3.9.3.3   Limitation of the ROC Analysis

While area under the ROC curve is useful measure, there is also an interest in determining the variation in the area under the ROC curve because of the limited size of the data. This aspect will be discussed in Chap. 5 when we will revisit the ROC curves and examine ways of improving the performance through modeling of the statistics of the two data sets.

The optimum threshold based on the shortest distance to the top left corner of the ROC curve given in Eq. (3.230) is not the only "optimum threshold." Another optimal threshold is based on Youden's index which is defined as the point on the ROC plot where the difference between the probabilities of detection and false alarm is the maximum (Youden 1950; Smits 2010; Roupp et al. 2018). It may be expressed as

$$[P_F, P_D]_{\text{optimum}} = [P_D + (1 - P_F) - 1]_{\text{max}} = [P_D - P_F]_{\text{max}} \qquad (3.237)$$

Neither one of the two optimal thresholds, one based on the proximity to the top left corner of the ROC curve and the other based on Youden's index, takes into account the actual prevalence. For example, if the occurrence of cancer in the general population is 1%, the a priori probabilities must reflect the actual values such as $p(H_1) = 0.01$ and $p(H_0) = 1-p(H_1)$. The threshold must also depend on the actual costs of the decision. These details are presented in Chap. 5 where the analysis is undertaken comparing optimal threshold and the corresponding positive predictive values based on all three approaches, closest to [0,1], Youden's index and prevalence and costs.

3.9.4 Mixture Densities

While we explored variables and densities used for modeling outcomes that are discrete, continuous, and mixed, such models may be inadequate in modeling the outcomes of some experiments. These experiments include examining responses from two or more chemicals mixed together, observing the image pixels from regions containing two or more distinct types of patterns in images, etc. This mixing aspect is not appropriately captured in the densities described so far even though Example 3.7 represented the case of mixture CDF associated with a mixture density (Papoulis and Pillai 2002; Shankar 2017).

To illustrate this issue, let us consider the case of an organization ordering a number of computers from three different manufacturers. Computers from each manufacturer have unique lifetimes. If A, B, and C are three manufacturers and 30% of the computers are from A, 20% are from B, and the remaining from C, we have

$$P(A) = p_1 = 0.3, \quad P(B) = p_2 = 0.2, \quad P(C) = p_3 = 0.5 \tag{3.238}$$

If a computer is randomly chosen, the probability that the computer fails within the first 5 years can be expressed as (X represents the lifetime) can be expressed by invoking Bayes' rule of total probability as

$$P(X \leq 5) = P(X \leq 5|A)P(A) + P(X \leq 5|B)P(B) + P(X \leq 5|C)P(C) \tag{3.239}$$

We can generalize the above equation to

$$P(X \leq x) = P(X \leq x|A)p_1 + P(X \leq x|B)p_2 + P(X \leq x|C)p_3 \tag{3.240}$$

Noting that $P(X \leq x)$ is the CDF, differentiation of Eq. (3.240) leads to

$$f_X(x) = f_X(x|A)p_1 + f_X(x|B)p_2 + f_X(x|C)p_3 \tag{3.241}$$

Equation (3.241) is a mixture density written as

$$f_X(x) = p_1 f_1(x) + p_2 f_2(x) + p_3 f_3(x) \tag{3.242}$$

Generalizing Eq. (3.242), we have

$$f_X(x) = \sum_{i=1}^{k} p_i f_i(x) \tag{3.243}$$

Equation (3.243) implies that

$$\sum_{i=1}^{k} p_i = 1 \tag{3.244}$$

$$\int_{-\infty}^{\infty} f_i(x)dx = 1, \quad i = 1, 2, \cdots, k \tag{3.245}$$

Equation (3.243) represents the density mixture and $f(x)$ on the left hand side is a mixture density (Shankar 2017).

The mixture model offers lots of possibilities of modeling outcomes because we can choose any valid pdf for the densities in Eq. (3.245) including non-identical ones. For $k = 3$, mixture models for the lifetime may be

$$f_X(x) = \sum_{i=1}^{3} p_i \frac{1}{\lambda_i} \exp\left(-\frac{x}{\lambda_i}\right) \tag{3.246}$$

$$f_X(x) = \sum_{i=1}^{3} p_i \frac{x}{b_i^2} \exp\left(-\frac{x^2}{2b_i^2}\right) \tag{3.247}$$

$$f_X(x) = p_1 \frac{1}{\lambda} \exp\left(-\frac{x}{\lambda}\right) + p_2 \frac{x}{b^2} \exp\left(-\frac{x^2}{2b^2}\right) + p_3 \frac{x^{a-1}}{c^a \Gamma(a)} \exp\left(-\frac{x}{c}\right) \tag{3.248}$$

While Eq. (3.246) is an exponential mixture, Eq. (3.247) is a Rayleigh mixture, and Eq. (3.248) has contributions from exponential, Rayleigh, and gamma densities.

A Rayleigh mixture with three components is displayed in Fig. 3.36.

A Gaussian mixture with three components is shown in Fig. 3.37.

3.9.5 Some Additional Aspects and Properties of Random Variables

3.9.5.1 Symmetric Random Variables

A random variable is called symmetric if

$$P\left(X \geq x\right) = P\left(X \leq -x\right) \tag{3.249}$$

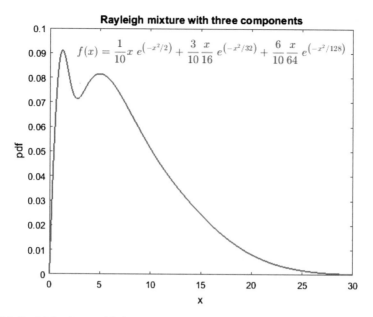

Fig. 3.36   Rayleigh mixture with three components

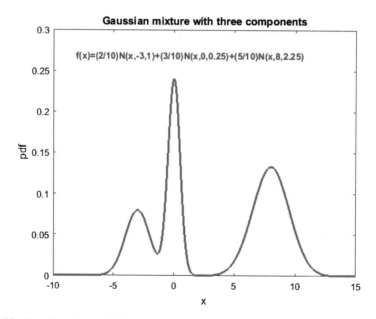

Fig. 3.37   Gaussian mixture with three components

This implies the following:

$$P(|X| \leq \alpha) = 2F(\alpha) - 1$$
$$P(|X| \geq \alpha) = 2[1 - F(\alpha)]$$
(3.250)

If the random variable is symmetric, the point of symmetry is the mean. For a symmetric random variable, the mean and median will be identical, where median x_m is defined as

$$\frac{1}{2} = \int_{-\infty}^{x_m} f(x)dx = \int_{x_m}^{\infty} f(x)dx$$
(3.251)

Equation (3.251) is identical to

$$\frac{1}{2} = P[X < x_m] = P[X > x_m]$$
(3.252)

3.9.5.2 Percentiles

Let α be such that $0 < \alpha < 1$. The $100(1-\alpha)^{th}$ percentile of a continuous random variable denoted by x_α is such that

$$F_X(x_\alpha) = P[X < x_\alpha] = 1 - \alpha$$
(3.253)

This means the following:

$\Rightarrow x_{0.5}$ is the median
$\Rightarrow x_{0.05}$ is the 95th percentile : separates the top 5%from the rest
$\Rightarrow x_{0.75}$ is the lower quartile
$\Rightarrow x_{0.25}$ is the upper quartile
(3.254)

3.9.5.3 Inequalities

The mean of a random variable (η) is often interpreted as the center of gravity, and the variance (σ^2) can be seen the moment of inertia with respect to the mean. For any arbitrary constant k, Tchebycheff's inequality is (Rohatgi and Saleh 2001; Papoulis and Pillai 2002)

$$P(\eta - k\sigma < X < \eta + k\sigma) \geq 1 - \frac{1}{k^2}$$
(3.255)

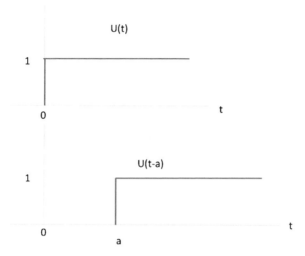

$$U(t) = \begin{cases} 1, & t > 0 \\ 0, & t < 0 \end{cases}$$

Fig. 3.38 Step functions

Equation (3.255) is also expressed as

$$P(|X - \eta| \geq k\sigma) \leq \frac{1}{k^2} \tag{3.256}$$

3.9.6 Step Functions and Delta Functions

1. A unit step function $U(t)$ is defined as (see Fig. 3.38)

$$U(t) = \begin{cases} 1, & t > 0 \\ 0, & t < 0 \end{cases} \tag{3.257}$$

2. A delta function is defined as (see Fig. 3.39)

$$\delta(t) = \begin{cases} 0, & t \neq 0 \\ \infty, & t = 0 \end{cases}$$

$$\int_{-\infty}^{\infty} \delta(t)dt = 1 \tag{3.258}$$

$$\delta(t) = \begin{cases} 0, & t \neq 0 \\ \infty, & t = 0 \end{cases}.$$

$$\int_{-\infty}^{\infty} \delta(t)\,dt = 1$$

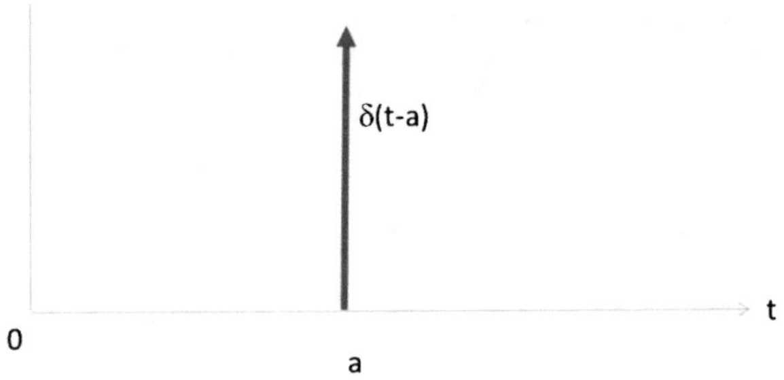

Fig. 3.39 Delta function

Equation (3.258) suggests that delta function integrates to unity even though it is infinite. Equation (3.258) further implies the following:

$$\int_{-3}^{5} \delta(t)\,dt = 1; \qquad \int_{-3}^{5} \delta(t-2)\,dt = 1 \qquad \int_{-3}^{0} \delta(t-2)\,dt = 0 \qquad (3.259)$$

In general, we have

$$\int_{A}^{B} \delta(t-C)\,dt = \begin{cases} 1, & A < C < B \\ 0, & \text{otherwise} \end{cases} \qquad (3.260)$$

Furthermore, Eq. (3.258) implies

$$\int_{t_1}^{t_2} f(t)\delta(t-a)\,dt = \begin{cases} f(a), & t_1 < a < t_2 \\ 0, & \text{otherwise} \end{cases} \qquad (3.261)$$

$$\int_{-\infty}^{\infty} f(t)\delta(t)\,dt = f(0) \qquad (3.262)$$

As an example, consider

$$f(t) = 10e^t \cos(t) \qquad (3.263)$$

$$\int_{-3}^{3} f(t)\delta(t)dt = \int_{-3}^{3} 10e^t \cos(t)\delta(t)dt = 10e^0 \cos(0) = 10 \qquad (3.264)$$

$$\int_{-3}^{3} f(t)\delta\left(t - \frac{\pi}{4}\right)dt = \int_{-3}^{3} 10e^t \cos(t)\delta\left(t - \frac{\pi}{4}\right)dt = 10e^{\frac{\pi}{4}} \cos\left(\frac{\pi}{4}\right) \qquad (3.265)$$

$$\int_{-3}^{3} f(t)\delta(t - 2\pi)dt = \int_{-3}^{3} 10e^t \cos(t)\delta(t - 2\pi)dt = 0 \qquad (3.266)$$

In Eq. (3.266), it can be seen that the value $2\pi = 6.28$ does not lie in the range $[-3,3]$, resulting in the integral being zero.

It should be noted that

$$\int_{0^-}^{0^+} \delta(t)dt = 1 \qquad (3.267)$$

Equation (3.267) implies that

$$\int_{0^-}^{\infty} \delta(t)dt = 1 \qquad \int_{0^+}^{\infty} \delta(t)dt = 1 \qquad (3.268)$$

This means 0 is actually 0^- starting from the lower end (left side) and the delta function is zero $[0^+,\infty]$. The delta function requires the lower limit be 0^-.

3. The derivative of the unit step function is a delta function,

$$U'(t) = \delta(t) \qquad (3.269)$$

4. The integral (*notice the limits of integration*) of a delta function is a step function,

$$U(t) = \int_{-\infty}^{t} \delta(y)dy \qquad (3.270)$$

3.9.7 *Normal Probability Table (Table 3.2)*

Table 3.2 Normal Probability Table

z	0	0.01	0.02	0.03	0.04	0.05	0.06	0.07	0.08	0.09
0	0.5	0.504	0.508	0.512	0.516	0.5199	0.5239	0.5279	0.5319	0.5359
0.1	0.5398	0.5438	0.5478	0.5517	0.5557	0.5596	0.5636	0.5675	0.5714	0.5753
0.2	0.5793	0.5832	0.5871	0.591	0.5948	0.5987	0.6026	0.6064	0.6103	0.6141
0.3	0.6179	0.6217	0.6255	0.6293	0.6331	0.6368	0.6406	0.6443	0.648	0.6517
0.4	0.6554	0.6591	0.6628	0.6664	0.67	0.6736	0.6772	0.6808	0.6844	0.6879
0.5	0.6915	0.695	0.6985	0.7019	0.7054	0.7088	0.7123	0.7157	0.719	0.7224
0.6	0.7257	0.7291	0.7324	0.7357	0.7389	0.7422	0.7454	0.7486	0.7517	0.7549
0.7	0.758	0.7611	0.7642	0.7673	0.7704	0.7734	0.7764	0.7794	0.7823	0.7852
0.8	0.7881	0.791	0.7939	0.7967	0.7995	0.8023	0.8051	0.8078	0.8106	0.8133
0.9	0.8159	0.8186	0.8212	0.8238	0.8264	0.8289	0.8315	0.834	0.8365	0.8389
1	0.8413	0.8438	0.8461	0.8485	0.8508	0.8531	0.8554	0.8577	0.8599	0.8621
1.1	0.8643	0.8665	0.8686	0.8708	0.8729	0.8749	0.877	0.879	0.881	0.883
1.2	0.8849	0.8869	0.8888	0.8907	0.8925	0.8944	0.8962	0.898	0.8997	0.9015
1.3	0.9032	0.9049	0.9066	0.9082	0.9099	0.9115	0.9131	0.9147	0.9162	0.9177
1.4	0.9192	0.9207	0.9222	0.9236	0.9251	0.9265	0.9279	0.9292	0.9306	0.9319
1.5	0.9332	0.9345	0.9357	0.937	0.9382	0.9394	0.9406	0.9418	0.9429	0.9441
1.6	0.9452	0.9463	0.9474	0.9484	0.9495	0.9505	0.9515	0.9525	0.9535	0.9545
1.7	0.9554	0.9564	0.9573	0.9582	0.9591	0.9599	0.9608	0.9616	0.9625	0.9633
1.8	0.9641	0.9649	0.9656	0.9664	0.9671	0.9678	0.9686	0.9693	0.9699	0.9706
1.9	0.9713	0.9719	0.9726	0.9732	0.9738	0.9744	0.975	0.9756	0.9761	0.9767
2	0.9772	0.9778	0.9783	0.9788	0.9793	0.9798	0.9803	0.9808	0.9812	0.9817
2.1	0.9821	0.9826	0.983	0.9834	0.9838	0.9842	0.9846	0.985	0.9854	0.9857
2.2	0.9861	0.9864	0.9868	0.9871	0.9875	0.9878	0.9881	0.9884	0.9887	0.989
2.3	0.9893	0.9896	0.9898	0.9901	0.9904	0.9906	0.9909	0.9911	0.9913	0.9916
2.4	0.9918	0.992	0.9922	0.9925	0.9927	0.9929	0.9931	0.9932	0.9934	0.9936
2.5	0.9938	0.994	0.9941	0.9943	0.9945	0.9946	0.9948	0.9949	0.9951	0.9952

2.6	0.9953	0.9955	0.9956	0.9957	0.9959	0.996	0.9961	0.9962	0.9963	0.9964
2.7	0.9965	0.9966	0.9967	0.9968	0.9969	0.997	0.9971	0.9972	0.9973	0.9974
2.8	0.9974	0.9975	0.9976	0.9977	0.9977	0.9978	0.9979	0.9979	0.998	0.9981
2.9	0.9981	0.9982	0.9982	0.9983	0.9984	0.9984	0.9985	0.9985	0.9986	0.9986
3	0.9987	0.9987	0.9987	0.9988	0.9988	0.9989	0.9989	0.9989	0.999	0.999
3.1	0.999	0.9991	0.9991	0.9991	0.9992	0.9992	0.9992	0.9992	0.9993	0.9993
3.2	0.9993	0.9993	0.9994	0.9994	0.9994	0.9994	0.9994	0.9995	0.9995	0.9995
3.3	0.9995	0.9995	0.9995	0.9996	0.9996	0.9996	0.9996	0.9996	0.9996	0.9997
3.4	0.9997	0.9997	0.9997	0.9997	0.9997	0.9997	0.9997	0.9997	0.9997	0.9997
3.5	0.9998	0.9998	0.9998	0.9998	0.9998	0.9998	0.9998	0.9998	0.9998	0.9998

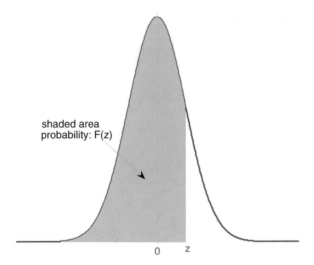

Fig. 3.40 Shaded area represents the CDF

$$F(z) = \frac{1}{\sqrt{2\pi}} \int_{-\infty}^{z} e^{-\frac{x^2}{2}} dx, \qquad z \geq 0 \tag{3.271}$$

What is the z-score? (Fig. 3.40)

$$z = \frac{x - \mu}{\sigma} = z_{\text{score}} \tag{3.272}$$

$$F(-z) = \frac{1}{\sqrt{2\pi}} \int_{-\infty}^{-z} e^{-\frac{x^2}{2}} dx = 1 - F(z), z > 0 \tag{3.273}$$

3.10 Summary

The concept of a random variable and its applications to engineering has been presented in detail. The various random variables (continuous, discrete, and mixed), associated densities, and cumulative distributions have been described in the proper context of issues of interest to engineering. The chapter also offered a preliminary peek into machine learning and machine vision problems through the introduction of the topics of error rates and receiver operating characteristics. Keeping up with the theme of the book, examples and exercises pertaining to data analytics are also offered. We revisited Bayes' rule within the context of extracting the probabilities of false alarm and miss with the availability of the density functions. A likelihood ratio test was also introduced to estimate the optimum threshold for the detection problems.

Three reports are presented below, one on random variables, one on transformations, and one on conditional densities and expected values.

3.10.1 Overview of Densities and Their Properties

An overview of the properties of several random variables that are generally used in engineering applications is given. All the results are created using the Symbolic toolbox in Matlab. In addition to the densities seen in earlier sections, gamma density and its relationships to other densities are described. Similarly, the relationships among Rayleigh to Nakagami, Rician, and Hoyt densities are explained. The descriptions of the properties of the random variables often contain citations of functions such as error functions, complimentary functions, Gaussian Q functions, imaginary error functions, gamma, upper and lower incomplete gamma functions, etc. These are catalogued. Similarly, the summary also provides the relationships among some of the discrete variables (Poisson, binomial, etc.) to Gaussian density.

Continuous Random Variables

CDF, pdf, moments and Characteristic Function

CDF →
$$F_X(x) = P(X \le x), \quad F_X(-\infty) = 0, \quad F_X(\infty) = 1$$

pdf →
$$f_X(x) = \frac{dF_X(x)}{dx}, \quad 0 \le f_X(x) \le \infty$$

CDF →
$$F_X(x) = \int_{-\infty}^{x} f_X(y)dy, \quad 0 \le F_X(x) \le 1$$

Probability →
$$P[x_1 \le X \le x_2] = \int_{x_1}^{x_2} f_X(x)dx = F_X(x_2) - F_X(x_1)$$

n^{th} moment →
$$E[X^n] = \int_{-\infty}^{\infty} x^n f_X(x)dx$$

Characteristic Function →
$$\phi_X(w) = E[e^{iXw}] = \int_{-\infty}^{\infty} e^{ixw} f_X(x)dx$$

Laplace Transform → $f_X(x), \; x>0$
$$L_X(s) = E[e^{-Xs}] = \phi_X(w)|_{(-jw=s)} = \int_0^{\infty} e^{-xs} f_X(x)dx$$

Properties → pdf and CDF
$$\begin{bmatrix} P[X \le x] \equiv P[X < x] = \int_{-\infty}^{x} f_X(\alpha)d\alpha \\ P[X = x] = 0, \quad \text{Continuous density} \\ P[X \ge x] \equiv P[X > x] = 1 - F_X(x) = \int_{x}^{\infty} f_X(\alpha)d\alpha \end{bmatrix}$$

Mean →
$$E[X] = \int_{-\infty}^{\infty} x f_X(x)dx$$

Variance →
$$var[X] = E[X^2] - (E[X])^2$$

Exponential pdf

pdf, CDF, Moments, Char. Function and Laplace Transform

pdf $\rightarrow$ $f_X(x) = \dfrac{e^{-\frac{x}{a}}}{a}$ $0 \le x \le \infty$

CDF $\rightarrow$ $F_X(x) = 1 - e^{-\frac{x}{a}}$

Mean $\rightarrow$ $E_X = a$

Variance $\rightarrow$ $\mathrm{var}_X = a^2$

Char. Fn $\rightarrow$ $G(w) = -\dfrac{1}{-1 + a\,w\,\mathrm{li}}$

Laplace $\rightarrow$ $L(s) = \dfrac{1}{a\,s + 1}$
Transform

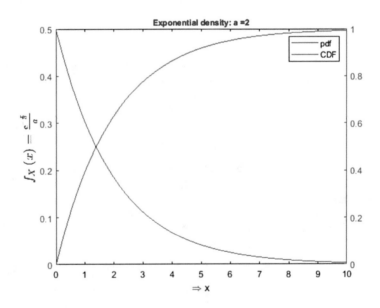

Gamma pdf

pdf, CDF, Moments, Char. Function and Laplace Transform

pdf $\rightarrow$ $f_X\left(x\right) = \frac{x^{a-1}\,e^{-\frac{x}{b}}}{b^a\,\Gamma(a)}$ $0 \le x \le \infty$

CDF $\rightarrow$ $F_X\left(x\right) = 1 - \frac{\Gamma\left(a,\frac{x}{b}\right)}{\Gamma(a)}$

Mean $\rightarrow$ $E_X = a\,b$

Variance $\rightarrow$ $\mathrm{var}_X = a\,b^2$

Char. Fn $\rightarrow$ $G\left(w\right) = \dfrac{\Gamma(a)-\left(\lim_{x\to\infty}\Gamma\left(a,-\frac{x(-1+b\,w\,\mathrm{1i})}{b}\right)\right)}{\Gamma(a)\,(1-b\,w\,\mathrm{1i})^a}$

Laplace $\rightarrow$ $L\left(s\right) = \frac{1}{(b\,s+1)^a}$
Transform

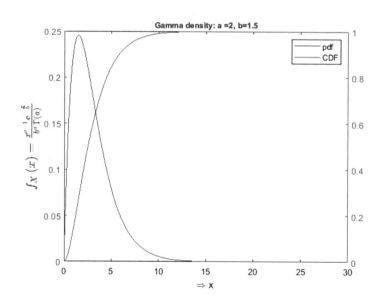

Gamma and related densities

---

gamma pdf →
G(a,b)

$$f_X(x) = \frac{x^{a-1} e^{-\frac{x}{b}}}{b^a \Gamma(a)}$$

$0 \le x \le \infty$

---

exp pdf →

$$f_X(x) = \frac{e^{-\frac{x}{b}}}{b}$$

$\rightarrow \quad G(1,b)$

χ^2 pdf →

$$f_X(x) = \frac{x^{\frac{c}{2}-1} e^{-\frac{x}{2}}}{2^{c/2} \Gamma\left(\frac{c}{2}\right)}$$

$\rightarrow \quad G\left(\frac{c}{2}, 2\right), \quad integer \ c$

Erlang pdf →

$$f_X(x) = \frac{x^{c-1} e^{-\frac{x}{b}}}{b^c \Gamma(c)}$$

$\rightarrow \quad G(c,b), \quad integer \ c$

Lognormal pdf

pdf, CDF, Moments, Char. Function and Laplace Transform

pdf   →
$$f_X(x) = \frac{\sqrt{2} e^{-\frac{(m-\ln(x))^2}{2s^2}}}{2\,s\,x\,\sqrt{\pi}}$$

$0 \le x \le \infty$

CDF   →
$$F_X(x) = \frac{1}{2} - \frac{\mathrm{erf}\left(\frac{\sqrt{2}m - \sqrt{2}\ln(x)}{2s}\right)}{2}$$

Mean   →
$$E_X = m\,e^{\frac{s^2}{2}}$$

Variance   →
$$\mathrm{var}_X = m^2 e^{s^2}\left(e^{s^2} - 1\right)$$

Char. Fn   →   NO Analytical expression

Laplace
Transform   →   No Analytical Expression

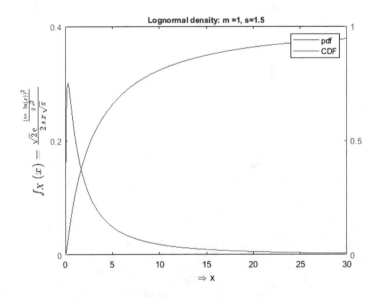

Lognormal density: m =1, s=1.5

Normal pdf

pdf, CDF, Moments, Char. Function and Laplace Transform

pdf   → $f_X(x) = \dfrac{\sqrt{2}\,e^{-\frac{(m-x)^2}{2s^2}}}{2s\sqrt{\pi}}$ $-\infty \le x \le \infty$

CDF   → $F_X(x) = \dfrac{1}{2} - \dfrac{\mathrm{erf}\left(\frac{\sqrt{2}(m-x)}{2s}\right)}{2}$

Mean   → $E_X = m$

Variance → $\mathrm{var}_X = s^2$

Char. Fn → $G(w) = e^{-\frac{s^2 w^2}{2} + m\,w\,1\mathrm{i}}$

Laplace →
Transform   Undefined $(-\infty < x < \infty)$

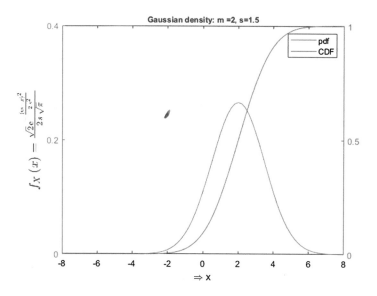

Rayleigh pdf

pdf, CDF, Moments, Char. Function and Laplace Transform

pdf $\rightarrow$ $f_X(x) = \dfrac{x e^{-\frac{x^2}{2b^2}}}{b^2}$ $\qquad 0 \le x \le \infty$

CDF $\rightarrow$ $F_X(x) = 1 - e^{-\frac{x^2}{2b^2}}$

Mean $\rightarrow$ $E_X = \dfrac{\sqrt{2}\,b\,\sqrt{\pi}}{2}$

Variance $\rightarrow$ $\mathrm{var}_X = -\dfrac{b^2(\pi-4)}{2}$

Char. Fn $\rightarrow$ $G(w) = \dfrac{b\,w\,e^{-\frac{b^2 w^2}{2}}\sqrt{-2\pi}}{2} - \dfrac{b\,w\,e^{-\frac{b^2 w^2}{2}}\,\mathrm{erfi}\left(\frac{\sqrt{2}\,b\,w}{2}\right)\sqrt{2\pi}}{2}$

Laplace Transform $\rightarrow$ $L(s) = 1 - \dfrac{\sqrt{2}\,b\,s\,\sqrt{\pi}\,e^{\frac{b^2 s^2}{2}}\,\mathrm{erfc}\left(\frac{\sqrt{2}\,b\,s}{2}\right)}{2}$

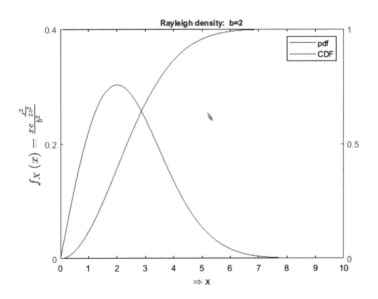

Rayleigh density: b=2

Weibull pdf

pdf, CDF, Moments, Char. Function and Laplace Transform

pdf $\rightarrow$ $f_X(x) = \dfrac{b\,x^{b-1}\,e^{-\left(\frac{x}{a}\right)^b}}{a^b}$ $\qquad\qquad$ $0 \le x \le \infty$

CDF $\rightarrow$ $F_X(x) = 1 - e^{-\left(\frac{x}{a}\right)^b}$

Mean $\rightarrow$ $E_X = \dfrac{a\,\Gamma\left(\frac{1}{b}\right)}{b}$

Variance $\rightarrow$ $\mathrm{var}_X = a^2\,\Gamma\left(\frac{2}{b}+1\right) - \dfrac{a^2\,\Gamma\left(\frac{1}{b}\right)^2}{b^2}$

Char. Fn $\rightarrow$ NO Analytical expression

Laplace Transform $\rightarrow$ No Analytical Expression

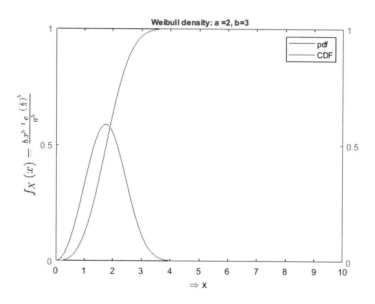

Weibull density: a = 2, b = 3

Uniform pdf

pdf, CDF, Moments, Char. Function and Laplace Transform

pdf $\rightarrow$ $f_X(x) = -\dfrac{1}{a-b}$ $a \leq x \leq b$

CDF $\rightarrow$ $F_X(x) = \dfrac{a-x}{a-b}$

Mean $\rightarrow$ $E_X = \dfrac{a}{2} + \dfrac{b}{2}$

Variance $\rightarrow$ $\mathrm{var}_X = \dfrac{(a-b)^2}{12}$

Char. Fn $\rightarrow$ $G(w) = -\dfrac{\left(e^{aw\,1i} - e^{bw\,1i}\right)1i}{w(a-b)}$

Laplace $\rightarrow$
Transform   Undefined $(-\infty < a,b < \infty)$

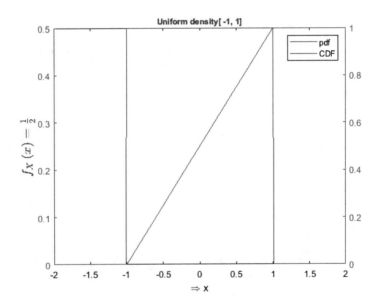

gamma functions

gamma function
$$\Gamma(x) = \int_0^\infty exp(-t)t^{x-1}dt$$

upper incomplete gamma function
$$\Gamma(a,x) = \int_x^\infty exp(-t)t^{a-1}dt$$

lower incomplete gamma function
$$\gamma(a,x) = \int_0^x exp(-t)t^{a-1}dt$$

$$\Gamma(x+1) = x\Gamma(x)$$
$$\Gamma(2) = \Gamma(1) = 1$$
$$\Gamma(\frac{1}{2}) = \sqrt{\pi}$$
$$\Gamma(-\frac{1}{2}) = -2\sqrt{\pi}$$

erf(.), erfc(.), Q(.) and erfi(.)

Error Function
$$erf(x) = \frac{2}{\sqrt{\pi}} \int_0^x e^{-z^2}dz$$

Complementary Error Function
$$erfc(x) = \frac{2}{\sqrt{\pi}} \int_x^\infty e^{-z^2}dz = 1 - erf(x)$$

Gaussian Q Function
$$Q(x) = \frac{1}{\sqrt{2\pi}} \int_x^\infty e^{-z^2/2}dz$$

Imaginary Error Function
$$erf\,i(x) = i\,erf(i\,x)$$

erfc(.) ⇒ Q(.)
$$erfc(x) = 2Q\left(x\sqrt{2}\right)$$

Q(.) ⇒ erfc(.)
$$Q(x) = \frac{1}{2}erfc\left(\frac{x}{\sqrt{2}}\right)$$

gamma function
$$\Gamma(x) = \int_0^\infty t^{x-1}e^{-t}dt$$

CDF, pdf, nth moment & Characteristic Function
Discrete Random Variables

Probability Mass Function $\rightarrow$ $P_X(k) = P(X = x_k)$

pdf $\rightarrow$

$$f_X(x) = \sum_k P_X(k)\delta(x - x_k)$$

CDF $\rightarrow$

$$F_X(x) = \sum_k P_X(k)U(x - x_k)$$

nth moment $\rightarrow$

$$E[X^n] = \sum_k [x_k]^n \, P_X(k)$$

Characteristic Function $\rightarrow$ $\phi_X(\omega) = E\left[e^{iX\omega}\right] = \sum_k e^{i\omega x_k} \, P_X(k)$

Poisson density

Prob. Mass Function $P_X(k) = \dfrac{\lambda^k e^{-\lambda}}{k!} \equiv f(k; \lambda), k = 0, 1, 2, .., \infty$

pdf

$$f_X(x) = \sum_{k=0}^{\infty} P_X(k)\delta(x - k)$$

CDF

$$F_X(x) = \sum_{k=0}^{\infty} P_X(k)U(x - k)$$

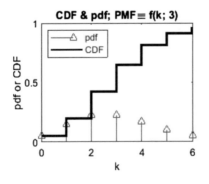

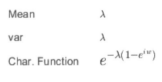

Mean	λ
var	λ
Char. Function	$e^{-\lambda(1-e^{i\omega})}$

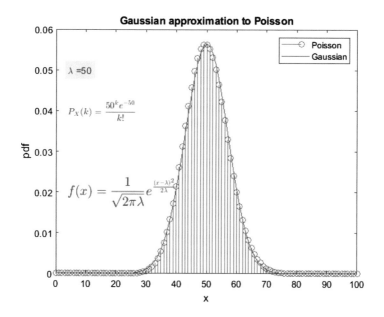

Geometric density

Prob. Mass Function $P_X(k) = p\,q^k \equiv f(k;p), k = 0, 2, .., \infty, (p+q) = 1$

pdf $$f_X(x) = \sum_{k=0}^{\infty} P_X(k)\delta(x-k)$$

CDF $$F_X(x) = \sum_{k=0}^{\infty} P_X(k)U(x-k)$$

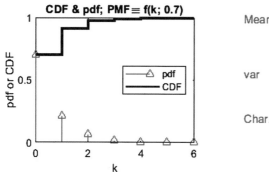

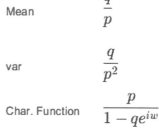

Mean $\dfrac{q}{p}$

var $\dfrac{q}{p^2}$

Char. Function $\dfrac{p}{1 - qe^{iw}}$

Geometric density (alternate form)

Prob. Mass Function $P_X(k) = p\,q^{k-1} \equiv f(k;p), k = 1, 2, .., \infty, (p+q) = 1$

pdf $$f_X(x) = \sum_{k=1}^{\infty} P_X(k)\delta(x - k)$$

CDF $$F_X(x) = \sum_{k=1}^{\infty} P_X(k)U(x - k)$$

Mean $$\frac{1}{p}$$

var $$\frac{q}{p^2}$$

Char. Function $$\frac{p}{e^{-iw} - q}$$

Binomial density

Prob. Mass Function $P_X(k) = \begin{pmatrix} n \\ k \end{pmatrix} p^k q^{n-k} \equiv f(k;n,p), k = 0, 1, .., n, (p+q) = 1$

pdf $$f_X(x) = \sum_{k=0}^{\infty} P_X(k)\delta(x - k)$$

CDF $$F_X(x) = \sum_{k=0}^{\infty} P_X(k)U(x - k)$$

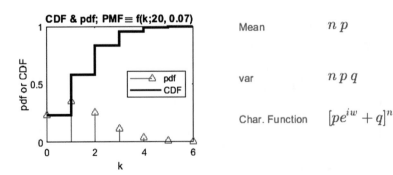

Mean $n\,p$

var $n\,p\,q$

Char. Function $[pe^{iw} + q]^n$

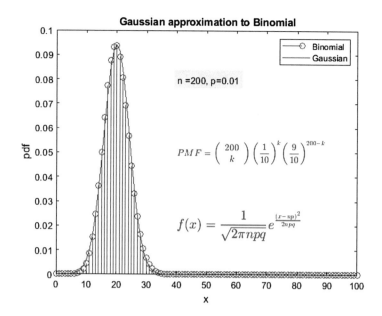

Negative Binomial (Pascal) density

Prob. Mass Function $P_X(k) = \begin{pmatrix} r+k-1 \\ k \end{pmatrix} p^r q^k, \quad k = 0, 1, .., \infty, (p+q) = 1$

pdf $\qquad f_X(x) = \sum_{k=0}^{\infty} P_X(k)\delta(x-k)$

CDF $\qquad F_X(x) = \sum_{k=0}^{\infty} P_X(k)U(x-k)$

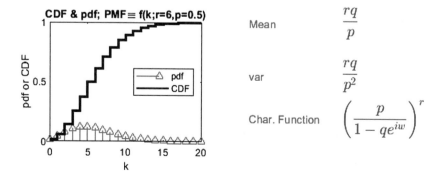

Mean $\qquad \dfrac{rq}{p}$

var $\qquad \dfrac{rq}{p^2}$

Char. Function $\qquad \left(\dfrac{p}{1 - qe^{iw}} \right)^r$

Negative Binomial (Pascal) density (alternate form)

Prob. Mass Func. $P_X(k) = \begin{pmatrix} k-1 \\ r-1 \end{pmatrix} p^r q^{k-r}, \ \ k = r, r+1, .., \infty, (p+q) = 1$

pdf $f_X(x) = \sum_{k=r}^{\infty} P_X(k)\delta(x-k)$

CDF $F_X(x) = \sum_{k=r}^{\infty} P_X(k)U(x-k)$

Mean $\dfrac{r}{p}$

var $\dfrac{rq}{p^2}$

Char. Function $\left(\dfrac{p}{e^{-iw} - q}\right)^r$

3.10.2 Moment Generating Function and Its Use: Exponential, Gamma, Rayleigh, Uniform, Gaussian, and Poisson Densities

The MGF has been generated using Matlab for a few densities. The moments are obtained by comparing the appropriate coefficients in MGF to those of the expected value exp.(Xt) obtained using the Taylor series expansion for exp.(.).

Establish the Relationship Between MGF and Moments (Matlab)

MGF of the random variable X is $E[e^{\wedge}(x*t)]$. Expanding $e^{\wedge}(X*t)$ in Taylor series, we get $e^{\wedge}(X*t) = 1 + X*t + (X^2*t^2)/2! + (X^3*t^3)/3! + ..$ If we take the expected value of $e^{\wedge}(X*t)$ using the Taylor series expansion, we get the expression for the MGF as $1 + tE(X) + E(X^2)(t^2/2) + E(X^3)(t^3/3!) + ...$ BY equating the MGF obtained directly to the MGF obtained from the series expansion, we see the relationship between the terms. Coefficient of t in MGF directly will be $E(X)$; coefficient of t^2 in MGF directly will be $E(X^2)/2$, coefficient of t^3 in MGF directly will be $E(X^3)/3!$, etc.

Exponential Density

```
pdf is ...>exp(-x/a)/a
Laplace transform of the density ..>1/(a*(w + 1/a))
MGF from Laplace by replacing s by -t  ...>-1/(a*(t - 1/a))
Taylor series expansion for MGF ...> a*t + a^2*t^2 + a^3*t^3 + 1
```

```
Coefficients of powerts of t in MGF ...>matrix([[1, a, a^2, a^3]])
Series expansion of exp(X*t) ..>X*t + (X^2*t^2)/2 + (X^3*t^3)/6 + 1
Coefficients of powerts of t in expansion of exp(X*t) ...>matrix([[1, X,
X^2/2, X^3/6]])
E_X0 = 1
E_X1 = a
E_X2 = 2*a^2
E_X3 = 6*a^3
var_X = a^2
Now obtain MGF directly instead of using the Laplace transform
MGF = -(a/(a*t - 1) - (a*limit(exp(x*t)*exp(-x/a), x, Inf))/(a*t - 1))/a
MGF = -1/(a*t - 1)
```

Gamma Density

```
pdf is ...>(x^(a - 1)*exp(-x/b))/(b^a*gamma(a))
MGF  ...>1/(b^a*(1/b - t)^a)
Taylor series expansion for MGF ...> a*b*t + (a*b^2*t^2*(a + 1))/2 +
(a*b^3*t^3*(a + 1)*(a + 2))/6 + 1
Coefficients of powerts of t in MGF ...>matrix([[1, a*b, (a*b^2*(a + 1))/
2, (a*b^3*(a + 1)*(a + 2))/6]])
Series expansion of exp(X*t) ..>X*t + (X^2*t^2)/2 + (X^3*t^3)/6 + 1
Coefficients of powerts of t in expansion of exp(X*t) ...>matrix([[1, X,
X^2/2, X^3/6]])


E_X0 = 1
E_X1 = a*b
E_X2 = a*b^2*(a + 1)
E_X3 = a*b^3*(a + 1)*(a + 2)
var_X = a*b^2
```

Rayleigh Density

```
pdf is ...>(x*exp(-x^2/(2*b^2)))/b^2
MGF  ...>(2^(1/2)*b*t*pi^(1/2)*exp((b^2*t^2)/2))/2 + (2^(1/2)
*b*t*pi^(1/2)*exp((b^2*t^2)/2)*erf((2^(1/2)*b*t)/2))/2 + 1
Taylor series expansion for MGF ...> b^2*t^2 + (2^(1/2)*b^3*t^3*pi^
(1/2))/4 + (2^(1/2)*b*t*pi^(1/2))/2 + 1
Coefficients of powerts of t in MGF ...>matrix([[1, (2^(1/2)*b*pi^
(1/2))/2, b^2, (2^(1/2)*b^3*pi^(1/2))/4]])
Series expansion of exp(X*t) ..>X*t + (X^2*t^2)/2 + (X^3*t^3)/6 + 1
Coefficients of powerts of t in expansion of exp(X*t) ...>matrix([[1, X,
X^2/2, X^3/6]])


E_X0 = 1
E_X1 = (2^(1/2)*b*pi^(1/2))/2
E_X2 = 2*b^2
E_X3 = (3*2^(1/2)*b^3*pi^(1/2))/2
var_X = 2*b^2 - (pi*b^2)/2
```

Uniform Density

```
pdf is ...>-1/(a - b), a<x<b
MGF  ...>(exp(a*t) - exp(b*t))/(t*(a - b))
Taylor series expansion for MGF ...> (t^2*(a*b + a^2 + b^2))/6 + (t*(a +
b))/2 + (t^3*(a + b)*(a^2 + b^2))/24 + 1
Coefficients of powerts of t in MGF ...>matrix([[1, a/2 + b/2, (a*b)/6 +
```

```
a^2/6 + b^2/6, ((a + b) * (a^2 + b^2))/24]])
Series expansion of exp(X*t) ..>X*t + (X^2*t^2)/2 + (X^3*t^3)/6 + 1
Coefficients of powerts of t in expansion of exp(X*t) ...>matrix([[1, X,
X^2/2, X^3/6]])
```

```
E_X0 = 1
E_X1 = a/2 + b/2
E_X2 = a^2/3 + (a*b)/3 + b^2/3
E_X3 = ((a + b) * (a^2 + b^2))/4
var_X = (a - b)^2/12
```

Gaussian Density

```
pdf is ...> (2^(1/2)*exp(-(m - x)^2/(2*sig^2)))/(2*sig*pi^(1/2))
MGF  ...>exp((sig^2*t^2)/2 + m*t)
Taylor series expansion for MGF ...> (t^2*(m^2 + sig^2))/2 + m*t +
(m*t^3*(m^2 + 3*sig^2))/6 + 1
Coefficients of powerts of t in MGF ...>matrix([[1, m, m^2/2 + sig^2/2,
(m*(m^2 + 3*sig^2))/6]])
Series expansion of exp(X*t) ..>X*t + (X^2*t^2)/2 + (X^3*t^3)/6 + 1
Coefficients of powerts of t in expansion of exp(X*t) ...>matrix([[1, X,
X^2/2, X^3/6]])
```

```
E_X0 = 1
E_X1 = m
E_X2 = m^2 + sig^2
E_X3 = m*(m^2 + 3*sig^2)
var_X = sig^2
```

Poisson

```
PMF of Poisson density ..>(lam^k*exp(-lam))/factorial(k)
MGF  ...>exp(lam*exp(t))*exp(-lam)
Taylor series expansion for MGF ...> lam*t + (lam*t^2*(lam + 1))/2 +
(lam*t^3*(3*lam + lam^2 + 1))/6 + 1
Coefficients of powerts of t in MGF ...>matrix([[1, lam, (lam*(lam + 1))/
2, (lam*(3*lam + lam^2 + 1))/6]])
Series expansion of exp(X*t) ..>X*t + (X^2*t^2)/2 + (X^3*t^3)/6 + 1
Coefficients of powerts of t in expansion of exp(X*t) ...>matrix([[1, X,
X^2/2, X^3/6]])
```

```
E_X0 = 1
E_X1 = lam
E_X2 = lam*(lam + 1)
E_X3 = lam*(lam^2 + 3*lam + 1)
var_X = lam
```

3.10.3   Overview of Transformations

Several transformations of random variables that are used in engineering applications are shown with detailed pedagogical steps.

Transformations of Exponential Variables

$$Y = X^{1/2}$$

$$f_X(x) = (1/a)e^{-x/a}, 0 \le x \le \infty \equiv EXP(a)$$

$$F_X(x) = 1 - e^{-x/a}, 0 \le x \le \infty$$

$$F_Y(y) = P(\sqrt{X} \le y) = P(X \le y^2)$$

$$F_Y(y) = F_X(y^2) = 1 - e^{-y^2/a}$$

$$f_Y(y) = dF/dy = (2y/a)e^{-y^2/a}, 0 \le y \le \infty$$

pdf of Y=X$^{1/2}$; f$_X$(x)$\equiv$ EXP(3)

$$f_X(x) = \frac{e^{-\frac{x}{3}}}{3} U(x)$$

$$F_Y(y) = 1 - e^{-\frac{y^2}{3}}$$

$$f_Y(y) = \frac{2y\,e^{-\frac{y^2}{3}}}{3} \; U(y)$$

Exponential $\Leftrightarrow$ Rayleigh

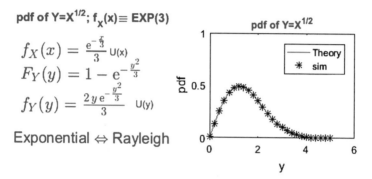

Exponential density: monotonic transformation Y = log$_e$(X)

$$f_X(x) = (1/a)e^{-x/a}, 0 \le x \le \infty \equiv EXP(a)$$

$$F_X(x) = 1 - e^{-x/a}, 0 \le x \le \infty$$

$$F_Y(y) = P(log_e(X) \le y) = P(X \le e^y)$$

$$F_Y(y) = F_X(e^y) = 1 - exp(\frac{-e^y}{a})$$

$$f_Y(y) = dF/dy = (\frac{e^y}{a})exp(\frac{-e^y}{a}), -\infty \le y \le \infty$$

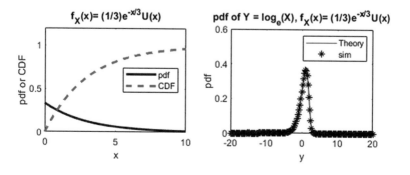

Exponential density: monotonic transformation Y = 1/X

$$f_X(x) = (1/a)e^{-x/a}, 0 \le x \le \infty \equiv EXP(a)$$

$$F_X(x) = 1 - e^{-x/a}, 0 \le x \le \infty$$

$$F_Y(y) = P(1/X \le y) = P(X \ge 1/y)$$

$$F_Y(y) = 1 - F_X(1/y) = exp\left(-\frac{1}{ay}\right)$$

$$f_Y(y) = dF/dy = \frac{1}{ay^2}exp\left(-\frac{1}{ay}\right), 0 \le y \le \infty$$

$f_X(x)$= (1/3)e$^{-x/3}$U(x)   **pdf of Y = 1/X, $f_X(x)$= (1/3)e$^{-x/3}$U(x)**

Exponential density: monotonic transformation Y = -cX, c>0

$$f_X(x) = \frac{1}{b}e^{-x/b}U(x), \; b > 0$$

$$F_X(x) = 1 - e^{-x/b}, 0 \le x \le \infty$$

$$Y = -cX, \; c > 0$$

$$F_Y(y) = P(Y < y) = P(-cX < y) = P\left(X > -\frac{y}{c}\right)$$

$$F_Y(y) = 1 - F_X\left(\frac{-y}{c}\right) = \begin{cases} exp\left(\frac{y}{bc}\right), & -\infty \le y \le 0 \\ 1, & y > 0 \end{cases}$$

$$f_Y(y) = dF/dy = \frac{1}{bc}exp\left(\frac{y}{bc}\right)U(-y)$$

$$E[Y] = -c\,E[X] = -bc$$

Exponential density: monotonic transformation Y = 1-exp(-X/b)

$$f_X(x) = \frac{1}{b}e^{-x/b}U(x), \; b > 0$$

$$F_X(x) = 1 - e^{-x/b}, 0 \le x \le \infty$$

$$Y = 1 - e^{-X/b}$$

$$F_Y(y) = P(Y < y) = P(1 - e^{-X/b} < y) = P\left(e^{X/b} < \frac{1}{1-y}\right)$$

$$F_Y(y) = P\left(X \le b\,log_e\frac{1}{1-y}\right) = F_X\left[b\,log_e\frac{1}{1-y}\right] = y$$

$$f_Y(y) = dF/dy = 1, \; 0 \le y \le 1 \Rightarrow U[0,1]$$

Exponential ⇔ Uniform

Exponential density: monotonic transformation Y=X$^{1/\alpha}$, $\alpha > 0$

$$f_X(x) = \frac{1}{b} e^{-x/b} U(x), \ b > 0$$

$$F_X(x) = 1 - e^{-x/b}, 0 \le x \le \infty$$

$$Y = X^{1/\alpha}, \ \alpha > 0$$

$$F_Y(y) = P(Y < y) = P(X^{1/\alpha} < y) = P[X \le y^\alpha]$$

$$F_Y(y) = F_X[y^\alpha] = 1 - e^{-y^\alpha/b}$$

$$f_Y(y) = dF/dy = \frac{\alpha}{b} y^{\alpha-1} e^{-y^\alpha/b} U(y)$$

Y has a Weibull distribution

Transformations of Rayleigh Variables

Rayleigh density: monotonic transformation Y = X^2

$$f_X(x) = \frac{x}{2b^2} exp\left(-\frac{x^2}{2b^2}\right), 0 \le x \le \infty \equiv R(b)$$

$$F_X(x) = 1 - exp\left(-\frac{x^2}{2b^2}\right), 0 \le x \le \infty$$

$$F_Y(y) = P(X^2 \le y) = P(-\sqrt{y} \le X \le \sqrt{y})$$

$$F_Y(y) = F_X(\sqrt{y}) - F_X(-\sqrt{y}) = F_X(\sqrt{y}) = 1 - exp\left(-\frac{y}{2b^2}\right)$$

$$f_Y(y) = dF/dy = \frac{1}{2b^2} exp\left(-\frac{y}{2b^2}\right), 0 \le y \le \infty$$

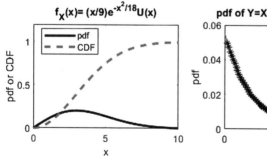

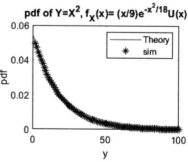

Rayleigh density: monotonic transformation Y = aX+c

$$f_X(x) = \frac{x}{2b^2} exp\left(-\frac{x^2}{2b^2}\right), 0 \le x \le \infty \equiv R(b)$$

$$F_X(x) = 1 - exp\left(-\frac{x^2}{2b^2}\right), 0 \le x \le \infty$$

$$F_Y(y) = P[aX + c \le y] = P[X \le (y-c)/a]$$

$$F_Y(y) = F_X([(y-c)/a) = 1 - exp\left(-\frac{(y-c)^2}{2a^2b^2}\right)$$

$$f_Y(y) = dF/dy = \frac{y-c}{a^2b^2} exp\left(-\frac{(y-c)^2}{2a^2b^2}\right), c \le y \le \infty, b > 0$$

$f_X(x)= (x/4)e^{-x^2/8}U(x)$ **pdf of Y = 3X+1, $f_X(x)$= (1/3)e$^{-x/3}$U(x)**

Rayleigh density: monotonic transformation Y = 1-exp(-X^2/2b^2)

$$f_X(x) = \frac{x}{b^2} e^{-x^2/2b^2} U(x), \ b > 0$$

$$F_X(x) = 1 - e^{-x^2/2b^2}, 0 \le x \le \infty$$

$$Y = 1 - e^{-x^2/2b^2}$$

$$F_Y(y) = P(Y < y) = P(1 - e^{-X^2/2b^2} < y) = P\left(e^{X^2/2b^2} < \frac{1}{1-y}\right)$$

$$F_Y(y) = P\left(X \le \sqrt{2b^2 \, log_e \frac{1}{1-y}}\right) = F_X\left[\sqrt{2b^2 \, log_e \frac{1}{1-y}}\right] = y$$

$$f_Y(y) = dF/dy = 1, \ 0 \le y \le 1 \implies U[0,1]$$

Rayleigh ⟺ Uniform

Transformations of Gaussian Variables

Monotonic transformation $Y = aX + b$

$$f_X(x) = \frac{1}{\sqrt{2\pi\sigma^2}} e^{-x^2/2\sigma^2}, -\infty \le x \le \infty \equiv N(0, \sigma^2)$$

$$F_Y(y) = P(aX + b \le y) = P\left[X \le \frac{y-b}{a}\right] = F_X\left[\frac{y-b}{a}\right]$$

$$f_Y(y) = dF/dy = \frac{1}{a} f_X\left(\frac{y-b}{a}\right) = \frac{1}{\sqrt{2\pi\sigma^2 a^2}} e^{-(y-b)^2/2\sigma^2 a^2}$$

$$f_Y(y) = \frac{f_X(x)}{\left|\frac{dy}{dx}\right|}\bigg|_{x = \frac{y-b}{a}} = \frac{1}{\sqrt{2\pi\sigma^2 a^2}} e^{-(y-b)^2/2\sigma^2 a^2}$$

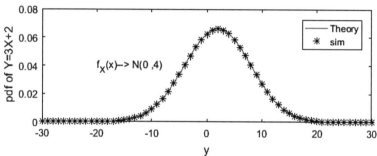

Non-monotonic transformation $Y = |X|$

$$f_X(x) = \frac{1}{\sqrt{2\pi\sigma^2}} e^{-x^2/2\sigma^2}, -\infty \le x \le \infty \equiv N(0, \sigma^2)$$

$$F_Y(y) = P(|X| \le y) = P(-y \le X \le y) = 2F_X(y) - 1 \quad \text{shaded areas bottom left figure}$$

$$f_Y(y) = \frac{dF}{dy} = 2f_X(y) = \frac{2}{\sqrt{2\pi\sigma^2}} e^{-y^2/2\sigma^2} U(y)$$

$$f_Y(y) = \sum_{k=1}^{2} \frac{f_X(x)}{\left|\frac{dy}{dx}\right|}\bigg|_{x = \pm y} = \frac{2}{\sqrt{2\pi\sigma^2}} e^{-y^2/2\sigma^2} U(y)$$

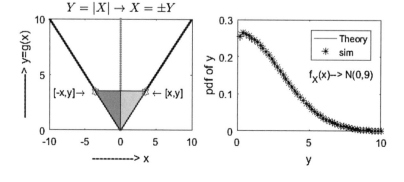

Non-monotonic transformation $Y = \sqrt{|X|}$

$$f_X(x) = \frac{1}{\sqrt{2\pi\sigma^2}} e^{-x^2/2\sigma^2}, \quad -\infty \le x \le \infty \equiv N(0,\sigma^2)$$

$$F_Y(y) = P(\sqrt{|X|} \le y) = P(-y^2 \le X \le y^2) = F_X(y^2) - F_X(-y^2) \quad \text{Areas shown below bottom left}$$

$$f_Y(y) = dF/dy = 4yf_X(y^2) = \frac{4y}{\sqrt{2\pi\sigma^2}} e^{-y^1/2\sigma^2} U(y)$$

$$f_Y(y) = \sum_{k=1}^{2} \frac{f_X(x)}{\left|\frac{dy}{dx}\right|}\bigg|_{x=\pm y^2} = \frac{4y}{\sqrt{2\pi\sigma^2}} e^{-y^1/2\sigma^2} U(y)$$

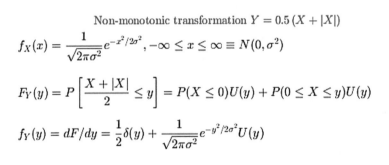

Non-monotonic transformation $Y = 0.5\,(X + |X|)$

$$f_X(x) = \frac{1}{\sqrt{2\pi\sigma^2}} e^{-x^2/2\sigma^2}, \quad -\infty \le x \le \infty \equiv N(0,\sigma^2)$$

$$F_Y(y) = P\left[\frac{X + |X|}{2} \le y\right] = P(X \le 0)U(y) + P(0 \le X \le y)U(y)$$

$$f_Y(y) = dF/dy = \frac{1}{2}\delta(y) + \frac{1}{\sqrt{2\pi\sigma^2}} e^{-y^2/2\sigma^2} U(y)$$

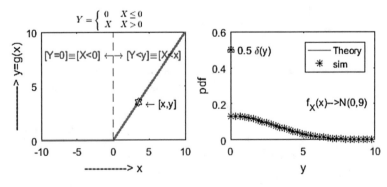

Mixed transformations $(b > 0, c > 0)$

$$f_X(x) = \frac{1}{\sqrt{2\pi\sigma^2}} e^{-(x-\mu)^2/2\sigma^2}, \quad -\infty \leq x \leq \infty \equiv N(\mu, \sigma^2)$$

$$Y = \begin{cases} X, & |X| < b \\ c, & \text{elsewhere} \end{cases} \qquad Z = \begin{cases} X, & |X| < b \\ -c, & x \leq -b \\ c, & x > b \end{cases}$$

$$f_Y(y) = p_1\delta(y - c) + f_X(y)[U(y + b) - U(y - b)],$$

$$p_1 = P(X < -b) + P(X > b)$$

$$f_Z(z) = q_1\delta(z + c) + q_2\delta(z - c) + f_X(z)[U(z + b) - U(z - b)],$$

$$q_1 = P(X < -b); \ q_2 = P(X > b)$$

$$W = \begin{cases} m, & |X| < b \\ -c, & x \leq -b \\ c, & x > b \end{cases}$$

$$f_W(w) = q_1\delta(w + c) + q_2\delta(w - c) + (1 - q_1 - q_2)\delta(w - m),$$

$$q_1 = P(X < -b); \ q_2 = P(X > b)$$

Transformations of Uniform Variables

Monotonic transformation Y = tan (X)

$$f_X(x) = 1/\pi, \quad -\pi/2 \leq x \leq \pi/2 \equiv U[-\pi/2, \pi/2]$$

$$F_X(x) = \begin{cases} 0, & x < -\pi/2 \\ x/\pi + 1/2, & -\pi/2 \leq x \leq \pi/2 \\ 1, & x > \pi/2 \end{cases}$$

$$F_Y(y) = P(\tan(X) \leq y) = P[X \leq \tan^{-1}(y)]$$

$$F_Y(y) = F_X[\tan^{-1}(y)] = \tan^{-1}(y)/\pi + 1/2$$

$$f_Y(y) = \frac{dF}{dy} = \frac{1}{\pi(1 + y^2)}, \quad -\infty \leq y \leq \infty$$

Uniform $\leftrightarrow$ Cauchy

Non-monotonic transformation Y = X^2

$$f_X(x) = 1/2a, -a \le x \le a \equiv U[-a, a]$$

$$F_X(x) = \begin{cases} 0, & x < -a \\ x/2a + 1/2, & -a \le x \le a \\ 1, & x > a \end{cases}$$

$$F_Y(y) = P(X^2 \le y) = P(-\sqrt{y} \le X \le \sqrt{y})$$

$$F_Y(y) = F_X(\sqrt{y}) - F_X(-\sqrt{y}) = \sqrt{y}/a$$

$$f_Y(y) = dF/dy = 1/(2a\sqrt{y}), 0 < y < a^2$$

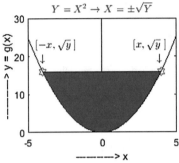
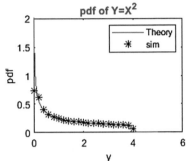

Monotonic transformation Y = X^2

$$f_X(x) = 1, 0 \le x \le 1 \equiv U[0, 1]$$

$$F_X(x) = \begin{cases} 0, & x < 0 \\ x, & 0 \le x \le 1 \\ 1, & x > 1 \end{cases}$$

$$F_Y(y) = P(X^2 \le y) = P(0 \le X \le \sqrt{y}) = F_X(\sqrt{y}) = \sqrt{y}$$

$$f_Y(y) = \frac{dF}{dy} = \frac{1}{2\sqrt{y}}, 0 < y < 1$$

Uniform ↔ Power function

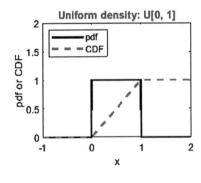
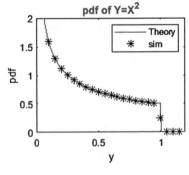

Non-monotonic transformation Y = cos(X)

$f_Y(y)$ from $f_X(x)$ using $\dfrac{dy}{dx}$

$f_X(x) = 1/2\pi, 0 \le x \le 2\pi \equiv U[0, 2\pi]$

$F_X(x) = \begin{cases} 0, & x < 0 \\ x/2\pi, & 0 \le x \le 2\pi \\ 1, & x > 2\pi \end{cases}$

$\dfrac{d}{dx}[cos(x)] = -sin(x) = -\sqrt{1 - y^2}$

$f_Y(y) = \dfrac{f_X(x)}{\left|\frac{dy}{dx}\right|}\Bigg|_{x=cos^{-1}y} + \dfrac{f_X(x)}{\left|\frac{dy}{dx}\right|}\Bigg|_{x=2\pi-cos^{-1}y}$

$f_Y(y) = \dfrac{1}{\pi\sqrt{1 - y^2}}, \quad -1 \le y \le 1$

$f_Y(y)$ from CDF of X , $F_X(x)$

$F_Y(y) = P[Y \le y] = P[cos(X) \le y] = P\left[cos^{-1}(y) \le X \le 2\pi - cos^{-1}(y)\right]$

$F_Y(y) = F_X\left[2\pi - cos^{-1}(y)\right] - F_X\left[cos^{-1}(y)\right] = \dfrac{2\pi - 2cos^{-1}(y)}{2\pi}$

$f_Y(y) = \dfrac{dF_Y(y)}{dy} = \dfrac{1}{\pi\sqrt{1 - y^2}}, \quad -1 \le y \le 1$

Uniform $\leftrightarrow$ Arcsine

Monotonic transformation Y = b $(-2\log_e X)^{1/2}$, b>0

$f_X(x) = 1, 0 \le x \le 1 \equiv U[0, 1]$

$F_X(x) = \begin{cases} 0, & x \le 0 \\ x, & 0 \le x \le 1 \\ 1, & x > 1 \end{cases}$

$F_Y(y) = P\left(b\sqrt{-2log_e(X)} < y\right) = P\left[X > e^{-\frac{y^2}{2b^2}}\right]$

$F_Y(y) = 1 - F_X(e^{-\frac{y^2}{2b^2}}) = 1 - e^{-\frac{y^2}{2b^2}}$

$f_Y(y) = \dfrac{dF}{dy} = \dfrac{y}{b^2} e^{-\frac{y^2}{2b^2}}, 0 \le y \le \infty$

Uniform $\leftrightarrow$ Rayleigh

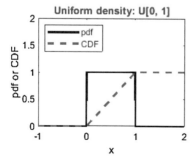

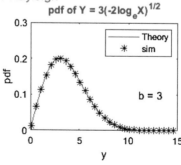

Mixed transformations (0 < b < a)

$$f_X(x) = \frac{1}{2a}, -a \le x \le a \equiv U[-a,a]$$

$$Y = \begin{cases} X, & |X| < b, \ 0 < b < a \\ c, & \text{elsewhere} \end{cases}$$

$$f_Y(y) = p_1\delta(y-c) + f_X(y)[U(y+b) - U(y-b)]$$

$$p_1 = P(X < -b) + P(X > b) = \frac{a-b}{a}$$

$$f_Y(y) = \left[\frac{a-b}{a}\right]\delta(y-c) + \frac{1}{2a}[U(y+b) - U(y-b)]$$

$$Z = \begin{cases} \alpha, & |X| < b \\ -\beta, & X \le -b \\ \beta, & X > b \end{cases}$$

$$f_Z(z) = p_\alpha\delta(z-\alpha) + p_{-\beta}\delta(z+\beta) + p_\beta\delta(z-\beta)$$

$$p_\alpha = P(|X| < b) = \frac{2b}{2a}; \ p_{-\beta} = P(X < -b) = p_\beta = P(X > b) = \frac{a-b}{2a}$$

$$f_Z(z) = \left[\frac{2b}{2a}\right]\delta(z-\alpha) + \left[\frac{a-b}{2a}\right]\delta(z+\beta) + \left[\frac{a-b}{2a}\right]\delta(z-\beta)$$

Monotonic transformation Y = X/b - 1, b>0

$$f_X(x) = \frac{1}{b}, 0 \le x \le b, \ b > 0$$

$$F_X(x) = \begin{cases} 0, & x < 0 \\ x/b, & 0 \le x \le b \\ 1, & x > b \end{cases}$$

$$Y = \frac{X}{b} - 1$$

$$F_Y(y) = P(\frac{X}{b} - 1 \le y) = P[X < b(1+y)] = F_X[b(1+y)]$$

$$F_Y(y) = \begin{cases} 0, & y < -1 \\ 1+y, & -1 \le y \le 0 \\ 1, & y > 0 \end{cases}$$

$$f_Y(y) = dF/dy = 1, \ -1 \le y \le 0 \equiv U[-1,0]$$

Monotonic transformation Y = 1-X/b , b>0

$$f_X(x) = \frac{1}{b}, 0 \le x \le b, \ b > 0$$

$$F_X(x) = \begin{cases} 0, & x < 0 \\ x/b, & 0 \le x \le b \\ 1, & x > b \end{cases}$$

$$Y = 1 - \frac{X}{b}$$

$$F_Y(y) = P(1 - \frac{X}{b} \le y) = P[X > b(1-y)] = 1 - F_X[b(1-y)]$$

$$F_Y(y) = \begin{cases} 0, & y < 0 \\ y, & 0 \le y \le 1 \\ 1, & y > 1 \end{cases}$$

$$f_Y(y) = dF/dy = 1, \ 0 \le y \le 1 \equiv U[0,1]$$

Monotonic transformation Y = log$_e$(1-X/b), b>0

$$f_X(x) = \frac{1}{b}, 0 \le x \le b, \ b > 0$$

$$F_X(x) = \begin{cases} 0, & x < 0 \\ x/b, & 0 \le x \le b \\ 1, & x > b \end{cases}$$

$$Y = log_e\left(1 - \frac{X}{b}\right)$$

$$F_Y(y) = P\left[log_e\left(1 - \frac{X}{b}\right) \le y\right] = P[X > b(1 - e^y)]$$

$$F_Y(y) = 1 - F_X[b(1 - e^y] = 1 - [1 - e^y] = e^y, \quad -\infty \le y \le 0$$

$$f_Y(y) = e^y U(-y)$$

Monotonic transformation Y = 1/(1-X)

$$f_X(x) = 1, 0 \le x \le 1$$

$$F_X(x) = \begin{cases} 0, & x < 0 \\ x, & 0 \le x \le 1 \\ 1, & x > 1 \end{cases}$$

$$Y = \frac{1}{1 - X}$$

$$F_Y(y) = P\left[\frac{1}{1-X} \le y\right] = P\left[1 - X \ge \frac{1}{y}\right] = P\left[X \le 1 - \frac{1}{y}\right]$$

$$F_Y(y) = F_X\left[1 - \frac{1}{y}\right] = 1 - \frac{1}{y}$$

$$f_Y(y) = dF/dy = \frac{1}{y^2}, \quad 1 \le y \le \infty$$

Transformations of Discrete Variables

Discrete random variable: transformation Y=X²

$$f_X(x) = p_{-2}\delta(x+2) + p_{-1}\delta(x+1) + p_0\delta(x) + p_1\delta(x-1)$$
$$+ p_2\delta(x-2) + p_3\delta(x-3)$$
$$p_{-2} + p_{-1} + p_0 + p_1 + p_2 + p_3 = 1$$

$$Y = X^2 \rightarrow Y = [0, 1, 4, 9] \rightarrow \quad
\begin{aligned}
&[Y = 0] \leftarrow [X = 0] \\
&[Y = 1] \leftarrow [X = -1] \cup [X = 1] \\
&[Y = 4] \leftarrow [X = -2] \cup [X = 2] \\
&[Y = 9] \leftarrow [X = 3]
\end{aligned}$$

$$f_Y(y) = q_0\delta(y) + q_1\delta(y-1) + q_4\delta(y-4) + q_9\delta(y-9)$$

$$q_0 = p_0$$
$$q_1 = p_{-1} + p_1$$
$$q_4 = p_{-2} + p_2$$
$$q_9 = p_3$$
$$E[Y] = 0 \times q_0 + 1 \times q_1 + 4 \times q_4 + 9 \times q_9 = q_1 + 4q_4 + 9q_9$$
$$E\left[Y^2\right] = (0)^2 q_0 + (1)^2 q_1 + (4)^2 q_4 + (9)^2 q_9 = q_1 + 16q_4 + 81q_9$$
$$var(Y) = E\left[Y^2\right] - E[Y]^2$$

Poisson density:Transformation Y=X²

$$PMF(x) = P_X(k) = \frac{\lambda^k e^{-\lambda}}{k!} \equiv f(k; \lambda), k = 0, 1, 2, ..$$

$$f_X(x) = \sum_{k=0}^{\infty} P_X(k)\delta(x-k) = e^{-\lambda}\sum_{k=0}^{\infty}\frac{\lambda^k}{k!}\delta(x-k)$$

$$F_X(x) = e^{-\lambda}\sum_{k=0}^{\infty}\frac{\lambda^k}{k!}U(x-k)$$

$$Y = X^2, \quad f_Y(y) = e^{-\lambda}\sum_{k=0,1,4,9,...}^{\infty}\frac{\lambda^{\sqrt{k}}}{\sqrt{k}!}\delta(y-k)$$

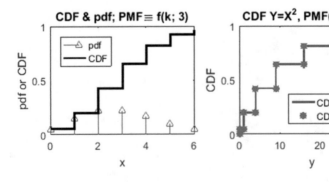

3.10.4 An Alternate View of Transformation of a Random Variable

Consider the transformation

$$Y = g(X) \tag{3.274}$$

The density function of X has the property

$$\int_{-\infty}^{\infty} f(x)dx = 1 \tag{3.275}$$

In terms of variables, Eq. (3.274) is expressed as

$$y = g(x) \tag{3.276}$$

Equation (3.276) implies that

$$x = g^{-1}(y) \tag{3.277}$$

Taking differentials on both sides of Eq. (3.276), we get

$$dy = g'(x)dx = g'\left(g^{-1}(y)\right)dx \tag{3.278}$$

In Eq. (3.278), $g'(x)$ is the derivative of $g(x)$ with respect to x, and we have used Eq. (3.277) to express x. We may now rewrite Eq. (3.275) as

$$\int_{-\infty}^{\infty} f(x)dx = \int_{-\infty}^{\infty} f\left(g^{-1}(y)\right)\frac{dy}{g'(x)} = \int_{-\infty}^{\infty} f\left(g^{-1}(y)\right)\frac{dy}{g'(g^{-1}(y))} = 1 \tag{3.279}$$

By virtue of the fact that Y is a random variable, the density of Y satisfies the property

$$\int_{-\infty}^{\infty} f(y)dy = 1 \tag{3.280}$$

Comparing Eqs. (3.279) and (3.280), we may write

$$1 = \int_{-\infty}^{\infty} f(g^{-1}(y)) \frac{dy}{g'(g^{-1}(y))} = \int_{-\infty}^{\infty} f(y) dy \qquad (3.281)$$

Rearranging the terms in Eq. (3.281), we get

$$\int_{-\infty}^{\infty} \frac{f(g^{-1}(y))}{g'(g^{-1}(y))} \, dy = \int_{-\infty}^{\infty} f(y) dy \qquad (3.282)$$

Comparing the integrands on both sides of Eq. (3.282), we may write

$$\frac{f(g^{-1}(y))}{g'(g^{-1}(y))} = f(y) \qquad (3.283)$$

Since the densities must always be positive, we may write

$$\frac{f(g^{-1}(y))}{|g'(g^{-1}(y))|} = f(y) \qquad (3.284)$$

The absolute value sign in Eq. (3.284) merely addresses the cases where the derivative may be negative (monotonically decreasing) to ensure that the density is always positive. We may rewrite Eq. (3.284) as

$$f(y) = \frac{f(g^{-1}(y))}{\left|\frac{dy}{dx}\right|_{x=g^{-1}(y)}} = \frac{f(x)}{\left|\frac{dy}{dx}\right|}\Bigg|_{x=g^{-1}(y)} \qquad (3.285)$$

Note that Eq. (3.285) is identical to the density obtained from the traditional approach.

3.10.5   Conditional Densities and Conditional Expectation Revisited

Earlier in this chapter, we had seen that that the expected value of a function of a random variable such as $g(X)$ can be found without obtaining the density of $Y = g(X)$. Eq. (3.55) is reproduced here for context

$$E[g(X)] = \int_{-\infty}^{\infty} g(x) f_X(x) dx \tag{3.286}$$

Consider the case of two functions of the random variable X defined as

$$g_1(X) = X, X > 6 \tag{3.287}$$

$$g_2(X) = \begin{cases} 0, X < 6 \\ X, X \geq 6 \end{cases} \tag{3.288}$$

To get a better perspective, let us assume that X as gamma distributed.

$$f_X(x) = \frac{x}{3^2} \exp\left(-\frac{x}{3}\right) U(x) \tag{3.289}$$

It can be seen that Eq. (3.289) represents the gamma density $G(2,3)$ with a mean of 6. Clearly

$$E[g_2(X)] = \int_0^{\infty} g_2(x) f_X(x) dx = \int_6^{\infty} x \frac{x}{3^2} \exp\left(-\frac{x}{3}\right) dx = 4.06 \tag{3.290}$$

The expected value of X is 6, and we expect the mean of $g_2(X)$ to be less than $E(X)$ because any value of X below 6 has been deemed to be 0. This would mean that the mean of $g_2(X)$ will be moving from 6 toward 0, and we have a value of 4.067.

If we now look at the mean of $g_1(X)$, we will see that

$$E[g_1(X)] \neq \int_0^{\infty} g_1(x) f_X(x) dx \tag{3.291}$$

The reason is that the sample size (range) of $g_1(X)$ is smaller than the sample size (range) of X which equals the sample (range) size of $g_2(X)$. This means that the density function needs to be weighted. In other words, to determine the mean of $g_1(X)$, we need to obtain the conditional density of $g_1(X)$. Using the results from Sect. 3.5, the conditional density will be

$$f_{X|X>6}(x) = \frac{\frac{x}{3^2} \exp\left(-\frac{x}{3}\right)}{1 - F_X(6)} U(x - 6) \tag{3.292}$$

The mean of $g_1(X)$

$$E[g_1(X)] = \int_0^\infty g_1(x) f_{X|X>6}(x) dx = \int_6^\infty x \frac{\frac{x}{3^2} \exp\left(-\frac{x}{3}\right)}{1 - F_X(6)} dx = 10 \qquad (3.293)$$

We expect the mean of $g_1(X)$ to exceed the mean of X, and we see that in Eq. (3.293). These results can be easily verified by obtaining the densities of $g_1(X)$ and $g_2(X)$. If we define two new random variables, Y and W as

$$Y = g_1(X) = [X|X > 6] \qquad (3.294)$$

$$W = g_2(X) = \begin{cases} 0, X < 6 \\ X, X \geq 6 \end{cases} \qquad (3.295)$$

Using the concepts of conditional densities,

$$f_Y(y) = \frac{f_X(y)}{1 - F_X(6)}, \quad x > 6 \qquad (3.296)$$

$$f_Y(y) = \frac{\frac{y}{3^2} \exp\left(-\frac{y}{3}\right)}{1 - F_X(6)} U(y - 6) \qquad (3.297)$$

$$f_W(w) = F_X(6)\delta(w) + \frac{w}{3^2} \exp\left(-\frac{w}{3}\right) U(w - 6) \qquad (3.298)$$

We can now obtain the expected values as

$$E[Y] = \int_0^\infty y f_Y(y) dy = \int_6^\infty y \frac{\frac{y}{3^2} \exp\left(-\frac{y}{3}\right)}{1 - F_X(6)} dy = 10 \qquad (3.299)$$

$$E[W] = \int_0^\infty w f_W(w) dw = \int_0^\infty w F_X(6)\delta(w) + \int_0^\infty \frac{w}{3^2} \exp\left(-\frac{w}{3}\right) U(w - 6) dw \quad (3.300)$$

$$E[W] = 0 + \int_6^\infty \frac{w}{3^2} \exp\left(-\frac{w}{3}\right) dw = 4.067 \qquad (3.301)$$

This case is also demonstrated through the use of Matlab where the analytical results are compared to simulation ones.

```
%expected_value_conditional
% There is confusion regarding the use of E(g(X))=int(g(x)f(x)dx) when g(x)
% does not span the whole range of X. This example shows ways of finding
% the expected values without first finding the density of g(X). The two
% case are g1(X)=X, X>6 and g2(X)= 0 if X<6 and X   if X>6.
clear;clc;close all
syms x
disp('-----------------------------------')
disp('Analytical results')
disp('-----------------------------------')
% X is G(2,3)
fx=x*exp(-x/3)/3^2;% gamma density
disp(['pdf of X is .... ',char(fx)])
Fx=int(fx,x,0,x);% CDF of X
disp(['CDF of X is .... ',char(Fx)])
Ex=double(int(x*fx,x,0,inf));% verify that the mean is 6
disp(['E(X) = ',num2str(Ex)])
% g1(X)=x,x>6
% g2(X)=0,x<6 and x, if x>6
% we can find E(g2(x)) using the standard defintion of E (.) of a function
Eg2x=double(int(x*fx,x,6,inf));
disp(['E(g2(X)) = ',num2str(Eg2x)])
% however, g1(x) does not cover the whole range of x. In g1(x), we are
% simply ignoring entries below 6. This means that we should weigh the
% density appropriately. In other words, we have to use the conditional
% density of x|x>6. It is expressed as f1(x) below. The denominator is
% 1-F(6)=P(X>6)
f1x=fx/subs(1-Fx,x,6);
disp(['weighted density for g1(X) = ',char(f1x)])
Eg1x=double(int(x*f1x,x,6,inf));
disp(['E(g1(X)) = ',num2str(Eg1x)])
```

Analytical results
----
pdf of X is .... (x*exp(-x/3))/9
CDF of X is .... 1 - (exp(-x/3)*(x + 3))/3
E(X) = 6
E(g2(X)) = 4.0601
weighted density for g1(X) = (x*exp(-x/3)*exp(2))/27
E(g1(X)) = 10
verification

```
disp('----------------------------------------')
disp('Results of random number simulation')
disp('----------------------------------------')
clear x
N=1e6; % one million samples
x=gamrnd(2,3,1,N);
ME= mean(x);
disp(['E(X) = ',num2str(ME)])
x1=x(x>6); % samples of g1(x)
ME1=mean(x1);
L=length(x1);
x2=[x1,zeros(1,N-L)]; % these are the samples corresponding to g2(x)
ME2=mean(x2);
disp(['E(g2(X)) = ',num2str(ME2)])
disp(['E(g1(X)) = ',num2str(ME1)])
% This presentes samples of x1 with remaining samples of x replaced by NaN
% so that we can see the histograms of the data for g1(x) and g2(x)
% side-by-side in Matlab
x11=[x1,NaN(1,N-L)];
y=[x11;x2]';
hist(y)
legend(['g_1(x), number of samples = ',num2str(L)],...
    ['g_2(x), number of samples = ',num2str(N)])
ylabel('frequency'),xlabel('values')
g1x='$$ g_1(x) = X, X\ge6 $$';
g2x='$$ g_2(x) = \left\{\begin{array}{lr} 0, & X<6 \\ X, & X\ge 6 \end{array}\right. $$';
text(20,4e5,g1x,'interpreter','latex')
text(20,3.5e5,g2x,'interpreter','latex')
```

```
----------------------------------------
Results of random number simulation
----------------------------------------
E(X) = 6.0012
E(g2(X)) = 4.0598
E(g1(X)) = 10.0062
```
Histograms of the samples for the two cases are shown. It can be seen that the number of samples $g_1(x)$ is smaller than the number of samples of $g_2(X)$.

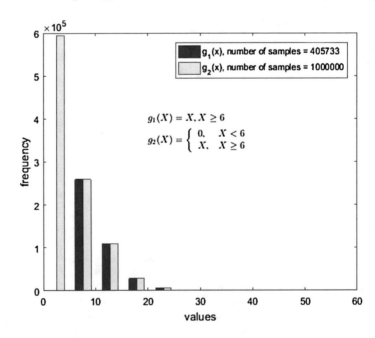

3.10.6 Proofs of the Moments of Gaussian and Rayleigh Densities

3.10.6.1 Proof of the Gaussian Integral

$$I = \int_{-\infty}^{\infty} \frac{1}{\sqrt{2\pi b^2}} \exp\left(-\frac{x^2}{2b^2}\right) dx$$

$$I^2 = \int_{-\infty}^{\infty} \frac{1}{\sqrt{2\pi b^2}} \exp\left(-\frac{x^2}{2b^2}\right) dx \int_{-\infty}^{\infty} \frac{1}{\sqrt{2\pi b^2}} \exp\left(-\frac{y^2}{2b^2}\right) dy$$

$$I^2 = \int_{-\infty}^{\infty} \int_{-\infty}^{\infty} \frac{1}{\sqrt{2\pi b^2}} \exp\left(-\frac{x^2}{2b^2}\right) \frac{1}{\sqrt{2\pi b^2}} \exp\left(-\frac{y^2}{2b^2}\right) dxdy$$

$$I^2 = \int_{-\infty}^{\infty} \int_{-\infty}^{\infty} \frac{1}{2\pi b^2} \exp\left(-\frac{x^2+y^2}{2b^2}\right) dxdy$$

Put

$$x = r\cos(\theta)$$
$$y = r\sin(\theta)$$

We have

$$dxdy = rdrd\theta$$

Changing the variables, we have

$$I^2 = \int_{-\infty}^{\infty} \int_{-\infty}^{\infty} \frac{1}{2\pi b^2} \exp\left(-\frac{x^2+y^2}{2b^2}\right) dxdy = \frac{1}{2\pi} \int_{0}^{2\pi} d\theta \int_{0}^{\infty} \frac{r}{b^2} \exp\left(-\frac{r^2}{2b^2}\right) dr$$

$$= \int_{0}^{\infty} \frac{r}{b^2} \exp\left(-\frac{r^2}{2b^2}\right) dr = 1$$

Note that the integral on the right is the integral of the Rayleigh density. Thus, we have

$$I^2 = 1 \Rightarrow I = 1 = \int_{-\infty}^{\infty} \frac{1}{\sqrt{2\pi b^2}} \exp\left(-\frac{x^2}{2b^2}\right) dx$$

3.10.6.2   Second Moment of the Gaussian

$$f(x) = \frac{1}{\sqrt{2\pi b^2}} \exp\left(-\frac{x^2}{2b^2}\right)$$

We would like to prove that

$$E(X^2) = b^2$$

$$E(X^2) = \int_{-\infty}^{\infty} x^2 \frac{1}{\sqrt{2\pi b^2}} \exp\left(-\frac{x^2}{2b^2}\right) dx = \frac{1}{\sqrt{2\pi b^2}} \int_{-\infty}^{\infty} x^2 \exp\left(-\frac{x^2}{2b^2}\right) dx$$

$$= \frac{1}{\sqrt{2\pi b^2}} I_1$$

In the equation above,

$$I_1 = \int_{-\infty}^{\infty} x^2 \exp\left(-\frac{x^2}{2b^2}\right) dx$$

Make the substitution

$$y = \frac{1}{b^2}$$

$$I_1 = \int_{-\infty}^{\infty} x^2 \exp\left(-\frac{yx^2}{2}\right) dx$$

We can write

$$x^2 \exp\left(-\frac{yx^2}{2}\right) = -2 \frac{\partial}{\partial y}\left[\exp\left(-\frac{yx^2}{2}\right)\right]$$

Substituting and reversing the order of differentiation and integration,

$$I_1 = -\int_{-\infty}^{\infty} 2 \frac{\partial}{\partial y}\left[\exp\left(-\frac{yx^2}{2}\right)\right] dx = -2 \frac{\partial}{\partial y}\left[\int_{-\infty}^{\infty} \exp\left(-\frac{yx^2}{2}\right) dx\right]$$

Note that the integral is the Gaussian integral and results in

$$\int_{-\infty}^{\infty} \exp\left(-\frac{yx^2}{2}\right) dx = \sqrt{\frac{2\pi}{y}}$$

This means that

$$I_1 = -2\frac{\partial}{\partial y}\left[\int_{-\infty}^{\infty} \exp\left(-\frac{yx^2}{2}\right)dx\right] = -2\frac{\partial}{\partial y}\left(\sqrt{\frac{2\pi}{y}}\right) = -2\sqrt{2\pi}\left(-\frac{1}{2y^{\frac{3}{2}}}\right) = \frac{\sqrt{2\pi}}{y^{\frac{3}{2}}}$$

Substituting for y, we have

$$I_1 = \frac{\sqrt{2\pi}}{y^{\frac{3}{2}}} = \sqrt{2\pi}b^3$$

The second moment is

$$E(X^2) = \frac{1}{\sqrt{2\pi b^2}}\sqrt{2\pi}b^3 = b^2$$

3.10.6.3   Mean and Second Moment of Rayleigh

The Rayleigh density is

$$f(x) = \frac{x}{b^2}\exp\left(-\frac{x^2}{2b^2}\right)U(x)$$

The mean is

$$E(X) = \int_0^{\infty} xf_X(x)dx = \int_0^{\infty} x\frac{x}{b^2}\exp\left(-\frac{x^2}{2b^2}\right)dx = \int_0^{\infty} \frac{x^2}{b^2}\exp\left(-\frac{x^2}{2b^2}\right)dx$$

The integral is obtained using the second moment of a zero mean Gaussian density

$$\int_{-\infty}^{\infty} \frac{x^2}{\sqrt{2\pi b^2}}\exp\left(-\frac{x^2}{2b^2}\right)dx = 2\int_0^{\infty} \frac{x^2}{\sqrt{2\pi b^2}}\exp\left(-\frac{x^2}{2b^2}\right)dx = b^2$$

We may express the mean of the Rayleigh as

$$E(X) = \int_0^\infty \frac{x^2}{b^2} \exp\left(-\frac{x^2}{2b^2}\right) dx = \frac{1}{b^2} \int_0^\infty x^2 \exp\left(-\frac{x^2}{2b^2}\right) dx$$

$$= \frac{1}{b^2} \sqrt{2\pi b^2} \int_0^\infty \frac{x^2}{\sqrt{2\pi b^2}} \exp\left(-\frac{x^2}{2b^2}\right) dx$$

Using the second moment of the zero mean Gaussian variable, we have

$$\int_0^\infty \frac{x^2}{\sqrt{2\pi b^2}} \exp\left(-\frac{x^2}{2b^2}\right) dx = \frac{b^2}{2}$$

Using the integral above, we can simplify the mean as

$$E(X) = \int_0^\infty \frac{x^2}{b^2} \exp\left(-\frac{x^2}{2b^2}\right) dx = \frac{1}{b^2} \sqrt{2\pi b^2} \int_0^\infty \frac{x^2}{\sqrt{2\pi b^2}} \exp\left(-\frac{x^2}{2b^2}\right) dx$$

$$= \frac{1}{b^2} \sqrt{2\pi b^2} \frac{b^2}{2} = b\sqrt{\frac{\pi}{2}}$$

The second moment is

$$E(X^2) = \int_0^\infty x^2 f_X(x) dx = \int_0^\infty x^2 \frac{x}{b^2} \exp\left(-\frac{x^2}{2b^2}\right) dx = \int_0^\infty \frac{x^3}{b^2} \exp\left(-\frac{x^2}{2b^2}\right) dx$$

Put

$$y = x^2 \Rightarrow$$
$$\frac{dy}{dx} 2x \Rightarrow xdx = \frac{dy}{2}$$

The second moment now becomes

$$E(X^2) = \int_0^\infty \frac{y}{2b^2} \exp\left(-\frac{y}{2b^2}\right) dy = \int_0^\infty \frac{y}{2b^2} \exp\left(-\frac{y}{2b^2}\right) dy = 2b^2$$

Notice that

$$\int_0^\infty \frac{y}{2b^2} \exp\left(-\frac{y}{2b^2}\right) dy = E(Y)$$

In the equation above,

$$E(Y) = \int_0^\infty y f_Y(y) dy$$

$$f_Y(y) = \frac{1}{2b^2} \exp\left(-\frac{y}{2b^2}\right) U(y) \Rightarrow \text{exponential density}$$

3.10.7   Moments of a Poisson Random Variable

The probability mass function of a Poisson random variable is

$$P(X = k) = \frac{\lambda^k}{k!} e^{-\lambda}$$

This means that

$$1 = \sum_{k=0}^\infty P(X = k) = e^{-\lambda} \sum_{k=0}^\infty \frac{\lambda^k}{k!}$$

The summation is nothing but the expansion of exp.(λ), and therefore,

$$e^{-\lambda} \sum_{k=0}^\infty \frac{\lambda^k}{k!} = e^{-\lambda} e^{\lambda} = 1$$

The first moment is

$$E(X) = \sum_{k=0}^\infty k P(X = k) = \sum_{k=0}^\infty k \frac{\lambda^k}{k!} e^{-\lambda}$$

The second moment is

$$E(X^2) = \sum_{k=0}^\infty k^2 P(X = k) = \sum_{k=0}^\infty k^2 \frac{\lambda^k}{k!} e^{-\lambda}$$

To obtain the first and second moments, we go back to the summation for the total probability,

$$e^{-\lambda} \sum_{k=0}^{\infty} \frac{\lambda^k}{k!} = 1$$

Rewriting in terms of the identity for the expansion of exp.(λ), we have

$$\sum_{k=0}^{\infty} \frac{\lambda^k}{k!} = e^{\lambda}$$

Differentiating once w. r. t to λ, we have

$$\sum_{k=0}^{\infty} k \frac{\lambda^{k-1}}{k!} = e^{\lambda}$$

Multiplying both sides by λ

$$\sum_{k=0}^{\infty} k\lambda \frac{\lambda^{k-1}}{k!} = \lambda e^{\lambda} \quad \Rightarrow \quad \sum_{k=0}^{\infty} k \frac{\lambda^k}{k!} = \lambda e^{\lambda}$$

Rearranging the terms, we have

$$e^{-\lambda} \sum_{k=0}^{\infty} k \frac{\lambda^k}{k!} = \lambda$$

The equation above matches the equation for the mean, and therefore, the mean is λ. Now differentiating twice, we have

$$\sum_{k=0}^{\infty} k(k-1) \frac{\lambda^{k-2}}{k!} = e^{\lambda}$$

Multiplying both sides by λ^2, we have

$$\sum_{k=0}^{\infty} k(k-1)\lambda^2 \frac{\lambda^{k-2}}{k!} = \lambda^2 e^{\lambda} \quad \Rightarrow \quad e^{-\lambda} \sum_{k=0}^{\infty} k(k-1) \frac{\lambda^k}{k!} = \lambda^2$$

Rewriting, we have

$$\sum_{k=0}^{\infty} (k^2 - k) \frac{\lambda^k}{k!} = \sum_{k=0}^{\infty} k^2 \frac{\lambda^k}{k!} - \sum_{k=0}^{\infty} k \frac{\lambda^k}{k!} = \lambda^2 e^{\lambda}$$

Rewriting in terms of moments, the equation above becomes

$$e^{-\lambda}\sum_{k=0}^{\infty}k^2\frac{\lambda^k}{k!}-e^{-\lambda}\sum_{k=0}^{\infty}k\frac{\lambda^k}{k!}=E(X^2)-E(X)=\lambda^2$$

Therefore,

$$E(X^2)=\lambda^2+E(X)=\lambda^2+\lambda$$

The variance is

$$\mathrm{var}(X)=E(X^2)-E^2(X)=\lambda^2+\lambda-\lambda^2=\lambda$$

Exercises

1. Test whether the function $h(x)$ below is a valid pdf.

$$h(x)=\frac{x}{2}\exp(-x)U(x)$$

2. Obtain the value of c so that the function $g(x)$ below is a valid pdf. Obtain its CDF.

$$g(x)=\frac{b}{[1-e^{-b}]}\exp(-xb)U(c-x),\qquad x,b,c>0$$

3. You are given $h(x)=k\exp(cx)$; c is positive. For $h(x)$ to be a valid pdf, what should be the range of validity of the variable and the relationship between c and k? If it is a valid pdf, obtain its CDF.

4. For the following pdf $f(x)$, obtain the relationship between p and q.

$$f(x)=\frac{p}{a}\exp\left(-\frac{x}{a}\right)+q\exp\left(-\frac{x}{b}\right),\qquad x>0,\quad p>0,q>0,a>0,b>0$$

5. A function $h(x)$ is given as

$$h(x)=\frac{1}{2x^2},\qquad a<x<\infty,\quad a>0$$

What value of a will make $h(x)$ a valid pdf? What is the CDF?

6. A function $h(x)$ is given as

$$h(x) = \frac{1}{\sqrt{x}}, \quad 0 < x < a, \quad a > 0$$

What value of a will make $h(x)$ a valid pdf? What is the CDF?

7. For the function $h(x)$ to be a valid CDF, what must be the range of the variable? What is the pdf?

$$h(x) = x^2, \quad a < x < b, \quad a \geq 0, \quad b > 0$$

8. For the following function $h(x)$, what is the range of the variable for its validity as a pdf? Obtain its CDF.

$$h(x) = \frac{4}{\pi\sqrt{1 - x^2}}, 0 < x < a, \quad 0 < a < 1$$

9. X is a number randomly picked from the interval [0,4]. What is the probability that $[X^2 - 6X + 8] > 0$?

Verify your results through random number simulation.

10. For the following triangular pdf, obtain the CDF and $P(|X| < 1)$.

$$f_x(x) = \frac{(2 - |x|)}{4}, \quad |x| < 2$$

Verify $P(|X| < 1)$ value through random number simulation.

11. A random variable X has the following pdf: Obtain the CDF. What is the probability that $X > 1$?

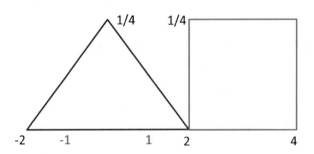

12. If X has the following pdf, obtain the value of the constant A and the CDF. What is the prob. $\{X < pi/4\}$?

$$f_X(x) = A \sin (x), 0 < x < \pi$$

13. A random variable X has the following pdf:

$$f_X(x) = \begin{cases} \dfrac{1}{4} e^{\frac{x}{2}}, & x < 0 \\ \dfrac{1}{2}, & 0 < x < 2 \\ 0, & x > 2 \end{cases}$$

Obtain the CDF. What is $P[X > -1]$?

14. A number X is picked randomly between 0 and 5.

What is the probability that $X > 3$?
What is the probability that $X > 2$ given that X has been observed to be always been larger than 1?
What is the probability that the first decimal place of X is 0.4?

15. X is uniform in $[0,\pi]$. What is the probability that $\sin(X)$ lies between 0.2 and 0.3?

16. A dart is thrown at a dartboard with concentric circles. If the radius of the circle is a Rayleigh variable with a parameter 0.5, what is the probability that the dart thrown lands in a disc of thickness 0.2 units with the center at 1 unit?

17. The voltage measured has been observed to follow a Rayleigh density. If the probability that the measured voltage exceeds 5 V is 0.3, what is the mean voltage?

18. X is uniform in $[-2,4]$ and Y is uniform in $[2,4]$. For non-zero real values of p and q, define

$$Z = pX^2 + qY^2$$

Obtain a relationship between p and q such that $E(Z) = 0$.

19. X is exponential with a mean of 3 and Y is gamma distributed G(3,4). Obtain a relationship between p and q (non-zero real values) such that $E(W)$ is 0 where

$$W = p(2X + 3Y^2) + q(4Y - X^2)$$

20. A string has a length of 5 units. A point is chosen on the string randomly, and the string is cut at a length X (to that point); what is the probability that this segment is *at least* 1.5 times longer than the remaining piece?

21. A call center receives telephone calls with an average duration of 3 minutes. What is the probability that the next call received will exceed the average duration given that calls generally last longer than 2 minutes? (Call duration is exponentially distributed.)

22. A highly sensitive optical receiver operates such that it shuts down the operation (to prevent damage to the sensitive electronics) when the optical power exceeds 1 mW. If the power is modeled as a gamma density with a mean power of (4/10) mW and order 2, what is the probability that the receiver shuts down?

23. The power received in mW by a wireless receiver is modeled in terms of a gamma distribution G(3,0.25e-6). What is the probability of outage (no service) if outage occurs when the received power falls below −75 dBm?

24. The lifetimes of computers are modeled in terms of Rayleigh density. Three manufacturers A, B, C are available. Their computers have average lifetimes of 5, 6, and 7 years, respectively. A company orders 20 computers from A, 30 from B, and 50 from C. What is the probability that a computer randomly examined is operating beyond 6 years?

25. An organization purchases 10 computers from a manufacturer with lifetimes modeled as exponential variables (mean of 5 years). What is the probability that at least 6 computers are functioning beyond 6 years?

26. In medical imaging, the amplitude of the image displayed is modeled as a Rayleigh random variable with an average brightness (intensity or power) of 4 units when dark suspicious regions in the image and average brightness 512 units when there are no dark regions in the image (normal). If a threshold is set at 16 units of brightness, (a) what is the probability that a suspicious region is misidentified as a normal region? (b) What is the probability that a normal region is considered to be a suspicious region? The probability in part (a) is identified as the false alarm rate, and the probability in part (b) is considered as the miss rate.

27. If in the previous image analysis experiment, an "image reader" is given 300 images containing suspicious regions and 700 regions containing no dark (normal) regions (all confirmed through other tests); what is the probability of correct decision dark identified as suspicious and bright identified as normal) being made as normal by the image reader? What is this rate if the threshold is changed from 2 units to 24 units of intensity? Obtain plots of the prob. False alarm (normal identified as suspicious) and prob. of miss (suspicious identified as normal) as threshold values change. Obtain plot of the probability of decision being "normal."

28. In a digital communication system, 1 s are transmitted 55% of the time and -1 s transmitted during the rest of the time. If noise is present in the channel (Gaussian noise N(0,1/25)), and the threshold is set to 0, what is the probability of error?

29. In a test with 10 Yes/No questions, the probability of being correct is 0.35. If the number of Y's is modeled as discrete random variable X, obtain expressions for the CDF and pdf of this random variable. What are the mean and variance of X?

30. The number of computers sold at a big box store during the start of the school period is modeled as a Poisson variable with an average of 5/day. What is the probability that on any given day, the number sold exceed the average? What is the probability that the number sold on any given day is exactly equal to the average?

31. X is a uniform random variable in the range [0,5]. If $g(X) = \exp.(-X)$, what is the average of $g(X)$? Verify the results through random number simulation.

32. X is an exponential random variable with a mean of 3. If a new random variable is defined as $Y = X + 3$, show that the variance of X and Y are the same by examining the characteristic function or MGF of X and Y.

33. X is a gamma random variable G(2,3). If $Y = 3X$, use the MGF to show that that the variance of Y is 9 times the variance of X.

34. X is a Poisson variable with a parameter 6. Using MGF, show that its mean and variance are equal.

35. X is a binomial random variable with $N = 5$ and $p = 0.6$. Obtain its mean and variance from its MGF.

36. X is uniform in [0,1]. $Y = -\log(1-X)$. Obtain the pdf, the CDF, and the mean of Y.

37. X is uniform in [0,1]. $Y = -\log(X)$. Obtain the pdf, the CDF, and the mean of Y.

38. X is uniform in [0,1]. $Y = -\log \sqrt{X}$. Obtain the pdf, the CDF, and the mean of Y.

39. X is uniform in [0,1]. $Y = \sqrt{-\log(X)}$. Obtain the pdf, the CDF, and the mean of Y.

40. X is uniform in [0,2π]. If $Y = |\sin(X)|$, obtain the pdf and CDF of Y.

41. X is uniform in [0,π/2]. If $Y = \tan(X)$, obtain the pdf and CDF of Y.

42. X is uniform in [0,1]. Obtain the pdf of $Y = X^n$, $n > 0$.

43. X is uniform in [−2,4]. Obtain the pdf and CDF of $Y = 1/X^2$.

44. X is uniform in [−2,1]. Obtain the pdf and CDF of $Y = 1/X$.

45. X is the exponentially distributed density of the power expressed in mW with an average of 10 mW. Obtain the pdf of the power expressed in dB_m units.

46. The current is Rayleigh distributed with an average of 2 mA. If the load is 5 Ohms, obtain an expression for the pdf of the power dissipated across the load.

47. The power in a wireless system X is gamma distributed with a pdf G(a,b). If the power is scaled such that $Y = X^k$, $k > 1$, obtain an expression for the pdf of Y. Verify your results through random number simulation for $k = 2$ and $a = 2$ and $b = 1$.

48. If X is Rayleigh distributed with an average power of 8 units (average power is the second moment), obtain the pdf and CDF of $Y = 2x^2 + 3$.

49. If X is Rayleigh distributed with an average power of 8 units (average power is the second moment), obtain the pdf and CDF of the power conditioned on being greater than 8 units.

50. If X exponential with an average of 5 units, obtain the pdf and CDF of $Y = \frac{1}{\sqrt{X}}$.

51. The power received at a radar station is modeled in terms of a gamma variable. When a target is present, the radar return is a gamma variable G(2.5,15), and when the target is absent, the radar return is a gamma variable G(1.5,4). If a threshold is set at 9 units, what is the probability of false alarm? What is the probability of miss? If a target is seen only 10% of the time, what is the error rate?

52. In the previous problem, if we expect a false alarm rate of 1e–2, what should be the threshold? At that threshold, what is the probability of miss?

53. X is Rayleigh distributed with a mean power of 8 units. A new random variable is defined as follows:

$$Y = \begin{cases} 3, X < 3 \\ X^2, X > 3 \end{cases}$$

Obtain and plot the pdf and CDF of Y.

54. X is an exponentially distributed random variable with a mean of 5. A new random variable is defined as

$$Z = \begin{cases} X, X < 5 \\ \frac{1}{X}, X > 5 \end{cases}$$

Obtain the pdf and CDF of Z.

55. Power received at the radar receiver can be described in terms of an exponential random variable. In the absence of any target, the power received is an exponential variable with a mean of 1 mW. When the target is present, power received is also an exponential variable with a mean of 16 mW. If the threshold for decision is set to 4 mW, obtain the probabilities of false alarm and miss. If the chances of the target being present are only 10%, what is the probability of error in this decision-making process?

56. X is Gaussian, $N(3,16)$. Obtain the pdf and CDF of $Y = X^2$.

57. X is Gaussian, $N(4,9)$. A new random variable is created such that

$$Z = \begin{cases} -3, X < -3 \\ X, \; -3 < X < 3 \\ 3, X > 3 \end{cases}$$

Obtain the pdf of Z.

58. X is Gaussian, $N(4,9)$. The noise modeled using this Gaussian is passed through an A/D converter such that any voltage less than -10 becomes -10 and any voltage greater than 10 becomes 10. The values between -10 and 10 are equally divided into 5 bins. Obtain the pdf of the digitized output.

59. X is Gaussian, $N(3,16)$. Obtain the pdf of

$$Y = 10^{\frac{X}{10}}$$

60. X is Gaussian, $N(0,16)$. Obtain the pdf of $Y = [X|X > 4]$.
61. X is Gaussian, $N(3,16)$. Obtain the pdf of $Y = [X|-4 < X < 4]$.
62. X is exponentially distributed with a parameter of unity. Obtain the pdf, CDF, mean, and variance of

$$Y = e^{-X}$$

63. If X is a continuous random variable with a CDF of $F(x)$, obtain the pdf, the CDF, and the mean of $Y = 2F(X) + 1$.
64. X is an exponential random variable with a mean of 5 units. A new random variable Y is defined as follows:

$$Y = \left[X \mid X > 5 \right]$$

Find the pdf of Y and the mean of Y.

Problems 3.65–3.74

> For the 10 data sets provided, obtain the ROC plots, area under the ROC curve, optimum threshold, and the positive predictive value at the optimum threshold (display results similar to shown in Table 3.1 and Fig. 3.17).

Problems 3.75–3.124

> For the 50 data sets provided from (target absent 70 samples and target present 30 samples), obtain the ROC plots, area under the ROC curve, optimum threshold, and the positive predictive value at the optimum threshold.

Problem 3.125 Use random number generation to verify the results of Problem 3.36.
Problem 3.126 Use random number generation to verify the results of Problem 3.37.
Problem 3.127 Use random number generation to verify the results of Problem 3.38.
Problem 3.128 Use random number generation to verify the results of Problem 3.39.
Problem 3.129 Use random number generation to verify the results of Problem 3.40.
Problem 3.130 Use random number generation to verify the results of Problem 3.41.
Problem 3.131 Use random number generation to verify results of Problem 3.43.
Problem 3.132 Use random number generation to verify results of Problem 3.44.
Problem 3.133 Use random number generation to verify results of Problem 3.45.
Problem 3.134 Use random number generation to verify results of Problem 3.46.
Problem 3.135 Use random number generation to verify results of Problem 3.47. Use $a = 2$, $b = 1$, and $k = 2$.

Problem 3.136 Use random number generation to verify the results of Problem 3.48.
Problem 3.137 Use random number generation to verify the results of Problem 3.49.
Problem 3.138 Use random number generation to verify the results of Problem 3.50.
For the following set of exercises, generate 1000 samples of data belonging to
hypothesis H_0 (target absent) and 900 samples of data belonging to hypothesis
H_1 (target present). In each case, obtain the density plots of the data,
corresponding ROC curves, and areas under the ROC curves (all from random
numbers). Obtain the ROC, area under the ROC curve, and optimum operating
point, using the theoretical densities. Compare the area under the ROC curve
(theory vs. simulation).
Problem 3.139.

Target absent: $N(-1,16)$.
Target present: $N(6,16)$.

Problem 3.140.

Target absent: $N(-1,25)$.
Target present: $N(8,9)$.

Problem 3.141.

Target absent: Rayleigh (3).
Target present: Rayleigh (8).

Problem 3.142.

Target absent: Rayleigh(1).
Target present: Nakagami(3,5).

Problem 3.143.

Target absent: Nakagami(2,1).
Target present: Nakagami(3,5).

Problem 3.144.

Target absent: gamma(2,4).
Target present: gamma(3,6).

Problem 3.145.

Target absent: Rayleigh(3).
Target present: Rician(2,6).

Problem 3.146.

Target absent: Rician(1.2,1).
Target present: Rician(2.5,2).

Problem 3.147.

Target absent: Nakagami(1.2,3).
Target present: Rician(2.5,4).

Problem 3.148.

Target absent: Nakagami(1.2,2).
Target present: Weibull(3,5).

Problem 3.149.

Target absent: Poisson(5).
Target present: Poisson(8).

Problem 3.150.

Target absent: Poisson(4).
Target present: Poisson(10).

Problems 3.151–260 For the data sets (target absent, target present: The first 40 samples correspond to the absence of the target and the rest represent the target present), determine the threshold at the intersection of the data sets, and calculate the corresponding operating point to the top left corner of the ROC plot. Obtain the confusion matrix.

Now, obtain the ROC plot and determine the optimum threshold. Mark the two operating points (threshold based on the intersection and optimal threshold) on the ROC plot. Obtain the confusion matrix, and check whether the distance to the top left corner of the ROC plot is less than the case where the threshold was chosen at the intersection of the densities (110 sets of data).

Problems 3.261–360 Repeat the ROC analysis for the data set given tasks (similar to the previous set). The set consists of 60 samples of target absent and 60 samples of target absent (100 sets of data).

Problems 361–460 Repeat the ROC analysis for the data set given tasks (similar to the previous set). The set consists of 60 samples of target absent and 50 samples of target absent.

References

Abramowitz M, Segun IA (1972) Handbook of mathematical functions with formulas, graphs, and mathematical tables. Dover Publications, New York

Block HW, Savits TH (1980) Laplace transforms for classes of life distributions. Ann Probab 8 (3):465–474

Bradley AP (1997) The use of the area under the ROC curve in the evaluation of machine learning algorithms. Pattern Recogn 30(7):1145–1159

Brown CD, Davis HT (2006) Receiver operating characteristics curves and related decision measures: a tutorial. Chemometr Intell Lab Syst 80:24–38

DeLong ER, DeLong DM, Clarke-Pearson DL (1988) Comparing the areas under two or more correlated receiver operating characteristic curves: a nonparametric approach. Biometrics 44:837–845

Eng J (2012) Teaching receiver operating characteristic analysis: an interactive laboratory exercise. Acad Radiol 19:1452–1456

Fawcett T (2006) An introduction to ROC analysis. Pattern Recogn Lett 27:861–874

Gradshteyn IS, Ryzhik IM (2000) Table of integrals, series, and products, 6th edn. Academic, New York

Hanley JA, McNeil B (1982) The meaning and use of the area under a receiver operating characteristic (ROC) curve. Radiology 143:29–36

Hanley JA, McNeil BJ (1983) A method for comparing areas under the receiver operating characteristic curves derived from the same cases. Radiology 148:839–843

Helstrom CW (1968) Statistical theory of signal detection. Pergamon Press, Oxford/New York

Lawton L (2009) An exercise for illustrating the logic of hypothesis testing. J Stat Educ 17:1–9

Limpert E, Stahel WA, Abbt M (2001) Log-normal distributions across the sciences: keys and clues: on the charms of statistics, and how mechanical models resembling gambling machines offer a link to a handy way to characterize log-normal distributions, which can provide deeper insight into variability and probability—normal or log-normal: that is the question. Bioscience 51(5):341–352

Metz CE (2006) Receiver operating characteristic analysis: a tool for the quantitative evaluation of observer performance and imaging systems. J Am Coll Radiol 3(6):413–422

Nakagami M (1960) The m-distribution—a general formula of intensity distribution of rapid fading. In: Hoffman WC (ed) Statistical methods in radio wave propagation. Pergamon, Elmsford

Papoulis A, Pillai U (2002) Probability, random variables, and stochastic processes. McGraw-Hill, New York

Rohatgi VK, Saleh AKME (2001) An introduction to probability and statistics. Wiley, New York

Rota M, Antolini L (2014) Finding the optimal cut-point for Gaussian and Gamma distributed biomarkers. Comput Stat Data Anal 69:1):1–1)14. https://doi.org/10.1016/j.csda.2013.07.015

Roupp MD, Perkins NJ, Whitcomb BW, Schisterman EF (2018) Youden index and optimal cut-point estimated from observations affected by a lower limit of detection. Biom J 50 (3):419–430

Shankar PM (2015) A composite shadowed fading model based on the McKay distribution and Meijer G functions. Wirel Pers Commun 81(3):1017–1030

Shankar PM (2017) Fading and shadowing in wireless systems, 2nd edn. Springer, Cham

Shankar PM (2019) Pedagogy of Bayes' rule, confusion matrix, transition matrix, and receiver operating characteristics. Comput Appl Eng Educ 27(2):510–518. https://doi.org/10.1002/cae.22093

Shankar PM (2020) Introduction of data analytics in the engineering probability course: Implementation and lessons learnt. Comput Appl Eng Educ 28(5):1072–1082

Smits N (2010) A note on Youden's J and its cost ratio. BMC Med Res Methodol 10(1):89–92

Smits N, Smit F, Cuipers P, De Graph R (2007) Using decision theory to derive optimal cut-off scores of screening instruments: an illustration explicating costs and benefits of mental health screening. Int J Methods Psychiatr Res 16(4):219–229. https://doi.org/10.1002/mpr.230

Stacy E (1962) A generalization of the gamma distribution. Ann Math Stat 33(3):1187–1192

Van Trees HL (1968) Detection, estimation, and modulation theory, part I. Wiley, New York

Youden WJ (1950) Index for rating diagnostic tests. Cancer 3(1):32–35

Zweig MH, Campbell G (1993) Receiver-operating characteristic (ROC) plots: a fundamental evaluation tool in clinical medicine. Clin Chem 39(4):561–577

Chapter 4
Multiple Random Variables and Their Characteristics

4.1 Introduction

Statistical analysis of engineering systems often relies on the outcomes of two or more experiments. Additionally further processing of these experimental results may be necessary for making decisions on the next step. Initially we examine applications involving the processing of outcomes of two experiments or trials constituting the study of two random variables. Later, we will explore the statistics of more than two random variables.

A few examples of interest dealing with two variables are cited below:

We have a signal with amplitude values that are complex. We are interested in understanding the statistics of the magnitude, power, or the phase.

We have two systems that are connected in parallel, series, or standby mode. We are interested in analyzing and comparing the performance of these systems.

We have collected the data from two sources. We are interested in finding out the best approach to process these by exploring whether we should take the arithmetic mean, geometric mean, product, ratio, maximum, or minimum of these data sets. We may expand the scope of the signal processing algorithms beyond the ones mentioned above.

We often conduct two stage experiments. This means that we perform one measurement first, and based on the outcome of this measurement, a second experiment is conducted. Our interest is in modeling such two stage experiments to gain knowledge about the statistics of outcomes of these experiments.

From the discussion on the statistics of a single random variable in Chap. 3, it can be understood that each of two sets of outcomes can be modeled as random

Electronic supplementary material: The online version of this chapter (https://doi.org/10.1007/978-3-030-56259-5_4) contains supplementary material, which is available to authorized users.

P. M. Shankar, *Probability, Random Variables, and Data Analytics with Engineering Applications*, https://doi.org/10.1007/978-3-030-56259-5_4

variables. The properties of these random variables can be understood by revisiting the concepts of joint probability introduced in Chap. 2.

4.2 Joint Distribution and Densities

The concept of the joint probability of two events A and B introduced in Chap. 2 will be the starting point of discussion. Following the study of a single random variable from Chap. 3, we know that events can be formed from random variables. If the measurements or outcomes are modeled as random variables X and Y, it is possible to define two events A and B as

$$
\begin{aligned}
A &= \{X \leq x\} \\
B &= \{Y \leq y\}
\end{aligned}
\tag{4.1}
$$

In Eq. (4.1), X and Y are random variables, and x and y, respectively, represent the value each takes. The joint probability is expressed as

$$
P(AB) = P[X \leq x, Y \leq y]
\tag{4.2}
$$

Using the concept of the CDF introduced in connection with a single random variable, the left hand side of Eq. (4.2) must represent the CDF. Rewriting Eq. (4.2), we get

$$
F_{X,Y}(x, y) = P[X \leq x, Y \leq y]
\tag{4.3}
$$

In Eq. (4.3), $F_{X,Y}(x, y)$ is the joint cumulative distribution function. The joint pdf will be

$$
f_{X,Y}(x, y) = \frac{\partial^2 F_{X,Y}(x, y)}{\partial x \partial y}
\tag{4.4}
$$

It is easy to observe that if X and Y are independent,

$$
F_{X,Y}(x, y) = P[X \leq x, Y \leq y] = P[X \leq x]P[Y \leq y] = F_X(x)F_Y(y)
\tag{4.5}
$$

$$
f_{X,Y}(x, y) = f_X(x)f_Y(y)
\tag{4.6}
$$

In other words, when two random variables are independent, the joint pdf is the product of the marginal pdfs $f_X(x)$ and $f_Y(y)$. Similarly for two independent random variables, the joint CDF is the product of the marginal CDFs.

Taking advantage of the properties of pdf and CDF associated with a single random variable, it is possible to state the properties of the joint pdf and the joint CDF. These are

$$
F_{X,Y}(-\infty, -\infty) = 0
\tag{4.7}
$$

$$
F_{X,Y}(-\infty, y) = F_{X,Y}(x, -\infty) = 0
\tag{4.8}
$$

$$0 \leq F_{X,Y}(x,y) \leq 1 \tag{4.9}$$

$$\frac{\partial F_{X,Y}(x,y)}{\partial x} \geq 0 \tag{4.10}$$

$$\frac{\partial F_{X,Y}(x,y)}{\partial y} \geq 0 \tag{4.11}$$

$$\frac{\partial^2 F_{X,Y}(x,y)}{\partial x \partial y} \geq 0 \tag{4.12}$$

$$F_{X,Y}(\infty, \infty) = 1 \tag{4.13}$$

$$F_{X,Y}(x,y) = \int_{-\infty}^{x} \int_{-\infty}^{y} f_{X,Y}(u,w) \, du \, dw \tag{4.14}$$

Equations (4.10), (4.11), and (4.12) imply that the joint cumulative distribution function is a monotonically increasing function of x and y. Equation (4.13) implies that the joint density integrates to unity as

$$F_{X,Y}(\infty, \infty) = \int_{-\infty}^{\infty} \int_{-\infty}^{\infty} f(x,y) \, dx \, dy = 1 \tag{4.15}$$

Additional attributes of the joint pdf and CDF can now be summarized in terms of the relationships among the joint and marginal densities.

$$f_X(x) = \frac{\partial F(x, \infty)}{\partial x} = \int_{-\infty}^{\infty} f_{X,Y}(x,w) \, dw \tag{4.16}$$

$$f_Y(y) = \frac{\partial F(\infty, y)}{\partial y} = \int_{-\infty}^{\infty} f_{X,Y}(u,y) \, du \tag{4.17}$$

Note that there is a difference between Eq. (4.15) and the integral of a single pdf being equal to unity,

$$\int f(x) \, dx = \int f(y) \, dy = 1 \tag{4.18}$$

In a one-dimensional case, Eq. (4.18) implies that the area under the curve is unity and in a two-dimensional case, the volume encompassed by the area is unity. This will now be shown by taking the case of a single Gaussian variable and the case of a joint density of two Gaussian variables. Figure 4.1 shows the density of a Gaussian random variable.

The joint pdf of two independent and identically distributed random variables, each being $N(0, \sigma^2)$ is

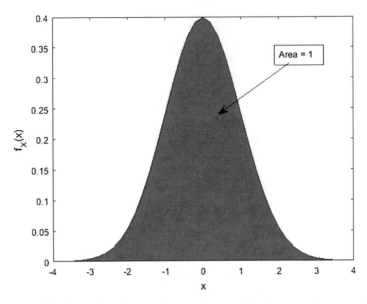

Fig. 4.1 Probability density function of a single random variable. Area under the curve is unity

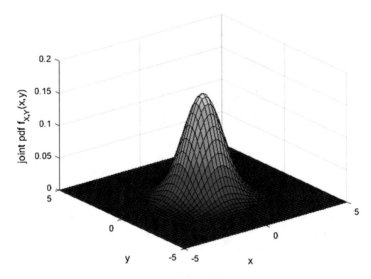

Fig. 4.2 Joint pdf in Eq. (4.19)

$$f_{X,Y}(x,y) = \frac{1}{\sqrt{2\pi\sigma^2}} \frac{1}{\sqrt{2\pi\sigma^2}} e^{-\frac{x^2+y^2}{2\sigma^2}} \tag{4.19}$$

The joint pdf is shown in Fig. 4.2.

Fig. 4.3 Joint pdf in Eq. (4.20)

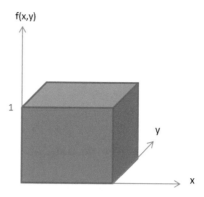

The integral in Eq. (4.15) in this case implies that the volume encompassed as shown is unity.

An example of a joint density associated with X and Y being uniform in the range [0, 1] is shown in Fig. 4.3. The joint density is given as

$$f(x, y) = 1, \quad 0 < x < 1, 0 < y < 1 \tag{4.20}$$

In this case, the volume of the cube with each side of unit length represents the integral in Eq. (4.15) and is equal to unity.

It is easy to summarize the following relationships of the joint CDF and the joint pdf.

$$F_{X,Y}(x, \infty) = F_X(x) = \int_{u=-\infty}^{x} \int_{-\infty}^{\infty} f_{X,Y}(u, v) du dv \tag{4.21}$$

$$F_{X,Y}(\infty, y) = F_Y(y) = \int_{-\infty}^{\infty} \int_{v=-\infty}^{y} f_{X,Y}(u, v) du dv \tag{4.22}$$

$$\frac{\partial F_{X,Y}(x, y)}{\partial x} = \frac{\partial}{\partial x} \left[\int_{-\infty}^{x} \int_{v=-\infty}^{y} f_{X,Y}(u, v) du dv \right] = \int_{-\infty}^{y} f_{X,Y}(x, v) dv \tag{4.23}$$

Note that the right hand side of Eq. (4.23) is obtained from the application of the Leibniz theorem (Sect. 4.9). Similarly, we get

$$\frac{\partial F_{X,Y}(x, y)}{\partial y} = \int_{-\infty}^{x} f_{X,Y}(u, y) du \tag{4.24}$$

$$f_X(x) = \int_{-\infty}^{\infty} f_{X,Y}(x, y)dy \qquad (4.25)$$

$$f_Y(y) = \int_{-\infty}^{\infty} f_{X,Y}(x, y)dx \qquad (4.26)$$

$$\text{Prob}[x_1 \le X \le x_2, y_1 \le Y \le y_2] = \int_{x_1}^{x_2} \int_{y_1}^{y_2} f_{X,Y}(x, y)dxdy \qquad (4.27)$$

The notion of joint statistics described so far was presented within the context of continuous random variables. One can extend this notion to the discrete random variables as well.

A joint CDF associated with a pair of discrete random variables can be written as

$$F_{X,Y}(x, y) = \sum_{n=1}^{N} \sum_{m=1}^{M} P(x_n, y_m) U(x - x_n) U(y - y_m) \qquad (4.28)$$

In Eq. (4.28), $P(x_n, y_m)$ is the joint PMF (probability mass function) associated with the event $\{X = x_n, Y = y_m\}$. The joint pdf is given as

$$f_{X,Y}(x, y) = \sum_{n=1}^{N} \sum_{m=1}^{M} P(x_n, y_m) \delta(x - x_n) \delta(y - y_m) \qquad (4.29)$$

Consider the case of an experiment consisting of a coin toss and a die roll. The random variable representing the coin toss has outcomes of 1 (Heads) with probability p and 0 (Tails) with probability $(1-p)$ implying a biased coin. If the die is a fair one, each outcome ranging from 1 to 6 will have a probability of $(1/6)$. X represents the random variable associated with the coin toss, and Y represents the random variable associated with the die roll. Since each of these experiments is independent, we have

$$F_{X,Y}(x, y) = \sum_{n=1}^{2} \sum_{m=1}^{6} P(x_n, y_m) U(x - x_n) U(y - y_m)$$

$$= \sum_{n=1}^{2} \sum_{m=1}^{6} P(x_n) P(y_m) U(x - x_n) U(y - y_m) \qquad (4.30)$$

In Eq. (4.30), the two probability mass functions (PMF) are given as

$$P(x_n) = \begin{bmatrix} P(X = 1) = p \\ P(X = 0) = 1 - p \end{bmatrix} \tag{4.31}$$

$$P(Y = m) = \frac{1}{6}, \quad m = 1, 2, \cdots, 6 \tag{4.32}$$

The joint PMF can be written as

$$P(0, m) = (1 - p)\frac{1}{6}, \quad m = 1, 2, \cdots, 6$$
$$P(1, m) = p\frac{1}{6}, \quad m = 1, 2, \cdots, 6 \tag{4.33}$$

It can be seen that

$$F(\infty, \infty) = \sum_{n=1}^{2} \sum_{m=1}^{6} P(x_n)P(y_m) = 6\left(\frac{1}{6}\right)(1 - p) + 6\left(\frac{1}{6}\right)p = 1 \tag{4.34}$$

When we have two random variables X and Y, we can define a complex random variable as

$$Z = X + jY \tag{4.35}$$

It can be seen that a complex random variable is made up of two real random variables, and therefore, the complex random variable is completely described by the joint density $f_{X, Y}(x, y)$ of the real and imaginary parts.

4.3   Conditional Densities

While the probability that a continuous random variable takes a specific value is zero (see Chap. 3), the events $\{X = x\}$ and $\{Y = y\}$ have special meaning and interpretation within the context of two random variables. To understand this, let us examine the event $[Y \leq y | x_1 \leq X \leq x_2]$ and associated conditional CDF and conditional pdf. Using Bayes' rule, we have the conditional probability as

$$P[Y < y | x_1 < X < x_2] = \frac{P[x_1 < X < x_2, Y < y]}{P[x_1 < X < x_2]} \tag{4.36}$$

Using the definition of the CDF, Eq. (4.36) becomes

$$F[y|x_1 < X < x_2] = \frac{F_{X,Y}[x_1 < X < x_2, y]}{F_X(x_2) - F_X(x_1)} = \frac{F_{X,Y}(x_2, y) - F_{X,Y}(x_1, y)}{F_X(x_2) - F_X(x_1)} \quad (4.37)$$

Taking the derivative of Eq. (4.37) w.r.t. y, we get

$$\frac{\partial}{\partial y} F[y|x_1 < X < x_2] = \frac{\frac{\partial}{\partial y} F_{X,Y}(x_2, y) - \frac{\partial}{\partial y} F_{X,Y}(x_1, y)}{F_X(x_2) - F_X(x_1)} \quad (4.38)$$

Using Eq. (4.24), Eq. (4.38) becomes

$$f[y|x_1 < X < x_2] = \frac{\int_{-\infty}^{x_2} f(x, y)dx - \int_{-\infty}^{x_1} f(x, y)dx}{F_X(x_2) - F_X(x_1)} = \frac{\int_{x_1}^{x_2} f(x, y)dx}{F_X(x_2) - F_X(x_1)} \quad (4.39)$$

Equation (4.39) provides an interesting result if we make

$$\begin{aligned} x_1 &= x \\ x_2 &= x + \Delta x \end{aligned} \quad (4.40)$$

If we now let $\Delta x \to 0$

$$\{x < X < x + \Delta x\}_{\Delta x \to 0} = \{X = x\} \quad (4.41)$$

Using Eqs. (4.40) and (4.41), Eq. (4.39) becomes

$$f[y|x < X < x + \Delta x]_{\Delta x \to 0} = \frac{\int_{x}^{x+\Delta x} f(x, y)dx}{F_X(x + \Delta x) - F_X(x)} = \frac{f(x, y)\Delta x}{f(x)\Delta x} = \frac{f(x, y)}{f(x)} \quad (4.42)$$

Rewriting the LHS of Eq. (4.42) leads to

$$f(y|x) = \frac{f(x, y)}{f(x)} \quad (4.43)$$

Notice that X and Y being independent lead to Eq. (4.43) to produce the expected result,

$$f(y|x) = \frac{f(x)f(y)}{f(x)} = f(y) \quad (4.44)$$

We can similarly define the conditional density of X given $Y = y$ as

$$f_{X|Y}(x|y) = \frac{f_{X,Y}(x,y)}{f_Y(y)} \tag{4.45}$$

It should be noted that Eq. (4.43) is valid only when $f(x) \neq 0$ and Eq. (4.45) is valid only when $f(y) \neq 0$.

An interesting case is obtained by examining the conditional joint CDF given by

$$F(x,y|x_1 \leq X \leq x_2) = \frac{P(X \leq x, Y \leq y, x_1 \leq X \leq x_2)}{P(x_1 \leq X \leq x_2)} \tag{4.46}$$

The joint probability in the numerator of Eq. (4.46) can be obtained by examining the overlapping region of the two events, $\{X < x, Y < y\}$ and $\{x_1 < X < x_2\}$, as shown in Fig. 4.4.

- When $x < x_1$, there is no overlap between the two events.
- When $x > x_2$, the region along X is fixed, and therefore, the overlap is determined by $\{Y < y\}$ alone.
- When $\{x_1 < X < x_2\}$, overlap exists when $x > x_1$ and $x < x_2$.

Based on this, the numerator of Eq. (4.46) simplifies to

$$F(x,y|x_1 \leq X \leq x_2) = \begin{cases} 0, x < x_1 \\ \dfrac{F(x_2,y) - F(x_1,y)}{F(x_2) - F(x_1)}, x > x_2 \\ \dfrac{F(x,y) - F(x_1,y)}{F(x_2) - F(x_1)}, x_1 \leq x \leq x_2 \end{cases} \tag{4.47}$$

Differentiating Eq. (4.47) w.r.t. x (x_1 and x_2 are fixed) and w.r.t. y, we get

Fig. 4.4 Regions of interest for conditioning

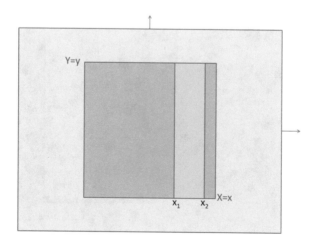

$$f(x, y | x_1 \leq X \leq x_2) = \begin{cases} \dfrac{f(x, y)}{F(x_2) - F(x_1)}, x_1 \leq x \leq x_2 \\ 0, \quad \text{elsewhere} \end{cases}$$

$$= \begin{cases} \dfrac{f(x, y)}{P(x_1 \leq X \leq x_2)}, x_1 \leq x \leq x_2 \\ 0, \quad \text{elsewhere} \end{cases} \qquad (4.48)$$

We may use Eq. (4.48) as a template to obtain the joint density conditioned on an event, similar to the results obtained in Chap. 3. In other words, *the conditional density is a scaled version of the joint density and it exists only in the range of the conditioning event.*

Interpretation of conditional densities defined with {$X = x$} or {$Y = y$}.
Even though X and Y are continuous random variables, the conditioned densities imply

$$\begin{aligned} f(x|y) &= f(x|Y = y) \\ f(y|x) &= f(y|X = x) \end{aligned} \qquad (4.49)$$

Equation (4.49) suggests that for the analysis of conditional densities, $Y = y$ means that *the value of Y is fixed at y* and $X = x$ means that *the value of X is fixed at x*. While the notion of the event {$X = x$} in the context of a single continuous random variables implies that the probability of such an event is zero, the physical interpretation of the conditional density in Eq. (4.43) is seen in Fig. 4.5. First we choose a specific value of $X = x$. The conditional density $f(y|X = x)$ allows us to examine the statistics of Y when the outcome associated with X is fixed at a specific value.

This understanding of the conditionality helps in solving engineering problems involving multiple random variables as shown later.

Fig. 4.5 Interpretation point of conditioning

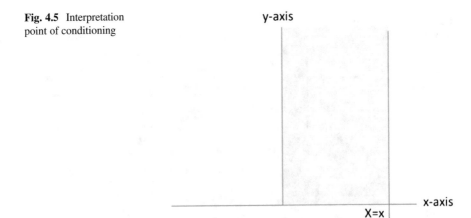

Example 4.1 The joint pdf of two random variables X and Y is

$$f_{X,Y}(x,y) = k(x^2 + y), 0 < x < 2, 0 < y < 1$$

Find the constant k. Obtain the marginal density functions of X and Y. Are X and Y independent?

Solution

$$\int_{x=0}^{2} \int_{y=0}^{1} k(x^2 + y)\,dxdy = 1 \Rightarrow k = \frac{3}{11}$$

$$f_X(x) = \int_0^1 \frac{3}{11}(x^2 + y)\,dy = \frac{3}{11}x^2 + \frac{3}{22}, \quad 0 < x < 2$$

$$f_Y(y) = \int_0^2 \frac{3}{11}(x^2 + y)\,dx = \frac{6}{11}y + \frac{8}{11}, \quad 0 < y < 1$$

It is easily seen that

$$f(x,y) \neq f(x)f(y)$$

Therefore X and Y are not independent.

Example 4.2 X and Y are not independent in Example 4.1. Obtain the conditional densities of X and Y conditioned on Y and X, respectively.

Solution

$$f(x|y) = \frac{f(x,y)}{f(y)} = \frac{3(x^2 + y)}{6y + 8}, 0 < x < 2, 0 < y < 1$$

$$f(y|x) = \frac{f(x,y)}{f(x)} = \frac{6(x^2 + y)}{6x^2 + 3}, 0 < x < 2, 0 < y < 1$$

Example 4.3 The joint density of X and Y is given as

$$f(x,y) = kxy, 0 < x < y < 2$$

What is the value of k? What are the marginal densities of X and Y?

Solution

$$\int_{y=0}^{2} \int_{x=0}^{y} kxy\,dx\,dy = 1 \Rightarrow 2k = 1 \Rightarrow k = \frac{1}{2}$$

$$f(x,y) = \frac{1}{2}xy, \quad 0 < x < y < 2$$

$$f_X(x) = \int_{x}^{2} f(x,y)dy = x - \frac{x^3}{4}, 0 < x < 2$$

$$f_Y(y) = \int_{0}^{y} f(x,y)dx = \frac{y^3}{4}, 0 < y < 2$$

Example 4.4 X and Y are random variables with a joint pdf

$$f(x,y) = \frac{1}{\pi r^2}, \quad 0 < \sqrt{x^2 + y^2} < r$$

Obtain the marginal densities of X and Y.

Solution

$$f(x) = \int_{y=-\sqrt{r^2-x^2}}^{y=\sqrt{r^2-x^2}} f(x,y)dy = \int_{y=-\sqrt{r^2-x^2}}^{y=\sqrt{r^2-x^2}} \frac{1}{\pi r^2}dy = \frac{2}{\pi r^2} \int_{0}^{y=\sqrt{r^2-x^2}} dy$$

$$f(y) = \int_{x=-\sqrt{r^2-y^2}}^{x=\sqrt{r^2-y^2}} f(x,y)dx = \int_{x=-\sqrt{r^2-y^2}}^{x=\sqrt{r^2-y^2}} \frac{1}{\pi r^2}dx = \frac{2}{\pi r^2} \int_{0}^{x=\sqrt{r^2-y^2}} dx$$

$$f(x) = \begin{cases} \dfrac{2\sqrt{r^2 - x^2}}{\pi r^2}, & -r < x < r \\ 0, \text{elsewhere} \end{cases} \qquad f(y) = \begin{cases} \dfrac{2\sqrt{r^2 - y^2}}{\pi r^2}, & -r < y < r \\ 0, \text{elsewhere} \end{cases}$$

Example 4.5 X and Y are independent random variables with densities $f_X(x)$ and $f_Y(y)$. Obtain the density of X given that $X + Y < b$ where $b > 0$.

Solution The conditional density is (see Eq. (4.48))

$$f(x|X + Y < b) = \frac{f_X(x)}{P[X + Y < b]}, x + y < b$$

The denominator is

$$P(X+Y < b) = \int\limits_{y=-\infty}^{\infty} \left[\int\limits_{-\infty}^{b-y} f_X(x)dx \right] f_Y(y)dy$$

4.4 Two-Stage Experiments

One of the interesting applications of the concept of two random variables pertains to its use in two- stage experiments or measurements. Two-stage experiments fall in two categories, experiments deliberately set up in two stages or measurements being interpreted and best modeled as two-stage experiments.

We will now explore the notion of two-stage experiment through a few examples by invoking the concept of conditioning described in Sect. 4.3. Note that two-stage experiments consist of outcomes that may be continuous (both stages) or discrete (both stages). We may also have instances of one of the stages resulting in continuous outcomes, while the other one results in discrete outcomes.

Example 4.6 A number X is picked randomly in the interval $[0,b]$. Based on the observed value of X, we conduct a second experiment. The second one consists of picking a number Y randomly between $[0,X]$. What is the pdf of Y?

Solution Invoking the concept of point conditioning presented earlier, we note that the second experiment is conducted based on the observations of the first one. This means that the second experiment is best modeled with the outcome as a conditioned variable with the conditioning event $X = x$. This leads to the conditional density of Y given $X = x$ as

$$f(y|X = x) = f(y|x) = \frac{1}{x}, 0 < y < x$$

The density of X is clearly uniform and is given by

$$f(x) = \frac{1}{b}, \quad 0 < x < b$$

The joint density of X and Y can now be written using Eq. (4.43) as

$$f(x,y) = f(y|x)f(x) = \frac{1}{bx}, 0 < y < x < b$$

The marginal density of Y is obtained as

$$f_Y(y) = \int f(x, y)dx = \int_y^b \frac{1}{bx}dx = \frac{1}{b}\left[\log_e\left(\frac{b}{y}\right)\right], \qquad 0 < y < b$$

Example 4.7 Consider an experiment in which a die is rolled. Depending on the outcome of the roll, a sample of an exponential random variable Y is chosen such that its mean is the outcome of the roll of the die. Obtain the density of Y.

Solution Based on the information provided, the pdf of Y will be conditioned on the outcome of the roll of the die, and it is expressed as

$$f(y|X = k) = f(y|k) = \frac{1}{k}\exp\left(-\frac{y}{k}\right)U(y)$$

If the number observed (between 1 and 6) is k, the PMF of X can be written as

$$P(X = k) = \frac{1}{6}, \quad k = 1, 2, \cdots, 6$$

The pdf of X is

$$f_X(x) = \frac{1}{6}\sum_{k=1}^{6}\delta(x - k)$$

Therefore, the joint density of X and Y is

$$f(x, y) = f(y|x)f(x) = \frac{1}{6}\sum_{k=1}^{6}\delta(x - k)\frac{1}{k}\exp\left(-\frac{y}{k}\right)$$

The marginal density of Y is

$$f_Y(y) = \int f(x, y)dx = \frac{1}{6}\sum_{k=1}^{6}\frac{1}{k}\exp\left(-\frac{y}{k}\right)U(y)$$

Example 4.8 Instead of the experiments described above, let us assume that we are measuring the RF signal strength (power) in a room and examining its statistics. Our measurements lead to two interesting observations. The power (X) at any given location appears to fit the description of an exponential random variable. The second observation is that the average power measured at several locations is not constant. When the statistics of the average power from these locations is examined, it also turns out to be best described as an exponential random variable. What is the pdf of X? Use Y to represent the average power.

Solution If X is the power measured, model for the power X now expressed as

$$f(x|y) = \frac{1}{y}e^{-\frac{x}{y}}U(x)$$

The conditioning shows our a priori information about the average power being a random variable with a density (assume that the mean of Y is b),

$$f(y) = \frac{1}{b}e^{-\frac{y}{b}}U(y)$$

The joint density now becomes

$$f(x,y) = f_X(x|y)f_Y(y) = \frac{1}{by}e^{-\left(\frac{x}{y}+\frac{y}{b}\right)}U(x)U(y)$$

The marginal density of X is obtained as

$$f_X(x) = \int_0^\infty f(x,y)dy = \int_0^\infty \frac{1}{by}e^{-\left(\frac{x}{y}+\frac{y}{b}\right)}dy = \frac{2}{b}K_0\left(2\sqrt{\frac{x}{b}}\right)U(x)$$

The pdf of X is obtained from the Table of integrals (Gradshteyn and Ryzhik 2000).

While Examples 4.6 and 4.7 illustrate the cases of experiments deliberately set up in two stages, Example 4.8 represents observations best modeled as a two-stage experiment (Shankar 2017).

While point conditioning is useful in modeling two-stage experiments (whether it is based on continuous or discrete outcomes), it is also useful in obtaining the pdf and CDF of a function of two random variables. These cases are discussed next.

4.5 Function of Two Random Variables

In Chap. 2, we examined ways of obtaining the densities of transformed variables created from a single random variable. Similarly, there is an interest in the statistics of functions of two random variables. For example, if X and Y represent the real and imaginary parts of the measured voltage, we may be interested in obtaining the density of $Z = X^2 + Y^2$. In this case, Z represents the power. Other functions could be $\sqrt{X^2 + Y^2}$ and $\tan^{-1}\left(\frac{Y}{X}\right)$.

Additionally, we may be interested in finding the densities of $X + Y$, $X - Y$, XY, $\max(X, Y)$, $\min(X, Y)$, etc. The methodology of obtaining the densities of the transformed variables is presented below.

4.5.1   Sum of Two Random Variables

While we will explore formal ways of deriving the pdf of the function of two random variables later in this chapter, we will examine the use of the point conditioning approach first. Let us obtain the density of the sum of two variables, X and Y.

$$g(X, Y) = Z = X + Y \tag{4.50}$$

The joint density function of X and Y is $f(x,y)$ and the joint CDF is $F(x,y)$. Let us assume that X and Y are independent. We will look at the case of dependent variables later.

The CDF of Z can be derived using the definition of the cumulative distribution function of any random variable (Chap. 3) as

$$F_Z(z) = P(Z \le z) = P(X + Y \le z) = P(X \le z - Y) \tag{4.51}$$

At this point, we use the point conditioning and fix Y as

$$Y = y \tag{4.52}$$

The conditioning in Eq. (4.52) leads to replacement of Y with its value y in Eq. (4.51) resulting in the left hand side of Eq. (4.51) becoming the conditional CDF. We may now rewrite Eq. (4.51) as

$$F_{Z|Y}(z|y) = P(X \le z - y) \tag{4.53}$$

Using the definition of the CDF of X, we may evaluate the right hand side of Eq. (4.53). The Eq. (4.53) now becomes

$$F_{Z|Y}(z|y) = F_X(z - y) \tag{4.54}$$

Differentiating both sides of Eq. (4.54) with respect to z, we have

$$f_{Z|Y}(z|y) = f_X(z - y) \tag{4.55}$$

The joint density of Z and Y is obtained by multiplying Eq. (4.55) by $f_Y(y)$ resulting in

$$f(z, y) = f_{Z|Y}(z|y)f_Y(y) = f_X(z - y)f_Y(y) \tag{4.56}$$

The density of X is now obtained as

$$f_Z(z) = \int_{y_1}^{y_2} f(z, y)dy = \int_{y_1}^{y_2} f_Y(y)f_X(z - y)dy \tag{4.57}$$

The limits of integration will be determined by the range of values of X and Y. It is seen that if X and Y are independent, the density function of the sum is the convolution of the densities of X and Y.

If X and Y only take positive values, Eq. (4.57) becomes

$$f_Z(z) = \int_{y_1}^{y_2} f_Y(y)U(y)f_X(z-y)U(z-y)dy = \int_0^z f_Y(y)f_X(z-y)dy \qquad (4.58)$$

The limits of integration are obtained by noting that the presence of $U(y)$ in Eq. (4.58) results in the lower limit of $y = 0$ and the presence of $U(z-y)$ limits the maximum value of y to z so that $z - y > 0$.

If X and Y take values in the range $-\infty$ to ∞, Eq. (4.57) becomes

$$f(z) = \int_{-\infty}^{\infty} f_Y(y)f_X(z-y)dy \qquad (4.59)$$

If we now consider the case when X and Y are not independent, Eq. (4.59) can be expressed as (see Sects. 4.5.4 and 4.5.6)

$$f(z) = \int_{-\infty}^{\infty} f_{X,Y}(z-y,y)dy \qquad (4.60)$$

It should be obvious that

$$f(z) = \int_{-\infty}^{\infty} f_{X,Y}(x,z-x)dx \qquad (4.61)$$

Interestingly, the CDF of Z may be obtained directly from Eq. (4.54) by applying Bayes' rule of total probability associated with densities (Chap. 3). The expression for CDF of Z becomes

$$F_Z(z) = \int_{-\infty}^{\infty} F_{Z|Y}(z|y)f_Y(y)dy = \int_{-\infty}^{\infty} F_X(z-y)f_Y(y)dy \qquad (4.62)$$

It is obvious that the differentiation of Eq. (4.62) w.r.t. z leads to the density of Z in Eq. (4.57).

We may also obtain the density of the sum using the concept of a single variable transformation as well. Using Eq. (4.52), the transformation in Eq. (4.50) becomes

$$Z = X + y \qquad (4.63)$$

Equation (4.63) represents a monotonic transformation, and therefore, the conditional density of Z given $Y = y$ becomes

$$f_{Z|y}(z|y) = \frac{f_X(x)}{\left|\frac{dz}{dx}\right|}\Bigg|_{x=z-y} = f_X(z-y) \tag{4.64}$$

Equation (4.64) implies the general case of X and Y being dependent. Left hand side of Eq. (4.64) recognizes the conditioning for the purposes of this process. The joint density of Z and Y becomes

$$f(z,y) = f_{Z|Y}(z|y)f_Y(y) = f_{X,Y}(z-y,y) \tag{4.65}$$

Using Eq. (4.65), the marginal density of Z becomes

$$f(z) = \int\limits_{-\infty}^{\infty} f_{X,Y}(z-y,y)dy \tag{4.66}$$

4.5.2   Ratio of Two Random Variables

Let us define W as the ratio of two random variables X and Y (assume that both X and Y exist only for positive values)

$$W = \frac{X}{Y} \tag{4.67}$$

The CDF of W can be expressed as

$$F_W(w) = P(W \leq w) = P\left(\frac{X}{Y} \leq w\right) = P(X \leq Yw) \tag{4.68}$$

Treating $Y = y$ as a conditioning event, Eq. (4.68) becomes

$$F_{W|Y}(w|y) = P(X \leq yw) = F_X(wy) \tag{4.69}$$

Differentiating Eq. (4.69) w.r.t. w, we have

$$f_{W|Y}(w|y) = |y|f_X(wy) \tag{4.70}$$

The absolute value sign in Eq. (4.70) is necessary to ensure that the pdf is always positive (even though in the present case, both X and Y are positive, and the absolute value sign is unnecessary). Treating X and Y to be independent, the joint density of W and Y now becomes

$$f_{W,Y}(w, y) = f_Y(y)f_{W|Y}(w|y) = |y|f_X(wy)f_Y(y) \tag{4.71}$$

The marginal density of W is obtained as

$$f_W(w) = \int f_{W,Y}(w, y)dy = \int |y|f_X(wy)f_Y(y)dy \tag{4.72}$$

If X and Y are not independent, Eq. (4.72) becomes

$$f_W(w) = \int |y|f_{X,Y}(wy, y)dy \tag{4.73}$$

When we are finding the ratio of two Gaussian variables, Eq. (4.68) needs to be modified to include the case when Y takes negative values. A simpler approach is to treat the transformation as a monotonic one of a single random variable X to W. By fixing $Y = y$, the transformation in Eq. (4.67) becomes

$$W = \frac{X}{y} \tag{4.74}$$

The conditional density of W given $Y = y$ now becomes

$$f_{W|y}(w|y) = \frac{f_X(x)}{\left|\frac{dw}{dx}\right|}\Bigg|_{x=wy} = |y|f_X(wy) \tag{4.75}$$

The joint density of W and Y is

$$f_{W,Y}(w, y) = f_{W|y}(w|y)f_Y(y) = |y| f_X(wy)f_Y(y) \tag{4.76}$$

The density of W is

$$f_W(w) = \int_{-\infty}^{\infty} f_{W,Y}(w, y)dy = \int_{-\infty}^{\infty} |y| f_X(wy)f_Y(y)dy \tag{4.77}$$

If X and Y are not independent,

$$f_W(w) = \int_{-\infty}^{\infty} |y| f_{X,Y}(wy, y)dy \tag{4.78}$$

When X and Y exist only for positive values it can be seen that Eq. (4.78) becomes

$$f_W(w) = \int_{0}^{\infty} yf_{X,Y}(wy, y)dy \tag{4.79}$$

4.5.3   *Product of Two Random Variables*

Let us define Z as the product of two random variables X and Y (assume that X and Y exist only for positive values).

$$Z = XY \tag{4.80}$$

Proceeding as before,

$$F_Z(z) = P(Z \leq z) = P(XY \leq z) = P\left(X \leq \frac{z}{Y}\right) \tag{4.81}$$

Treating $Y = y$ as a conditioning event, Eq. (4.81) becomes

$$F_{Z|Y}(z|y) = P\left(X \leq \frac{z}{y}\right) = F_X\left(\frac{z}{y}\right) \tag{4.82}$$

Differentiating Eq. (4.82) w.r.t. z, we have

$$f_{Z|Y}(z|y) = \frac{1}{|y|} f_X\left(\frac{z}{y}\right) \tag{4.83}$$

The absolute value sign around y is necessary to ensure that the pdf is always positive even though in the present case, X and Y are both positive.

Treating X and Y as independent, the joint density of Z and Y is obtained as

$$f_{Z,Y}(z,y) = f_Y(y)f_{Z|Y}(z|y) = \frac{1}{|y|} f_X\left(\frac{z}{y}\right) f_Y(y) \tag{4.84}$$

The marginal density of the product of two random variables is

$$f_Z(z) = \int f(z,y)dy = \int \frac{1}{|y|} f_X\left(\frac{z}{y}\right) f_Y(y)dy \tag{4.85}$$

If X and Y are not independent, Eq. (4.85) becomes

$$f_Z(z) = \int \frac{1}{|y|} f_{X,Y}\left(\frac{z}{y},y\right)dy \tag{4.86}$$

As it was done with the case of the ratio, the density of the product when X and Y exist for both positive and negative values may be obtained by treating the transformation as a monotonic one. Fixing $Y = y$, we have

$$f_{Z|y}(z|y) = \frac{f_X(x)}{\left|\frac{dz}{dx}\right|}\Bigg|_{x=\frac{z}{y}} = \frac{f_X\left(\frac{z}{y}\right)}{|y|} = \frac{1}{|y|}f_X\left(\frac{z}{y}\right) \tag{4.87}$$

The joint density of Z and Y becomes

$$f_{Z,Y}(z,y) = f_{Z|y}(z|y)f_Y(y) = \frac{1}{|y|}f_X\left(\frac{z}{y}\right)f_Y(y) \tag{4.88}$$

The density of Z becomes

$$f_Z(z) = \int_{-\infty}^{\infty} f_{Z,Y}(z,y)dy = \int_{-\infty}^{\infty} \frac{1}{|y|}f_X\left(\frac{z}{y}\right)f_Y(y)dy \tag{4.89}$$

If X and Y are not independent,

$$f_Z(z) = \int_{-\infty}^{\infty} \frac{1}{|y|}f_{X,Y}\left(\frac{z}{y},y\right)dy \tag{4.90}$$

A close inspection of Example 4.8 will show that a two-stage experiment leads to results matching the solution in Eq. (4.90).

We will now look at a few additional examples where the concept of point conditioning is used to obtain the pdf of the function of two random variables. While the results presented above were of general nature devoid of specific values for the limits of integration, these examples will illustrate the choice of the limits.

Example 4.9 If X and Y are independent identically distributed (i. i. d) Gaussian random variables, each represented as $N(0,\sigma^2)$, obtain the pdf of

$$Z = X^2 + Y^2$$

Solution Using the definition of the CDF, the CDF of Z is

$$F_Z(z) = P(X^2 + Y^2 \leq z) = P\left(-\sqrt{z - Y^2} \leq X \leq \sqrt{z - Y^2}\right)$$

Treating $Y = y$ as a conditioning event, the CDF becomes

$$F_{Z|y}(z|y) = P\left(-\sqrt{z - y^2} \leq X \leq \sqrt{z - y^2}\right) = F_X\left(\sqrt{z - y^2}\right) - F_X\left(-\sqrt{z - y^2}\right)$$

Notice that by choosing $Y = y$, the CDF has become a conditional one. Differentiation of the CDF w.r.t. z leads to

$$f_{Z|y}(z|y) = \frac{d}{dz}F_{Z|y}(z|y) = \frac{1}{2\sqrt{z-y^2}}\left[f_X\left(\sqrt{z-y^2}\right) + f_X\left(-\sqrt{z-y^2}\right)\right]$$

The joint density of Z and Y becomes

$$f(z,y) = f_{Z|y}(z|y)f_Y(y) = \frac{f_Y(y)}{2\sqrt{z-y^2}}\left[f_X\left(\sqrt{z-y^2}\right) + f_X\left(-\sqrt{z-y^2}\right)\right]$$

We have

$$f_X(x) = \frac{1}{\sqrt{2\pi\sigma^2}}\exp\left(-\frac{x^2}{2\sigma^2}\right) \qquad f_Y(y) = \frac{1}{\sqrt{2\pi\sigma^2}}\exp\left(-\frac{y^2}{2\sigma^2}\right)$$

Because of symmetry of $f_X(x)$, the joint pdf of Z and Y becomes

$$f(z,y) = \frac{2f_Y(y)f_X\left(\sqrt{z-y^2}\right)}{2\sqrt{z-y^2}} = \frac{f_Y(y)f_X\left(\sqrt{z-y^2}\right)}{\sqrt{z-y^2}}$$

$$= \frac{1}{2\pi\sigma^2\sqrt{z-y^2}}\exp\left(-\frac{z}{2\sigma^2}\right)$$

The marginal density of Z is

$$f_Z(z) = \int_{-\infty}^{\infty} f(z,y)dy = \int_{-\sqrt{z}}^{\sqrt{z}} \frac{1}{2\pi\sigma^2\sqrt{z-y^2}}\exp\left(-\frac{z}{2\sigma^2}\right)dy$$

$$= \frac{1}{2\pi\sigma^2}\exp\left(-\frac{z}{2\sigma^2}\right)\int_{-\sqrt{z}}^{\sqrt{z}} \frac{1}{\sqrt{z-y^2}}dy$$

Notice that the limits of integration are obtained by examining that

$$x^2 + y^2 \leq z \quad \Rightarrow \quad \sqrt{z-y^2} > 0 \quad \Rightarrow \quad -\sqrt{z} \leq y \leq \sqrt{z}$$

Using the limits and noting that the integral in y is symmetric, the pdf of Z becomes

$$f_Z(z) = \frac{2}{2\pi\sigma^2} \exp\left(-\frac{z}{2\sigma^2}\right) \int_0^{\sqrt{z}} \frac{1}{\sqrt{z-y^2}} dy = \frac{1}{2\sigma^2} \exp\left(-\frac{z}{2\sigma^2}\right) U(z)$$

Notice that Z is exponentially distributed.

While we considered i. i. d. Gaussian variables with zero mean, a case of special interest is when the random variable Y has a non-zero mean,

$$f_Y(y) = \frac{1}{\sqrt{2\pi\sigma^2}} \exp\left(-\frac{(y-\mu)^2}{2\sigma^2}\right)$$

The density of Z when Y is of non-zero mean becomes

$$f_Z(z) = \frac{1}{2\pi\sigma^2} \exp\left(-\frac{z+\mu^2}{2\sigma^2}\right) \int_{-\sqrt{z}}^{\sqrt{z}} \frac{1}{\sqrt{z-y^2}} \exp\left(\frac{\mu y}{\sigma^2}\right) dy$$

$$= \frac{1}{2\sigma^2} \exp\left(-\frac{z+\mu^2}{2\sigma^2}\right) I_0\left(\frac{\mu}{\sigma^2}\sqrt{z}\right) U(z)$$

The function $I_0()$ is the modified Bessel function of the first kind (Abramowitz and Segun 1972; Gradshteyn and Ryzhik 2000) and the density $f(z)$ is identified by its moniker, the Rician density (Nakagami 1960; Shankar 2017). By symmetry, if X is $N(\mu, \sigma^2)$ and Y is $N(0, \sigma^2)$, the density of Z will be same. If X is $N(\mu_1, \sigma^2)$ and Y is $N(\mu_2, \sigma^2)$, the density of Z will be Rician (Papoulis and Pillai 2002; Shankar 2017). We may now obtain the pdf of the amplitude

$$W = \sqrt{X^2 + Y^2}.$$

We may use the concept of a single variable transformation by treating

$$W = \sqrt{X^2 + Y^2} = \sqrt{Z}$$

Since Z is exponentially distributed as shown above (when X and Y are i. i. d. and zero mean), we expect W to be Rayleigh distributed as shown in Chap. 3. We may derive this result using the conditioning approach followed by the transformation of a single variable.

$$W = \sqrt{X^2 + Y^2}$$

We chose a specific value for Y as y. This means that

$$Y = y$$

By virtue of fixing Y, W also becomes a conditioned variable as

$$W = \sqrt{X^2 + y^2}$$

Note that the transformation from X to W is non-monotonic, and therefore,

$$f_{W|y}(w|y) = \frac{f_X(x)}{\left|\frac{dw}{dx}\right|}\Bigg|_{x=-\sqrt{w^2-y^2}} + \frac{f_X(x)}{\left|\frac{dw}{dx}\right|}\Bigg|_{x=\sqrt{w^2-y^2}}$$

$$\left|\frac{dw}{dx}\right| = \left|\frac{x}{\sqrt{x^2+y^2}}\right| = \frac{\sqrt{w^2-y^2}}{w}$$

If X and Y are independent and identically distributed, each with zero mean, the conditional density becomes

$$f_{W|y}(w|y) = \frac{w}{\sqrt{w^2-y^2}} f_X\left(-\sqrt{w^2-y^2}\right) + \frac{w}{\sqrt{w^2-y^2}} f_X\left(\sqrt{w^2-y^2}\right)$$

$$= \frac{2w}{\sqrt{w^2-y^2}} f_X\left(\sqrt{w^2-y^2}\right)$$

The joint density of Z and Y becomes

$$f_{W,Y}(w,y) = f(w|y)f_Y(y) = \frac{2w}{\sqrt{w^2-y^2}} f_X\left(\sqrt{w^2-y^2}\right) f_Y(y)$$

The marginal density of W is

$$f_W(w) = \int f(w,y)dy = \int_{-\infty}^{\infty} \frac{2w}{\sqrt{w^2-y^2}} f_X\left(\sqrt{w^2-y^2}\right) f_Y(y)dy$$

$$f_W(w) = \int_{-w}^{w} \frac{w}{\sqrt{w^2-y^2}} 2 \frac{1}{\sqrt{2\pi\sigma^2}} \exp\left(-\frac{w^2-y^2}{2\sigma^2}\right) \frac{1}{\sqrt{2\pi\sigma^2}} \exp\left(-\frac{y^2}{2\sigma^2}\right) dy$$

Simplifying, we have

$$f_W(w) = \frac{w}{\pi\sigma^2} \exp\left(-\frac{w^2}{2\sigma^2}\right) \int_{-w}^{w} \frac{1}{\sqrt{w^2-y^2}} dy = \frac{2w}{\pi\sigma^2} \exp\left(-\frac{w^2}{2\sigma^2}\right) \int_{0}^{w} \frac{1}{\sqrt{w^2-y^2}} dy$$

$$f_W(w) = \frac{w}{\sigma^2} \exp\left(-\frac{w^2}{2\sigma^2}\right) U(w)$$

Note that W is Rayleigh distributed. While we had seen the density of Z is Rician when the means are not equal (identical variances), the density of W will also be Rician. With Z, it will be Rician distribution of power; it is identified as the Rician distribution of the envelope for the case of W. The Rician envelope or amplitude density can be obtained using the transformation of variables as

$$W = \sqrt{Z}$$

$$f_W(w) = \frac{f_Z(z)}{\left|\frac{dw}{dz}\right|}\Bigg|_{z=w^2} = \frac{w}{\sigma^2} \exp\left(-\frac{w^2 + \mu^2}{2\sigma^2}\right) I_0\left(\frac{\mu}{\sigma^2} w\right) U(w)$$

When $\mu \to 0$, the Rician density of the envelope becomes the Rayleigh density.

Example 4.10 If X and Y are independent identically distributed Gaussian variables, each represented as $N(0, \sigma^2)$, obtain the pdf of

$$\Theta = \tan^{-1}\left(\frac{Y}{X}\right)$$

Solution The CDF of Θ is

$$F_\Theta(\theta) = P\left[\tan^{-1}\left(\frac{Y}{X}\right) \le \theta\right] = P[Y \le X \tan(\theta)]$$

Defining $X = x$ as the conditioning event, the expression for the CDF becomes

$$F_{\Theta|x}(\theta|x) = P[Y \le x \tan(\theta)] = F_Y[x \tan(\theta)]$$

Differentiating the CDF w.r.t. θ leads to (important to use the absolute value sign with x)

$$f_{\Theta|x}(\theta|x) = |x| \sec^2(\theta) f_Y[x \tan(\theta)]$$

The joint density of Θ and X is

$$f_{\Theta,X}(\theta, x) = f_{\Theta|x}(\theta|x) f_X(x) = |x| \sec^2(\theta) f_Y[x \tan(\theta)] f_X(x)$$

The marginal pdf of the phase Θ is

$$f_\Theta(\theta) = \int_{-\infty}^{\infty} |x| \sec^2(\theta) f_Y[x \tan(\theta)] f_X(x) dx$$

$$= \int_{-\infty}^{\infty} \frac{|x| \sec^2(\theta)}{2\pi\sigma^2} \exp\left(-\frac{x^2 \tan^2(\theta)}{2\sigma^2}\right) \exp\left(-\frac{x^2}{2\sigma^2}\right) dx$$

$$f_\Theta(\theta) = \int\limits_{-\infty}^{\infty} \frac{|x| \sec^2(\theta)}{2\pi\sigma^2} \exp\left(-\frac{x^2 \sec^2(\theta)}{2\sigma^2}\right) dx$$

The integrand above is an even function of x and therefore it becomes

$$f_\Theta(\theta) = 2 \int\limits_{0}^{\infty} \frac{x \sec^2(\theta)}{2\pi\sigma^2} \exp\left(-\frac{x^2 \sec^2(\theta)}{2\sigma^2}\right) dx$$

The integral can be easily integrated resulting in

$$f_\Theta(\theta) = \frac{1}{\pi}, \quad -\frac{\pi}{2} \leq \theta \leq \frac{\pi}{2}$$

Since X and Y span $-\infty$ to ∞, the phase must span 0 to 2π or $-\pi$ to π. Taking this into account, the uniform density in the range $[-\pi/2, \pi/2]$ above needs to be changed to a uniform density in $-\pi$ to π as

$$f_\Theta(\theta) = \frac{1}{2\pi}, \quad -\pi \leq \theta \leq \pi$$

Another interesting result is seen by exploring the density of

$$W = \frac{X}{Y}$$

Using Eq. (4.77), the density of W becomes

$$f_W(w) = \int |y| f_{X,Y}(wy, y) dy = \int\limits_{-\infty}^{\infty} \frac{|y|}{2\pi\sigma^2} \exp\left(-\frac{w^2 y^2}{2\sigma^2}\right) \exp\left(-\frac{y^2}{2\sigma^2}\right) dy$$

Because of the symmetry of the integrand, the marginal density of W becomes

$$f_W(w) = 2 \int\limits_{0}^{\infty} \frac{y}{2\pi\sigma^2} \exp\left(-\frac{\left[1 + w^2\right] y^2}{2\sigma^2}\right) dy = \frac{1}{\pi\left(1 + w^2\right)}, \quad -\infty < w < \infty$$

The density of $W = X/Y$ is therefore a Cauchy density described in Chap. 3. By virtue of symmetry, the density of Y/X will also be identical to the density of X/Y, and the density of the ratio of two independent identically distributed Gaussian variables of zero mean is the Cauchy density.

We may now get the density of Θ defined as $\tan^{-1}\left(\frac{Y}{X}\right)$ earlier. Note that X and Y being identical, $\tan^{-1}\left(\frac{Y}{X}\right)$ and $\tan^{-1}\left(\frac{X}{Y}\right)$ will have identical densities. This means that

$$F_\Theta(\theta) = P\left(\tan^{-1}\frac{X}{Y} < \theta\right) = P\left(\tan^{-1}W < \theta\right) = P(W < \tan[\theta]) = F_W(\tan[\theta])$$

Differentiating w. r. t. θ, we have

$$f_\Theta(\theta) = \frac{d}{dw}F_W(\tan[\theta]) = \sec^2(w)f_W(\tan[\theta]) = \frac{\sec^2(w)}{\pi[1 + \tan^2(w)]} = \frac{1}{\pi}, \quad -\frac{\pi}{2} \le \theta \le \frac{\pi}{2}$$

Since X and Y span $-\infty$ to ∞, the phase must span 0 to 2π or $-\pi$ to π. Taking this into account, the uniform density in the range $[-\pi/2, \pi/2]$ above needs to be changed to a uniform density in $-\pi$ to π as

$$f_\Theta(\theta) = \frac{1}{2\pi}, \quad -\pi \le \theta \le \pi$$

Example 4.11 X is exponential with a mean of 4 and Y is exponential with a mean of 3. If X and Y are independent, what is the probability that $X > 2Y$?

Solution The solution may be obtained using the method of point conditioning or directly. Let us look at the method based on point conditioning. If W is X/Y, the density of W is

$$f(w) = \int |y| f_{X,Y}(wy, y)dy = \int_0^\infty y\frac{1}{4}\exp\left(-\frac{wy}{4}\right)\frac{1}{3}\exp\left(-\frac{y}{3}\right)dy$$

$$f_W(w) = \frac{12}{(3w + 4)^2}U(w)$$

$$F_W(w) = 1 - \frac{4}{(3w + 4)}$$

$$P(X > 2Y) = P\left(\frac{X}{Y} > 2\right) = 1 - F_W(2) = \frac{4}{10}$$

From Figure Ex.4.11, the probability of $X > 2Y$ is the volume contained by the area below the line $X = 2Y$.

$$P(X > 2Y) = \int_{x=0}^\infty \int_{y=0}^{\frac{x}{2}} \frac{1}{12}\exp\left(-\frac{x}{4} - \frac{y}{3}\right)dydx = \frac{4}{10}$$

Figure Ex.4.11

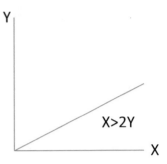

The solution may also be obtained as follows:

$$P(X > 2Y) = \int\limits_{x=0}^{\infty} f_X(x) \int\limits_{y=0}^{\frac{x}{2}} f_Y(y)dydx = \int\limits_{0}^{\infty} f_X(x)F_Y\left(\frac{x}{2}\right)dx$$

$$= \int\limits_{0}^{\infty} f_X(x)\left[1 - \exp\left(-\frac{x}{6}\right)\right]dx$$

$$P(X > 2Y) = \int\limits_{0}^{\infty} f_X(x)dx - \int\limits_{0}^{\infty} f_X(x)\exp\left(-\frac{x}{6}\right)dx = 1 - \frac{1}{4}\int\limits_{0}^{\infty} \exp\left(-\frac{x}{4} - \frac{x}{6}\right)dx$$

$$P(X > 2Y) = 1 - \frac{1}{4}\int\limits_{0}^{\infty} \exp\left(-\frac{10x}{24}\right)dx$$

Rewriting, we get

$$P(X > 2Y) = 1 - \frac{1}{4}\left(\frac{24}{10}\right)\int\limits_{0}^{\infty} \frac{10}{24}\exp\left(-\frac{10x}{24}\right)dx = 1 - \frac{1}{4}\left(\frac{24}{10}\right) = \frac{4}{10}$$

Note that the integral above equals unity because the integrand is an exponential density with a parameter of (24/10).

Example 4.12 If X and Y are two independent and identically distributed Rayleigh variables of second moment 8, obtain the joint pdf of X and Y under the condition that $(X + Y) > 6$.

Solution We are given two Rayleigh variables with $b = 2$ (based on the second moment). The joint pdf is

$$f(x, y) = \frac{xy}{16}\exp\left(-\frac{x^2 + y^2}{8}\right)$$

Using the pdf, we have

$$P(X+Y>6) = 1 - \int\limits_{x=0}^{6} \int\limits_{y=0}^{6-x} f(x,y)dydx = 0.2818 = P_0$$

Note that there is no analytical expression for the integral and the result is easily obtained using Matlab or Maple. Using Eq. (4.48) for the joint conditional density becomes

$$f(x,y|X+Y>6) = \begin{cases} 0, & x+y<6 \\ \dfrac{f(x,y)}{P_0}, & x+y>6 \end{cases}$$

Example 4.13 X and Y are lifetimes of two computers in an office. X and Y are independent and identically distributed exponential variables with mean lifetimes of 5 years. What is the probability that both computers will become inoperable within 4 years?

Solution

$$f(x,y) = \frac{1}{25} \exp\left(-\frac{x+y}{5}\right)$$

The probability that both of stop functioning within 4 years is

$$P(X<4, Y<4) = F_{X,Y}(4,4) = F_X(4)F_Y(4) = \left[1 - \exp\left(-\frac{4}{5}\right)\right]^2$$

Example 4.14 A manufacturer is testing computer chips to see how long they perform effectively. It is known that the lifetime is modeled as a gamma variable of order 2. After testing several, it was determined that the mean appears to be uniform in the range [4, 6]. Model the lifetimes and obtain expression for the pdf of the lifetime of the chips.

Solution   If $G(a,b)$ represents a gamma density with order a and scaling parameter b, the mean of the gamma density is ab. Since a is deterministic, the mean must be conditioned on the parameter b being a random variable,

$$E(X|b) = ab = 2b$$

The mean of the chips is seen to be uniform in [4, 6]. This implies that $W = 2b$ is uniform in [4,6]. We conclude that b is uniform in [2,3]. Therefore,

$$f(b) = 1, \quad 2 < b < 3$$

The density (conditional) associated with the lifetime is

$$f(x|b) = \frac{x^{a-1}}{b^a \Gamma(a)} \exp\left(-\frac{x}{b}\right) U(x)$$

The joint density of X and B is

$$f(x, b) = f(x|b) f(b)$$

Therefore, the density of X is given by

$$f(x) = \int f(x, b)db = \int_0^\infty \frac{x^{a-1}}{b^a \Gamma(a)} \exp\left(-\frac{x}{b}\right) f(b)db = \int_2^3 \frac{x^{a-1}}{b^a \Gamma(a)} \exp\left(-\frac{x}{b}\right) db$$

Example 4.15 X is uniform in $[0,8]$ and Y is uniform in $[0,6]$. X and Y are also independent. Find

(i) prob. $\{X + Y > 4\}$
(ii) prob. $\{XY > 2\}$
(iii) prob. $\{X + Y < 4 | X > 2\}$
(iv) prob. $\{X > 4 \,|\, X + Y < 6\}$

Solution   We are given that (Figure Ex.4.15)

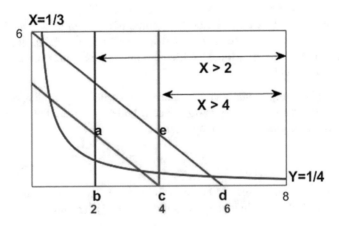

Figure Ex.4.15

$$f(x, y) = \frac{1}{48}, \quad 0 < x < 8, \quad 0 < y < 6$$

$$P\{X + Y > 4\} = 1 - P\{X + Y < 4\} = 1 - \text{Volume(triangle sides 4 and 4}\}$$
$$= 1 - (4 * 4/2)/48 = 5/6.$$

$$P(XY > 2) = \int\limits_{1/4}^{6} \int\limits_{x=2/y}^{8} \frac{1}{48} dx dy = 0.8259$$

$$P[X + Y < 4 \mid X > 2] = \frac{P[X + Y < 4, X > 2]}{P[X > 2]}$$

$$= \frac{\text{volume of triangle abc}}{(6/8)} = \frac{(2*2)\frac{1}{2} * \frac{1}{48}}{6/8} = \frac{(1/24)}{(6/8)} = 0.0556$$

$$P(X + Y < 6) = \frac{6 * 6 * \frac{1}{2}}{48} = \frac{3}{8}$$

$$P[X > 4 \mid X + Y < 6] = \frac{P[X + Y < 6, X > 4]}{P[X + Y < 6]}$$

$$= \frac{\text{volume of triangle ecd}}{(3/8)} = \frac{(2*2)\frac{1}{2} * \frac{1}{48}}{3/8} = 0.1111$$

Example 4.16 X and Y are independent and identically distributed exponential variables with a mean 4. Obtain the probability that $X^2 + Y^2 < 16$.

Solution

$$f(x, y) = \frac{1}{16} \exp\left(-\frac{x+y}{4}\right) U(x) U(y)$$

$$P(X^2 + Y^2 < 16) = \int\limits_{y=0}^{4} \int\limits_{x=0}^{\sqrt{16-y^2}} \frac{1}{16} \exp\left(-\frac{x+y}{4}\right) dx dy$$

$$P(X^2 + Y^2 < 16) = \int\limits_{y=0}^{4} \left[1 - \exp\left(-\frac{\sqrt{16 - y^2}}{4}\right)\right] \frac{1}{4} \exp\left(-\frac{y}{4}\right) dy = 0.3535$$

Example 4.17 The joint pdf of two random variables is

$$f(x,y) = k, \quad 0 < |y| < x < 5$$

What is the value of k? Obtain the marginal density functions of X and Y. What is $P(|Y| < \frac{6}{10}X)$?

Solution The region of validity of the joint density is shown in Figure Ex.4.17. It will be a triangle with corners at [0,0], [5,5], and [5,−5].

$$1 = \int\limits_{x=0}^{5} \int\limits_{y=-x}^{x} k \ dydx = \int\limits_{0}^{5} k2xdx = 25k \quad \Rightarrow \quad k = \frac{1}{25}$$

You may also obtain the value of k directly by summing the area of the two isosceles triangles, each of area 5*5/2.

$$f_X(x) = \int\limits_{-x}^{x} \frac{1}{25} dy = \frac{2}{25}x, 0 < x < 5$$

$$f_Y(y) = \int\limits_{|y|}^{5} \frac{1}{25} dx = \frac{5 - |y|}{25}, \quad -5 < y < 5$$

$$P\left(|Y| < \frac{6}{10}X\right) = 2[\text{Volume in area of trianagle : height} = 3, \text{base} = 5]$$

Figure Ex.4.17

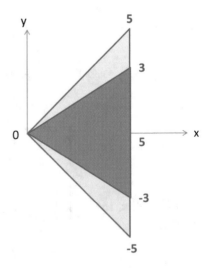

$$P\left(|Y| < \frac{6}{10}X\right) = \frac{1}{25}2\left(\frac{15}{2}\right) = \frac{15}{25}$$

In the previous section, we explored the use of point conditioning (choosing $X = x$ or $Y = y$) to obtain the density of the function of two random variables. Another approach relies on the basic definition of the CDF and obtains the CDF of the transformation as in the case of max (X,Y) and min (X,Y). While one "formal" approach relies on graphical methods, the one is based on the Jacobian of the transformation. In the latter case, we first find the joint density of two functions of two random variables. We use the new joint density to obtain the density of a single function of two random variables.

4.5.4   Density of g(X,Y) Graphically

It is possible to get the pdf and CDF by examining the region described by $Z = g(X,Y)$ and estimating the probability. Let us reexamine the pdf of the sum X and Y.

From Fig. 4.6, the CDF of Z can be written as

$$F_Z(z) = \iint_{\text{shaded area}} f(x,y)dxdy = \int_{x=-\infty}^{\infty} \int_{y=-\infty}^{z-x} f(x,y)dydx \qquad (4.91)$$

Fig. 4.6 Region of interest for $X + Y \le z$

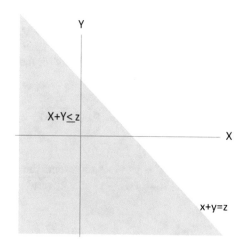

The pdf is obtained by differentiating Eq. (4.91) w.r.t. z.

$$f_Z(z) = \frac{d}{dz} \int_{x=-\infty}^{\infty} \int_{y=-\infty}^{z-x} f(x,y)dydx = \int_{x=-\infty}^{\infty} \frac{d}{dz}\left[\int_{y=-\infty}^{z-x} f(x,y)dy\right]dx \qquad (4.92)$$

The derivative in Eq. (4.92) is obtained by applying the Leibniz rule (see Sect. 4.9) to the inner integral in Eq. (4.92). Following the application of the Leibniz rule (Gradshteyn and Ryzhik 2000), Eq. (4.92) becomes

$$f_Z(z) = \int_{x=-\infty}^{\infty} f_{x,y}(x, z-x)dx \qquad (4.93)$$

If X and Y are independent, Eq. (4.93) becomes

$$f_Z(z) = \int_{x=-\infty}^{\infty} f_x(x)f_y(z-x)dx \qquad (4.94)$$

Additionally, if X and Y exist only for positive values, Eq. (4.94) becomes

$$f_Z(z) = \int_{x=0}^{z} f_x(x)f_y(z-x)dx \qquad (4.95)$$

Let us also find out the pdf of

$$W = \sqrt{X^2 + Y^2} \qquad (4.96)$$

The CDF of W is the probability (or the volume) contained in the shaded region in Fig. 4.7

$$F_W(w) = P\left[\sqrt{X^2 + Y^2} < w\right] = \iint_{\text{shaded area}} f(x,y)dxdy \qquad (4.97)$$

Note that $x^2 + y^2 \leq w^2$ represents the shaded area of radius w. The range of the shaded area is $-\sqrt{w^2 - y^2} < x < \sqrt{w^2 - y^2}$ and $-w < y < w$. This means that the CDF is

$$F_W(w) = \iint_{\text{shaded area}} f(x,y)dxdy = \int_{y=-w}^{w} \int_{x=-\sqrt{w^2-y^2}}^{\sqrt{w^2-y^2}} f(x,y)dxdy \qquad (4.98)$$

Fig. 4.7 Shaded region corresponding to $\{W < w]$

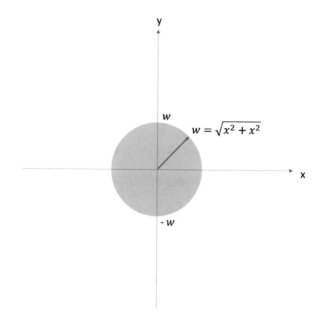

The pdf is obtained by differentiating Eq. (4.98) resulting in

$$f_W(w) = \frac{d}{dw} F_W(w) = \frac{d}{dw} \left[\int_{y=-w}^{w} \int_{x=-\sqrt{w^2-y^2}}^{\sqrt{w^2-y^2}} f(x,y)dxdy \right] \qquad (4.99)$$

Applying Leibniz theorem for the double integral (see Sect. 4.9), Eq. (4.99) becomes

$$f_W(w) = \int_{-w}^{w} \frac{w}{\sqrt{w^2-y^2}} \left[f\left(\sqrt{w^2-y^2},y\right) + f\left(-\sqrt{w^2-y^2},y\right) \right] dy \qquad (4.100)$$

4.5.5 Other Useful Transformations: max{X,Y} and min{X,Y}

If X and Y are two random variables with a joint CDF $F(x,y)$, we are interested in obtaining the pdf of

$$Z = \max(X, Y) \qquad (4.101)$$

Using the definition of the CDF,

$$F_Z(z) = P(Z \leq z) = P(\max(X, Y) \leq z) \qquad (4.102)$$

The max (X,Y) being less than z leads to

$$\max(X, Y) \leq z \quad \Rightarrow \quad \{X \text{ and } Y \leq z\} \qquad (4.103)$$

Equation (4.103) leads to

$$F_Z(z) = P(\max(X, Y) \leq z) = P(X \leq z, X \leq z) = F_{X,Y}(z, z) \qquad (4.104)$$

If X and Y are independent, Eq. (4.104) leads to

$$F_Z(z) = F_X(z)F_Y(z) \qquad (4.105)$$

The pdf of Z is obtained by differentiating Eq. (4.105) resulting in

$$f_Z(z) = f_X(z)F_Y(z) + F_X(z)f_Y(z) \qquad (4.106)$$

If X and Y are two random variables with a joint CDF $F(x,y)$, we are interested in obtaining the pdf of

$$W = \min(X, Y) \qquad (4.107)$$

If min (X,Y) is less than W, it implies that

$$W = \begin{cases} X, X \leq w \\ Y, Y \leq w \end{cases} \qquad (4.108)$$

Equation (4.108) implies

$$F_W(w) = P(W \leq w) = P(\min(X, Y) \leq w) = P(X \leq w \cup Y \leq w) \qquad (4.109)$$

Noting that $\cup$ in Eq. (4.109) represents **union** of two events; the CDF in Eq. (4.109) becomes

$$\begin{aligned} F_W(w) &= P(X \leq w) + P(Y \leq w) - P(X \leq w)P(Y \leq w) \\ &= F_X(w) + F_Y(w) - F_X(w)F_Y(w) \end{aligned} \qquad (4.110)$$

We have assumed that X and Y are independent. Differentiating Eq. (4.110) leads to the expression for the pdf as

$$f_W(w) = f_X(w) + f_Y(w) - f_X(w)F_Y(w) - F_X(w)f_Y(w) \qquad (4.111)$$

Another way to obtain the density is to rewrite Eq. (4.108) in terms of events

$$\{W \le w\} = \begin{cases} X \le w, Y > w \\ Y \le w, X > w \\ X \le w, Y \le w \end{cases} \tag{4.112}$$

Equation (4.112) implies that the event $\{W < w\}$ means any one of the three events. The three "events" (#1: when X is smaller than w, Y must be larger than W, # 2: when Y is smaller than w, X must be larger than W and #3: X and Y both may be smaller than w) in Eq. (4.112) are mutually exclusive, and this leads to the CDF of the minimum as

$$F_W(w) = F_X(w)[1 - F_Y(w)] + F_Y(w)[1 - F_X(w)] + F_X(w)F_Y(w) \tag{4.113}$$

Simplifying, Eq. (4.113) becomes

$$F_W(w) = F_X(w) + F_Y(w) - F_X(w)F_Y(w) \tag{4.114}$$

An indirect means of obtaining the pdf of the minimum is described below.

$$F_W(w) = P(W \le w) = P(\min(X,Y) \le w) = 1 - P(\min(X,Y) > w) \tag{4.115}$$

If we treat X and Y as independent,

$$\{\min(X,Y) > w\} \Rightarrow \{X > w, Y > w\} \tag{4.116}$$

Making use of Eq. (4.116), the CDF of the minimum in Eq. (4.115) becomes

$$F_W(w) = 1 - P(X > w, Y > w) = 1 - [1 - F_X(w)][1 - F_Y(w)] \tag{4.117}$$

Note that the right hand side of Eq. (4.117) matches the right hand side of Eq. (4.110).

Example 4.18 X is Rayleigh distributed with a parameter b and y is exponentially distributed with a parameter a. Obtain the following

(i) $P(X > Y)$: Choose $a = 2$; $b = 4$;
(ii) Prob $(XY > ab)$

Solution $P(X > Y)$: Choose $a = 2$; $b = 4$;

$$f(x) = \frac{x}{b^2} \exp\left(-\frac{x^2}{2b^2}\right)$$

$$f(y) = \frac{1}{a} \exp\left(-\frac{y}{a}\right)$$

$$P(X > Y) = \int_0^\infty \frac{1}{a} \exp\left(-\frac{y}{a}\right) \int_y^\infty \frac{x}{b^2} \exp\left(-\frac{x^2}{2b^2}\right) dx \quad dy$$

$$P(X > Y) = \int_0^\infty \frac{1}{a} \exp\left(-\frac{y}{a}\right) \exp\left(-\frac{y^2}{2b^2}\right) dy$$

$P(X > Y) = -(2^{\wedge}(1/2)*b*pi^{\wedge}(1/2)*exp(b^{\wedge}2/(2*a^{\wedge}2))*(erf((2^{\wedge}(1/2)*b)/(2*a)) - 1))/(2*a)$
$=0.8427$ (with $a = 2$ and $b = 4$)
Verification:
>> $x1 = raylrnd(4,1,1e6)$;
>> $y1 = exprnd(2,1,1e6)$;
>> $P(X > Y) = sum(x1 > y1)/1e6 = 0.8425$
Prob($XY > ab$)

$$P(XY > ab) = \int_0^\infty \frac{1}{a} \exp\left(-\frac{y}{a}\right) \int_{\frac{ab}{y}}^\infty \frac{x}{b^2} \exp\left(-\frac{x^2}{2b^2}\right) dx \quad dy$$

$$P(XY > ab) = \int_0^\infty \frac{1}{a} \exp\left(-\frac{y}{a}\right) \exp\left(-\frac{[ab]^2}{2b^2 y^2}\right) dy$$

$P(XY > ab) = \text{meijerG}([[], []], [[0, 1/2, 1], []], 1/8)/pi^{\wedge}(1/2)$.

```
(1/sqrt(pi))*double(evalin(symengine,sprintf('meijerG([[], [],
[[0, 1/2, 1], [], 1/8)'))) = 0.3997
```

It is also possible to obtain the result as

```
double(int(fy*int(fx,x,a*b/y,inf),y,0,inf))
```

Verification:

sum($x1.*y1 > 8$)/1e6: Note that this probability does not depend on a and b
$P(XY > ab) = 0.3997$

Note that Matlab automatically generates solution in terms of Meijer G functions (Mathai and Haubold 2008; Shankar 2015, 2016, 2017). The properties of Meijer G functions are given in Sect. 4.9.3.

Example 4.19 Using the densities in Example 4.18, obtain

(i) pdf of $Z = \max (X,Y)$
(ii) pdf of $W = \min (X,Y)$

Solution The pdf

$$f_Z(z) = f_X(z)F_Y(z) + f_Y(z)F_X(z)$$

$$f_Z(z) = \frac{z}{b^2} \exp\left(-\frac{z^2}{2b^2}\right)\left[1 - \exp\left(-\frac{z}{a}\right)\right] + \frac{1}{a}\exp\left(-\frac{z}{a}\right)\left[1 - \exp\left(-\frac{z^2}{2b^2}\right)\right]$$

Pdf of $W = \min (X,Y)$
The pdf

$$f_W(w) = f_X(w) + f_y(w) - f_X(w)F_Y(w) - f_Y(w)F_X(w)$$

$$f_W(w) = \frac{w}{b^2}\exp\left(-\frac{w^2}{2b^2}\right) + \frac{1}{a}\exp\left(-\frac{w}{a}\right) + f_y(w)$$

$$-\frac{w}{b^2}\exp\left(-\frac{w^2}{2b^2}\right)\left[1 - \exp\left(-\frac{w}{a}\right)\right] - \frac{1}{a}\exp\left(-\frac{w}{a}\right)\left[1 - \exp\left(-\frac{w^2}{2b^2}\right)\right]$$

Example 4.20 X is exponentially distributed with parameter 3 and Y is Rayleigh distributed with parameter 4. What is the probability that $X^2 + Y^2 > 25$?

Solution We are given

$$f_X(x) = \frac{1}{3}\exp\left(-\frac{x}{3}\right)U(x)$$

$$f_Y(y) = \frac{y}{4^2}\exp\left(-\frac{y^2}{32}\right)U(y)$$

$$P(X^2 + Y^2 > 25) = 1 - P\left(\sqrt{X^2 + Y^2} < 5\right)$$

$$= 1 - \int_0^5 \frac{y}{4^2}\exp\left(-\frac{y^2}{32}\right)\int_0^{\sqrt{25-y^2}} \frac{1}{3}\exp\left(-\frac{x}{3}\right)dxdy$$

$$P(X^2 + Y^2 > 25) = 1 - \int_0^5 \frac{y}{4^2}\exp\left(-\frac{y^2}{32}\right)\left[1 - \exp\left(-\frac{\sqrt{25-y^2}}{3}\right)\right]dy$$

$$= 0.6349$$

Verification

```
>> syms x y
>> fx = (1/3)*exp(-x/3);
>> fy = (y/16)*exp(-y^2/32);
>> F = int(fx,x,0,sqrt(25 - y^2)) = 1 - exp(-(25 - y^2)^(1/2)/3)
>> P = 1 - int(fy*F,y,0,5) = 1 - int(-(y*exp(-y^2/32)*(exp(-(25 - y^2)^(1/2)/
   3) - 1))/16, y, 0, 5)
>> double(P) = 0.6349
```

Simulation

```
x1 = exprnd(3,1,1e6);y1 = raylrnd(4,1,1e6);
sum(sqrt(x1.^2 + y1.^2) > 5)/1e6 = 0.6352
```

Example 4.21 When electrical or optical fiber cables are fabricated, there is an outer sheath (thicker than the core). If Y represents the radius of the outer sheath and X represents the radius of the fiber, the randomness is expressed in terms of the joint density as

$$f_{X,Y}(x, y) = \frac{k}{y}, 0 < x < y < 1.$$

What is the value of k?

What is the probability that $X + Y > 0.75$?

Solution (Figure Ex.4.21)

$$1 = \int_{y=0}^{1} \int_{x=0}^{y} \frac{k}{y} dx dy = k \int_{0}^{1} dy = 1 \Rightarrow k = 1$$

Note the region of interest corresponding to the joint density is the triangle above the diagonal line $(x < y < 1)$, and the probability of interest falls in the region outside the dark triangle **abc** of height 0.75.

Figure Ex.4.21

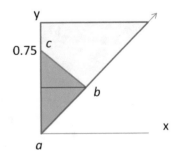

$$P\left(X + Y > \frac{3}{4}\right) = 1 - \iint_{abc} f(x, y)dxdy = 1 - \int_{y=0}^{\frac{3}{8}} \int_{x=0}^{y} \frac{1}{y} dxdy - \int_{y=\frac{3}{8}}^{\frac{3}{4}} \int_{x=0}^{\frac{3}{4}-y} \frac{1}{y} dxdy.$$

$$P\left(X + Y > \frac{3}{4}\right) = 1 - 0.5199 = 0.4801.$$

Example 4.22 Two competing computer service shops repair computers and the number of computers dropped off for service per day follow a Poisson distribution. Shop A gets an average of 5 computers a day, and shop B gets 7 computers a day. On any given day, find the probability that

1. A receives at least the same number of computers for service as B.
2. A and B receive exactly equal number.
3. A exceeds B.

Solution Let X represents the number of computers received by A, and Y represents the number of computers received by B. We are given that

$$P(X = k) = \frac{5^k}{k!} \exp(-5), k = 0, 1, 2, 3, ..$$

$$P(Y = j) = \frac{7^j}{j!} \exp(-7), j = 0, 1, 2, 3, ..$$

To obtain answer to part (1), we need to find $P(X \geq Y)$. We are dealing with discrete random variables, and therefore the presence of the equal sign does matter.

$$P(X \geq Y) = \sum_{k=0}^{\infty} \left[\sum_{j=0}^{k} \frac{7^j}{j!} \exp(-7) \right] \frac{5^k}{k!} \exp(-5) = 0.3328$$

For part (2), we are seeking the $P(X = Y)$

$$P(X = Y) = \sum_{k=0}^{\infty} \left[\sum_{j=k}^{k} \frac{7^j}{j!} \exp(-7) \right] \frac{5^k}{k!} \exp(-5) = 0.0991$$

For part (3), we are seeking $P(X > Y) = P(X \geq Y) - P(X = Y) = 0.3328 - 0.0991 = 0.2337$.

The summations can easily be carried out in Matlab using the symbolic toolbox. Results can also be verified through random number simulation.

```
clear;clc
syms k j integer
lam1=5; lam2=7;
```

```
x=poissrnd(5,1,1e6); y=poissrnd(7,1,1e6);
f1=exp(-lam1)*lam1^k/factorial(k);
f2=exp(-lam2)*lam2^j/factorial(j);
% the probability that RV 1 (parameter 5)>= RV 2 (parameter 7)
double(symsum(f1*symsum(f2,j,0,k),k,0,100)) % 100 is more than enough
% verification
sum(x>=y)/1e6
% the probability that RV 1 (parameter 5)> RV 2 (parameter 7)
double(symsum(f1*symsum(f2,j,0,k-1),k,1,100))
% verification
sum(x>y)/1e6
% prob that the outcomes are equal
double(symsum(f1*symsum(f2,j,k,k),k,0,100))
sum(x==y)/1e6
```

Example 4.23 Two students take a test consisting of 10 questions requiring
YES/NO answers. Student A expects to pick the correct answer 70% of the time,
and student B expects to pick the correct answer 60% of the time. What is the
probability that student B will score equal to or higher than student A?

Solution It is clear that the choice of the correct answer are modeled a Binomial
random numbers. For student A if X represents the random variable,

$$P(X = k) = C_k^{10}(0.7)^k(0.3)^{10-k}, k = 0, 1, 2, 3, \ldots, 10$$

If Y represents the outcomes for student B, we have

$$P(Y = j) = C_j^{10}(0.6)^j(0.4)^{10-j}, j = 0, 1, 2, 3, \ldots, 10$$

We are seeking $P(Y \geq X)$.

$$P(Y \geq X) = \sum_{j=0}^{10} \left[\sum_{k=0}^{j} C_k^{10}(0.7)^k(0.3)^{10-k} \right] C_j^{10}(0.6)^j(0.4)^{10-j} = 0.4046$$

```
syms j k integer
p1=0.7; p2=0.6;
n=20;
f1=nchoosek(n,k)*p1^k*(1-p1)^(n-k);
f2=nchoosek(n,j)*p2^j*(1-p2)^(n-j);
% the probability that RV 2 (p2=0.6)>= RV 2 (p1=0.7)
double(symsum(f2*symsum(f1,k,0,j),j,0,10)) %

x=binornd(10,0.7,1,1e6);
y=binornd(10,0.6,1,1e6);
sum(y>=x)/1e6
```

4.5.6   Joint Density of Two Functions of a Pair of Random Variables

So far, we explored ways of obtaining the density of one function of two random variables. In engineering applications, there is also an interest in obtaining the joint density of two new variables, each of which is a function of two random variables X and Y.

Given the joint density of X and Y, the interest is in obtaining the joint density of

$$
\begin{aligned}
Z &= g(X, Y) \\
W &= h(X, Y)
\end{aligned}
\tag{4.118}
$$

In other words, the joint pdf $f_{Z,W}(z,w)$ is to be obtained from $f_{X,Y}(x,y)$. Rather than delve into detailed calculations based on integral and differential calculus to accomplish this task, let us revisit the method developed to obtain the pdf of the function of a random variable in Chap. 3 and examine how the results could be imported to a pair of variables and a pair of functions of two variables.

In Chap. 3, it was shown that the density of $Y = g(X)$ for a continuous to continuous transformation can be written as

$$
f_Y(y) = \frac{f_X(x)}{\left|\frac{dy}{dx}\right|}
\tag{4.119}
$$

Ignoring the presence of the absolute value sign, Eq. (4.119) can be rewritten as

$$
f_Y(y)dy = f_X(x)dx
\tag{4.120}
$$

Equation (4.120) implies that

$$
\int f_Y(y)dy = \int f_X(x)dx = 1
\tag{4.121}
$$

If we now extend this notion to a pair of random variables, Eq. (4.121) takes the form

$$
\iint f_{Z,W}(z, w)dzdw = \iint f_{X,Y}(x, y)dxdy = 1
\tag{4.122}
$$

The process of finding the joint density of two functions of two random variables has been simplified to Eq. (4.122). To make this process clearer, let us go back to the case of a double integral such as the one involving the joint Gaussian density, $f(x,y)$ as

$$f(x, y) = \frac{1}{2\pi\sigma^2} \exp\left(-\frac{x^2 + y^2}{2\sigma^2}\right) \tag{4.123}$$

The double integral pertaining to the validity of the joint density is

$$I = \int_{-\infty}^{\infty} \int_{-\infty}^{\infty} f(x, y) dx dy = \int_{-\infty}^{\infty} \int_{-\infty}^{\infty} \frac{1}{2\pi\sigma^2} \exp\left(-\frac{x^2 + y^2}{2\sigma^2}\right) dx dy = 1 \tag{4.124}$$

The integral in Eq. (4.124) is solved by defining two new variables (conversion to polar coordinates)

$$\begin{aligned} x &= r\cos(\theta) \\ y &= r\sin(\theta) \end{aligned} \tag{4.125}$$

We are familiar with the identity,

$$dx dy = r dr d\theta \tag{4.126}$$

In other words, we have used the simple identity that the infinitesimal area in rectangular coordinates $dx dy$ can be equated to the corresponding area in polar coordinates $r dr d\theta$. Using the transformation in Eq. (4.126), Eq. (4.124) becomes

$$\int_{-\infty}^{\infty} \int_{-\infty}^{\infty} \frac{1}{2\pi\sigma^2} \exp\left(-\frac{x^2 + y^2}{2\sigma^2}\right) dx dy == \int_{0}^{\infty} \int_{0}^{2\pi} \frac{1}{2\pi\sigma^2} \exp\left(-r^2\right) r dr d\theta \tag{4.127}$$

Let us rewrite Eq. (4.125) as

$$\begin{aligned} r &= \sqrt{x^2 + y^2} \\ \theta &= \tan^{-1}\left(\frac{y}{x}\right) \end{aligned} \tag{4.128}$$

Since X and Y are random variables, R and Θ are also random variables expressed as

$$\begin{aligned} R &= \sqrt{X^2 + Y^2} \\ \Theta &= \tan^{-1}\left(\frac{Y}{X}\right) \end{aligned} \tag{4.129}$$

The ranges of the two random variables in Eq. (4.129) (see Chap. 3) are

$$R: \quad 0 < r < \infty \tag{4.130}$$

$$\Theta: \quad 0 < \theta < 2\pi \tag{4.131}$$

Closely following Eq. (4.122), Eq. (4.127) can be rewritten as

$$\int_{-\infty}^{\infty} \int_{-\infty}^{\infty} f_{X,Y}(x,y)dxdy = \int_{0}^{\infty} \int_{0}^{2\pi} f_{R,\Theta}(r,\theta)\,drd\theta \tag{4.132}$$

Note that both sides of Eq. (4.132) represent the normalization associated with the densities, $f_{X,Y}(x,y)$ and $f_{r,\theta}(r,\theta)$. Furthermore, by examining Eq. (4.127), for Eq. (4.132), we get

$$f_{R,\Theta}(r,\theta) = \frac{r}{2\pi\sigma^2} \exp\left(-\frac{r^2}{2\sigma^2}\right), 0 < r < \infty, 0 < \theta < 2\pi \tag{4.133}$$

Equation (4.132) clearly demonstrates that $f_{R,\,\Theta}(r,\theta)$ is a joint density with $0 < r < \infty$ and $0 < \theta < 2\pi$. Let us now establish the association between the two joint densities, $f_{X,Y}(x,y)$ and $f_{r,\theta}(r,\theta)$ by obtaining the Jacobian. The Jacobian associated with Eq. (4.128) is

$$J(x,y) = \begin{vmatrix} \dfrac{\partial r}{\partial x} & \dfrac{\partial r}{\partial y} \\ \dfrac{\partial \theta}{\partial x} & \dfrac{\partial \theta}{\partial y} \end{vmatrix} = \begin{vmatrix} \dfrac{x}{\sqrt{x^2+y^2}} & \dfrac{y}{\sqrt{x^2+y^2}} \\ \dfrac{-y}{x^2+y^2} & \dfrac{x}{x^2+y^2} \end{vmatrix} = \frac{1}{\sqrt{x^2+y^2}} = \frac{1}{r} \tag{4.134}$$

Using Eq. (4.134), we can express the relationship between the two densities as

$$f(r,\theta) = \frac{f(x,y)}{|J(x,y)|}\bigg|_{x=r\cos\theta,y=r\sin\theta} \tag{4.135}$$

In other words, it is possible to obtain the joint density of two functions of two random variables using the Jacobian, and Eq. (4.135) is very similar to Eq. (4.119). The Jacobian can also be defined in terms of the derivatives w. r. t. r and θ as

$$J(r,\theta) = \begin{vmatrix} \dfrac{\partial r}{\partial x} & \dfrac{\partial r}{\partial y} \\ \dfrac{\partial \theta}{\partial x} & \dfrac{\partial \theta}{\partial y} \end{vmatrix} = \begin{vmatrix} \cos(\theta) & \sin(\theta) \\ -r\sin(\theta) & r\cos(\theta) \end{vmatrix} = r = \frac{1}{J(x,y)} \tag{4.136}$$

We can now generalize the result as follows. If Z and W are functions of X and Y, the joint density of Z and W is

$$f(z, w) = \left(f(x, y) |J(z, w)| \right) \big|_{X, Y \Rightarrow g^{-1}(Z, W), h^{-1}(Z, W)}$$

$$= \frac{f(x, y)}{|J(x, y)|} \bigg|_{X, Y \Rightarrow g^{-1}(Z, W), h^{-1}(Z, W)} \tag{4.137}$$

Throughout this work, the density of Z and W will be expressed as

$$f(z, w) = \frac{f(x, y)}{|J(x, y)|} \bigg|_{X, Y \Rightarrow g^{-1}(Z, W), h^{-1}(Z, W)} \tag{4.138}$$

If multiple roots exist (as in the case of a one-to-many transformation of a single variable), Eq. (4.138) is expressed as a sum.

We will now look at a few examples of joint densities of functions of two random variables and examine how the approach can be tailored to obtain the density of a single function of two variables.

Example 4.24 If X is Rayleigh distributed with parameter b and Y is uniform in $[0, 2\pi]$, obtain the joint and marginal densities of $Z = X\cos(Y)$ and $W = X\sin(Y)$. Note that X and Y are independent.

Solution The marginal densities of X and Y are

$$f_X(x) = \frac{x}{b^2} \exp\left(-\frac{x^2}{2b^2}\right) U(x) \qquad f_Y(y) = \frac{1}{2\pi}, 0 < y < 2\pi$$

$$Z = X \cos(Y)$$
$$W = X \sin(Y)$$

The Jacobian J is

$$J(x, y) = \begin{vmatrix} \dfrac{\partial z}{\partial x} & \dfrac{\partial z}{\partial y} \\[2mm] \dfrac{\partial w}{\partial x} & \dfrac{\partial w}{\partial y} \end{vmatrix} = \begin{vmatrix} \cos(y) & -x\sin(y) \\[1mm] \sin(y) & x\cos(y) \end{vmatrix} = x$$

The joint density of Z and W is

$$f(z, w) = \frac{f(x, y)}{|J|} = \frac{f(x)f(y)}{|J|} = \frac{\frac{x}{b^2} \exp\left(-\frac{x^2}{2b^2}\right)\left(\frac{1}{2\pi}\right)}{x} = \frac{1}{2\pi b^2} \exp\left(-\frac{x^2}{2b^2}\right).$$

In terms of Z and W, X is given as

$$X^2 = Z^2 + W^2$$

Substituting for x, the joint density becomes

$$f(z, w) = \frac{1}{2\pi b^2} \exp\left(-\frac{z^2 + w^2}{2b^2}\right) = \frac{1}{\sqrt{2\pi b^2}} \exp\left(-\frac{z^2}{2b^2}\right) \frac{1}{\sqrt{2\pi b^2}} \exp\left(-\frac{w^2}{2b^2}\right)$$

We see that Z and W are jointly Gaussian. They are also identical and independent.

Example 4.25 The joint density of X and Y is given by

$$f(x, y) = \exp\left(-x - y\right) U(x) U(y)$$

Obtain the pdf of $Z = X + Y$.

Solution It can be easily seen that X and Y are independent and identically distributed random variables. In this example, we only have a single function of the two random variables. Even under these circumstances, we can use the concept of the Jacobian by defining an auxiliary or a dummy variable W as

$$W = X$$

With two functions Z and W, the Jacobian becomes

$$J(x, y) = \begin{vmatrix} \dfrac{\partial z}{\partial x} & \dfrac{\partial z}{\partial y} \\ \dfrac{\partial w}{\partial x} & \dfrac{\partial w}{\partial y} \end{vmatrix} = \begin{vmatrix} 1 & 1 \\ 1 & 0 \end{vmatrix} = 1$$

The joint density of Z and W is

$$f(z, w) = \frac{f(x, y)}{|J|} = \exp\left([-w - (z - w)]\right) = \exp\left(-z\right), 0 < w < z$$

Notice the range of the variable W. It represents X, and therefore, it will always be less than Z. The marginal density of Z is obtained as

$$f(z) = \int_{-\infty}^{\infty} f(z, w) dw = \int_{0}^{z} \exp\left(-z\right) dw = z \exp\left(-z\right) U(z)$$

The method used in this example is referred to as the **Method of Auxiliary Variable**. Thus, the method based on the Jacobian can be used successfully even when only a single function of two variables is of interest.

Using the standard expression for the density of the sum of two variables, the density of Z also can be expressed as

$$f(z) = \int f(x, z - x)dx = \int_0^z \exp\left(-x - [z - x]\right)dx = \int_0^z \exp\left(-z\right)dx = z\exp\left(-z\right)U(z)$$

The limits of integration are 0 and z because X and Y are always positive.

Example 4.26 This example demonstrates why general approaches may not be the most efficient means of solving problems. The problem specific approach based on fundamental concepts is the best-suited. This example has two parts starting with independent random variables, followed by the creation of a joint density when the variables are no longer independent. The example concludes with methods for finding the pdf of the sum of the two variables.

Let X and Y be two independent and identically exponentially distributed random variables, each with a mean of unity. This means that

$$f(x, y) = \exp\left(-x\right)U(x)\exp\left(-y\right)U(y) = \exp\left(-x - y\right)U(x)U(y)$$

(i) Obtain the pdf of X and Y with the constraint that $Y < X$.

The shaded region (extends to infinity) in Figure Ex.4.26a corresponds to $\{Y < X\}$. This means that

$$P(X < x, Y < y|Y < X) = \frac{P(X < x, Y < y, Y < X)}{P(Y < X)} = \frac{P(X < x, Y < y)}{P(Y < X)}, 0 < y < x < \infty$$

From the shaded area indicated,

$$P(Y < X) = \int_{y=0}^{\infty} \int_{x=y}^{\infty} \exp\left(-x - y\right)dxdy = \frac{1}{2}$$

Figure Ex.4.26a Region where Y≤ x is shown

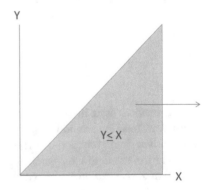

The conditional joint pdf is obtained using Eq. (4.48) as

$$f(x, y|Y < X) = \frac{f(x, y)}{P(Y < X)} = 2 \exp(-x - y), \quad 0 < y < x < \infty$$

Note that equation above is a joint density, and once the limits are stated explicitly, it is possible to view the density simply as a joint density of two variables X and Y that are no longer independent. In other words, we have a valid joint pdf,

$$f(x, y) = 2 \exp(-x - y), \quad 0 < y < x < \infty$$

(ii) Given the joint density obtained in the previous example, obtain the pdf of

$$W = X + Y$$

Our starting point is the joint density

$$f(x, y) = 2 \exp(-x - y), \quad 0 < y < x < \infty$$

The sample space for obtaining the CDF of W is the region below the diagonal line of $x = y$ and $x + y = w$ shown in Figure Ex.4.26b. Before we explore the approach of the convolution of the densities, it should be noted that with the nature of the joint density with restriction $0 < y < x < \infty$, it does not fit the two general set of results, one with X and Y, each being limited to 0 to ∞ without any restrictions on the relationship between x and y and the other one with both x and y with limits of $-\infty$ to ∞ without any restrictions on the relationship between x and y.

Let us go back to the basic definition of the CDF of W.

$$F_W(w) = P(W \le w) = P(\text{volume contained in the triangle ABC})$$

Figure Ex.4.26b Regions #1 and #2 consist of the region of interest for obtaining the CDF

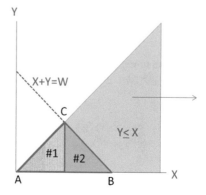

The area of the triangle needs to be broken into two, #1 and #2. This means that

$$F_W(w) = F_{W\#1}(w) + F_{W\#2}(w)$$

$$F_{W\#1}(w) = \int\limits_{x=0}^{\frac{w}{2}} \int\limits_{y=0}^{x} 2\exp(-x-y)dydx = \exp(-w)\left[\exp\left(\frac{w}{2}\right) - 1\right]^2$$

$$F_{W\#2}(w) = \int\limits_{x=\frac{w}{2}}^{w} \int\limits_{y=0}^{w-x} 2\exp(-x-y)dydx = \exp(-w)\left[2\exp\left(\frac{w}{2}\right) - 2 - w\right]$$

Adding the two, we have

$$F_W(w) = 1 - \exp(-w) - w\exp(-w)$$

Differentiating the CDF, the pdf of W becomes

$$f_W(w) = \frac{d}{dw}F_W(w) = w\exp(-w)U(w)$$

We will now revisit the method based on the convolution described earlier resulting in

$$f(w) = \int\limits_{-\infty}^{\infty} f_{X,Y}(x, w - x)dx$$

The difficulties with putting the appropriate limits become evident as soon as the actual pdf $f(x,y)$ is inserted into equation above because the region of interest comprises of regions #1 and #2 identified in Figure Ex.4.26b. Rewriting the equation in terms of actual densities, we have

$$f(w) = \int\limits_{-\infty}^{\infty} 2\exp(-x)\exp(-[w-x])U(x)U(w-x)U(x-y)dx$$

In the integral above, the $U(x-y)$ represents the conditioning that $X > Y$. Since $W = X + Y$, we replace y with $(w - x)$.

$$f(w) = 2\exp(-w) \int\limits_{0}^{\infty} U(w-x)U(2x-w)dx$$

The lower limit is set to 0 to satisfy $U(x)$. But, we also have $U(2x - w)$. This requires that $x > w/2$ requiring the change of the lower limit to $w/2$. The upper limit is w because of the presence of $U(w - x)$. The pdf now becomes

$$f(w) = 2 \exp(-w) \int_{\frac{w}{2}}^{w} dx = w \exp(-w)U(w)$$

Instead of using the concept of convolution, we may use CDF directly and differentiate the integral itself,

$$F_W(w) = P(W \leq w) = \int_{x=0}^{\frac{w}{2}} \int_{y=0}^{x} f(x,y)dydx + \int_{x=\frac{w}{2}}^{w} \int_{y=0}^{w-x} f(x,y)dydx$$

The pdf is obtained by differentiating the CDF above w. r. t to w. This requires invoking the Leibniz rule in two dimensions leading to

$$f_W(w) = \frac{1}{2} \int_{0}^{\frac{w}{2}} f_{X,Y}\left(\frac{w}{2}, y\right) dy + \int_{\frac{w}{2}}^{w} f_{X,Y}(x, w - x)dx - \frac{1}{2} \int_{0}^{\frac{w}{2}} f_{X,Y}\left(\frac{w}{2}, y\right) dy$$

Simplifying further, the pdf becomes

$$f_W(w) = \int_{\frac{w}{2}}^{w} f_{X,Y}(x, w - x)dx$$

The equation above offers a general expression for the pdf of the sum for cases where the joint pdf exists in the range $0 < y < x < \infty$.

Substituting the joint density, the pdf of the sum becomes

$$f_W(w) = 2 \int_{\frac{w}{2}}^{w} \exp(-w)dx = w \exp(-w)U(w)$$

Example 4.27 If X and Y are independent and identically distributed exponential random variables, obtain the pdf of $Z = X - Y$.

Solution Method 1

We can use the method available for the sum of two random variables to obtain the pdf of $X - Y$. Let us redefine

$$Z = X - Y = X + V$$

In the expression for Z above,

$$V = -Y$$

Using the transformation of random variables presented in Chap. 3, the pdf of v is

$$f_V(v) = e^v, \quad U(-v)$$

Note that

$$f_{X,Y}(x, y) = \exp(-x - y)U(x)U(y)$$

But, the joint density of X and V is

$$f_{X,V}(x, v) = \exp(-x)\exp(v)U(x)U(-v)$$

Note that X and Y are independent and hence, X and V are also independent. The density of function of the difference $X - Y$ is identical to the density of the sum $X + V$, and therefore, it will be the convolution of the densities of X and V. The pdf of Z can therefore be written as

$$f_Z(z) = \int f_X(x)f_V(z-x)\,dx = \int_{-\infty}^{\infty} \exp(-x)\exp(z-x)U(x)U(x-z)\,dx$$

Simplifying further, we get

$$f_Z(z) = \exp(z) \int_{-\infty}^{\infty} \exp(-2x)U(x)U(x-z)\,dx$$

Notice the step function associated with the variable V. Since $X - Y$ can be positive or negative, the limits of the integral will change because of the step functions in the integral.

Case 1: $z > 0$

When $z > 0$, we examine what happens to $U(x - z)$ because $U(x)$ always exists. $U(x - z)$ will exist only when $x > z$. Therefore, the lower limit will be z. The upper limit will be ∞. The pdf now becomes

$$f_Z(z) = \exp(z) \int_z^{\infty} \exp(-2x)dx = \frac{1}{2}\exp(-z)$$

Case 2: $z < 0$

In this case, the limit is determined by $U(x)$ alone; $(x - z)$ is always positive because z is negative. Therefore, the pdf becomes

$$f_Z(z) = \exp(z) \int_0^\infty \exp(-2x)dx = \frac{1}{2}\exp(z)$$

Combining the two cases, the density of Z can be expressed as

$$f_Z(z) = \frac{1}{2}\exp(|z|), \; -\infty \le z \le \infty$$

Method 2

The density of Z may also be obtained graphically. In Figure Ex.4.27, the region above the diagonal at $X = Y$ represents $Z < 0$, and region below represents $Z > 0$. Because of symmetry (X and Y are identical), we know that

$$P(Z < 0) = P(Z > 0) = \frac{1}{2}$$

For $Z > 0$, the CDF is

$$F_Z(z) = P(Z < 0) + \frac{1}{2} - P(Z > z) = 1 - \int_{y=0}^\infty \int_{x=y+z}^\infty \exp(-x - y)dxdy$$

$$= 1 - \frac{1}{2}\exp(-z)$$

By symmetry, for $Z < 0$, the CDF must equal the value of $P(Z > z)$ when z is positive.

Figure Ex.4.27

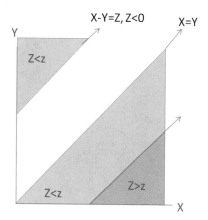

$$F_Z(z) = \frac{1}{2} \exp(z), \quad -\infty < z < 0$$

The CDF for the case when $X - Y$ is negative may also be obtained directly, (shaded region above the diagonal $X = Y$), as the integral

$$F_Z(z) = P(Z < z) \int\limits_{y=-z}^{\infty} \int\limits_{x=0}^{y+z} \exp(-x-y)dxdy = \frac{1}{2} \exp(z), \ z < 0$$

Combining the results, the CDF is

$$F_Z(z) = \begin{cases} \dfrac{1}{2} \exp(z), & -\infty < z < 0 \\ 1 - \dfrac{1}{2} \exp(-z), & z > 0 \end{cases}$$

Differentiating the CDF, the pdf becomes

$$f_Z(z) = \begin{cases} \dfrac{1}{2} \exp(z), & z < 0 \\ \dfrac{1}{2} \exp(-z), & z > 0 \end{cases} = \frac{1}{2} \exp(-|z|)$$

Method 3

Let us define a new function of the two variables, X and Y as

$$W = X + Y$$

The Jacobian of the transformation is

$$J = \begin{vmatrix} \dfrac{\partial z}{\partial x} & \dfrac{\partial z}{\partial y} \\ \dfrac{\partial w}{\partial x} & \dfrac{\partial w}{\partial y} \end{vmatrix} = \begin{vmatrix} 1 & -1 \\ 1 & 1 \end{vmatrix} = 2$$

The joint density of Z and W is

$$f(z, w) = \frac{f(x, y)}{J} \bigg|_{x=\frac{z+w}{2}, y=\frac{w-z}{2}}$$

$$f(z, w) = \frac{1}{2} \exp(-w), \quad 0 \leq |z| \leq w \leq \infty$$

Notice that the range of the variables is expressed taking into account:

(i) Z takes both positive and negative values.
(ii) $X - Y$ will always be less than $X + Y$.
(iii) $X + Y$ only takes positive values resulting in the use of absolute value of z in the range.

The marginal density of Z is given as

$$f(z) = \int_{|z|}^{\infty} f(z, w)dw = \int_{|z|}^{\infty} \frac{1}{2} \exp(-w)dw = \frac{1}{2}\exp(-|z|)$$

Even though unnecessary for this example, the density of the sum may be obtained as

$$f(w) = \int_{-w}^{w} f(z, w)dz = \int_{-w}^{w} \frac{1}{2} \exp(-w)dz = w \exp(-w)U(w)$$

Method 4 We use an auxiliary variable, $W = X$. We have

$$f(x, y) = \exp(-x - y)U(x)U(y)$$
$$Z = X - Y$$
$$W = X$$

The Jacobian is

$$J = \begin{vmatrix} 1 & -1 \\ 1 & 0 \end{vmatrix} \Rightarrow |J| = 1$$

$$f_{Z,W}(z, w) = \frac{f(x, y)}{|J|} = \exp(-w - [w - z]) = \exp(-2w + z)$$

The determination of the range of z and w has to be done in two stages, $x > y$ and $x < y$.

When $x > y$, we also have $z > 0$ (in this case z and w are positive, and $z < w$)

$$f_{Z,W}(z, w) = \exp(-2w + z), 0 < z < w < \infty$$

When $y > x$, we have $z < 0$ (in this case, $w > 0$ and $|z|$ may be larger or smaller than w)

$$f_{Z,W}(z, w) = \exp(-2w + z), \; -\infty < z < 0, \quad 0 < w < \infty$$

The marginal density of $X - Y$ is

$$f_Z(z) = \int\limits_z^\infty f(z, w)dw = \frac{1}{2}\exp(-z), z > 0$$

$$f_Z(z) = \int\limits_0^\infty f(z, w)dw = \frac{1}{2}\exp(z), z < 0$$

Combining both ranges

$$f_Z(z) = \frac{1}{2}\exp(-|z|)$$

Example 4.28 X and Y are independent identically distributed random variables. Obtain the joint density of $Z = X - Y$ and $W = X + Y$ for the following two cases:

1. X and Y are exponential each with a parameter a.
2. X and Y are Gaussian random variables, $N(0,\sigma^2)$.

Solution Case (1)

$$f(x, y) = \frac{1}{a^2}\exp\left(-\frac{x+y}{a}\right)U(x)U(y)$$

$$J = \begin{vmatrix} \dfrac{\partial z}{\partial x} & \dfrac{\partial z}{\partial y} \\[2mm] \dfrac{\partial w}{\partial x} & \dfrac{\partial w}{\partial y} \end{vmatrix} = \begin{vmatrix} 1 & -1 \\ 1 & 1 \end{vmatrix} = 2$$

$$f(z, w) = \left.\frac{f(x, y)}{J}\right|_{x=\frac{z+w}{2}, y=\frac{w-z}{2}} = \frac{1}{2a^2}\exp\left(-\frac{w}{a}\right), \quad 0 \le |z| \le w \le \infty$$

$$f(z) = \int\limits_{|z|}^\infty f(z, w)dw = \int\limits_{|z|}^\infty \frac{1}{2a^2}\exp\left(-\frac{w}{a}\right)dw = \frac{1}{2a}\exp\left(-\frac{|z|}{a}\right)$$

$$f(w) = \int\limits_{-w}^w f(z, w)dz = \int\limits_{-w}^w \frac{1}{2a^2}\exp\left(-\frac{w}{a}\right)dz = \frac{w}{a^2}\exp\left(-\frac{w}{a}\right)U(w)$$

It can be seen that the random variables Z and W are not independent because

$$f(z, w) \ne f(z)f(w).$$

Now, consider the case of X and Y being Gaussian,

$$f(x, y) = \frac{1}{2\pi\sigma^2} \exp\left(-\frac{x^2 + y^2}{2\sigma^2}\right)$$

We have

$$x = \frac{z + w}{2} \quad y = \frac{w - z}{2}$$

$$x^2 + y^2 = \frac{z^2 + w^2}{2}$$

$$f(z, w) = \left.\frac{f(x, y)}{J}\right|_{x=\frac{z+w}{2}, y=\frac{w-z}{2}} = \frac{1}{4\pi\sigma^2} \exp\left(-\frac{z^2 + w^2}{4\sigma^2}\right)$$

$$f(z, w) = \frac{1}{\sqrt{4\pi\sigma^2}} \exp\left(-\frac{z^2}{4\sigma^2}\right) \frac{1}{\sqrt{4\pi\sigma^2}} \exp\left(-\frac{w^2}{4\sigma^2}\right) = f(w)f(z)$$

You will notice that Z and W are independent. Each is a Gaussian random variable with a variance equal to 2 times the variance of X. In other words, both sum and difference of two Gaussian variables are also Gaussian.

In general, we can state that the sum of two independent Gaussian variables is another Gaussian with a variance equal to the sum of the variances and a mean equal to the sum of the means. The difference is also a Gaussian with a variance equal to the sum of the variances and mean equal to the difference of the means.

4.6 Joint Moments, Correlation, Covariance, Etc.

The joint moment (k^{th} joint moment) of a pair of random variables is defined as

$$E\left[(XY)^k\right] = \int\int (xy)^k f(x, y)\,dx\,dy \tag{4.139}$$

If two random variables are independent,

$$E[XY] = E(X)E(Y) \tag{4.140}$$

It should be noted that if one finds that $E[(XY)] = E(X)E(Y)$, it does not mean that X and Y are independent. In other words, independence of X and Y ensures that the mean of the product is the product of the means; the converse is not TRUE.

The covariance of two random variables, $C(X,Y)$, is defined as

$$C(X, Y) = E(XY) - E(X)E(Y) \tag{4.141}$$

The correlation coefficient ρ is defined as

$$\rho = \frac{C(X, Y)}{\sigma_X \sigma_Y} \tag{4.142}$$

In Eq. (4.142), the denominator is the product of the standard deviations of X and Y, namely, σ_X and σ_Y. If $\rho = 0$, variables X and Y are uncorrelated. It can be seen that if X and Y are independent, they are always uncorrelated. Two random variables are orthogonal if

$$E(XY) = 0 \tag{4.143}$$

Bivariate Normal Distribution (Correlated Case)

$$f(x, y) = A \exp \left\{ -\frac{1}{2(1 - \rho^2)} \left[\frac{(x - \eta_1)^2}{\sigma_1^2} - \frac{2\rho(x - \eta_1)(y - \eta_2)}{\sigma_1 \sigma_2} + \frac{(y - \eta_2)^2}{\sigma_2^2} \right] \right\} \tag{4.144}$$

with

$$A = \frac{1}{2\pi\sigma_1\sigma_2\sqrt{1 - \rho^2}} \tag{4.145}$$

X and Y are Gaussian with means of η_1 and η_2 and standard deviations of σ_1 and σ_2, respectively, and ρ is the correlation coefficient. Note that when ρ is zero, the joint pdf becomes the product of the marginal density functions of X and Y.

In general, the expected value of a function $h(x,y)$ of two random variables is expressed as

$$E[h(X, Y)] = \int\int h(x, y)f(x, y)dxdy \tag{4.146}$$

We can also define conditional expectations as

$$E[g(y)|x] = \int g(y)f(y|x)dy \tag{4.147}$$

The conditional expected value in Eq. (4.147) is a random variable, and $E[g(y)]$ is obtained as

$$E[g(y)] = \int E[g(y)|x]f(x)dx = \int\int g(y)f(x,y)dxdy \tag{4.148}$$

We can generalize Eq. (4.148) as

$$E_X(E[g(x,y)|x]) = E[g(x,y)] \tag{4.149}$$

The subscript X with the expectation E represents the averaging w. r. t. variable X.

We will explore the relationship between being independent and correlated by revisiting one of the previous examples. If we have two functions of a pair of random variables expressed as the sum and difference, we will examine whether the two new variables are independent or not. For the case of two independent and identically distributed variables X and Y, let us define

$$\begin{aligned} Z &= X - Y \\ W &= X + Y \end{aligned} \tag{4.150}$$

The Jacobian of this transformation is

$$J = \begin{vmatrix} \dfrac{\partial z}{\partial x} & \dfrac{\partial z}{\partial y} \\ \dfrac{\partial w}{\partial x} & \dfrac{\partial w}{\partial y} \end{vmatrix} = \begin{vmatrix} 1 & -1 \\ 1 & 1 \end{vmatrix} = 2 \tag{4.151}$$

The joint density of Z and W will be

$$f(z,w) = \frac{f(x,y)}{2} \tag{4.152}$$

Let us first examine whether the variables, Z and W, are correlated or not by finding the joint moments and marginal moments.

$$E(ZW) = E[(X+Y)(X-Y)] = E(X^2 - Y^2) = E(X^2) - E(Y^2) \tag{4.153}$$

$$\begin{aligned} E(Z) &= E(X-Y) = E(X) - E(Y) \\ E(W) &= E(X+Y) = E(X) + E(Y) \end{aligned} \tag{4.154}$$

The covariance of Z and W is

$$\text{cov}(Z,W) = E(ZW) - E(Z)E(W) \tag{4.155}$$

Since X and Y are identical,

$$\begin{aligned} E(ZW) &= E(X^2) - E(Y^2) = 0 \\ E(Z)E(W) &= 0 \end{aligned} \tag{4.156}$$

By virtue of Eq. (4.156), the covariance in Eq. (4.155) becomes

$$C(Z, W) = 0 \tag{4.157}$$

This means that the random variables, Z and W, are uncorrelated. They are also orthogonal because

$$E(ZW) = E(X^2) - E(Y^2) = 0 \tag{4.158}$$

Testing for orthogonality and uncorrelatedness does not require the probability densities of Z and W. However, testing for independence requires explicit expression of the joint density.

The joint density of Z and W (when X and Y are independent identically distributed exponential variables of unity) is (see Example 4.28)

$$f(z, w) = \frac{1}{2} \exp(-w), \quad 0 \le |z| \le w \le \infty \tag{4.159}$$

The marginal densities of Z and we were obtained as

$$f(z) = \frac{1}{2} \exp(-|z|) \tag{4.160}$$

$$f(w) = w \exp(-w) U(w) \tag{4.161}$$

Clearly the variables Z and W are not independent because

$$f(z, w) \ne f(z) f(z) \tag{4.162}$$

Let see what happens if X and Y are independent identically distributed Gaussian variables, each being $N(0, \sigma^2)$. In this case, the joint pdf of Z and W is

$$f(z, w) = \frac{f(x, y)}{2} = \left. \frac{\exp\left(-\frac{x+y^2}{2\sigma^2}\right)}{2(2\pi\sigma^2)} \right|_{x^2+y^2=\frac{z^2+w^2}{2}} = \frac{\exp\left(-\frac{w+z^2}{4\sigma^2}\right)}{4\pi\sigma^2} \tag{4.163}$$

The joint density in Eq. (4.163) can be written as

$$f(z, w) = \frac{1}{\sqrt{2\pi(2\sigma^2)}} \exp\left(-\frac{z^2}{4\sigma^2}\right) \frac{1}{\sqrt{2\pi(2\sigma^2)}} \exp\left(-\frac{w^2}{4\sigma^2}\right) = f(w) f(z) \tag{4.164}$$

Note that $f(z, w)$ is jointly Gaussian and $f(z)$ and $f(w)$ are marginally Gaussian variables, each being $N(0, 2\sigma^2)$.

These results indicate that the variables, Z and W, are independent if X and Y are independent, identical and *Gaussian*. On the other hand, Z and W are **not independent** if X and Y are independent, identical and *exponential*, even though in both cases, the variables, Z *and* W, *are uncorrelated*.

4.6.1   Bayes' Rule, Conditionality and Correlation

Having seen the relationship of correlation and independence, we can revisit Bayes' rule presented in Chap. 2. Let us restate Bayes' rule relating any two events A and B with probabilities of $P(A)$ and $P(B)$, respectively,

$$P(A|B) = \frac{P(AB)}{P(B)}, \quad P(B) > 0 \tag{4.165}$$

To explore this further, we consider the case of two discrete random variables X and Y, each one being binary type. In other words, we have

$$\begin{aligned} f_X(x) &= P(X=0)\delta(x) + P(X=1)\delta(x-1) \\ f_Y(y) &= P(Y=0)\delta(y) + P(Y=1)\delta(y-1) \end{aligned} \tag{4.166}$$

If A and B represent the outcomes in X and Y, respectively, we may rewrite Eq. (4.166) as

$$\begin{aligned} f_X(x) &= P(\overline{A})\delta(x) + P(A)\delta(x-1) \\ f_Y(y) &= P(\overline{B})\delta(y) + P(B)\delta(y-1) \end{aligned} \tag{4.167}$$

Equation (4.167) implies that we identify the following events,

$$\begin{aligned} A &= \{X=1\} \quad \overline{A} = \{X=0\} \\ B &= \{Y=1\} \quad \overline{B} = \{Y=0\} \end{aligned} \tag{4.168}$$

The respective expected values of X and Y become

$$\begin{aligned} E(X) &= 1P(A) + 0P(\overline{A}) = P(A) \\ E(Y) &= 1P(B) + 0P(\overline{B}) = P(B) \end{aligned} \tag{4.169}$$

The second moments of X and Y are

$$\begin{aligned} E(X^2) &= 1^2 P(A) + 0^2 P(\overline{A}) = P(A) \\ E(Y^2) &= 1^2 P(B) + 0^2 P(\overline{B}) = P(B) \end{aligned} \tag{4.170}$$

The variances of X and Y are

$$\begin{aligned} \text{var}(X) &= P(A) - P^2(A) = P(A)[1 - P(A)] = P(A)P(\overline{A}) \\ \text{var}(Y) &= P(B) - P^2(B) = P(B)[1 - P(B)] = P(B)P(\overline{B}) \end{aligned} \tag{4.171}$$

The correlation coefficient of X and Y is

$$\rho(X,Y) = \frac{E(XY) - E(X)E(Y)}{\sqrt{\mathrm{var}(X)\mathrm{var}(Y)}} \tag{4.172}$$

The joint moment is

$$E(XY) = \sum_{j=1}^{2}\sum_{k=1}^{2} x_j y_k P(X = x_k, Y = y_j) \tag{4.173}$$

Rewriting Eq. (4.173), we have

$$E(XY) = 1 \times 0 P(X = 1, Y = 0) + 1 \times 1 P(X = 1, Y = 1) \\ + 0 \times 1 P(X = 0, Y = 1) + 0 \times 0 P(X = 0, Y = 0) \tag{4.174}$$

Equation (4.174) simplifies to

$$E(XY) = 1 \times 1 P(X = 1, Y = 1) = P(X = 1, Y = 1) = P(AB) \tag{4.175}$$

In terms of Eq. (4.175), Eq. (4.172) becomes

$$\rho(X,Y) = \frac{P(AB) - P(A)P(B)}{\sqrt{P(A)P(\overline{A})P(B)P(\overline{B})}} \tag{4.176}$$

Examining the numerator of Eq. (4.176), we see that the correlation coefficient is positive if

$$P(AB) - P(A)P(B) > 0 \tag{4.177}$$

The correlation coefficient is negative if

$$P(AB) - P(A)P(B) < 0 \tag{4.178}$$

The correlation coefficient is zero if

$$P(AB) - P(A)P(B) = 0 \tag{4.179}$$

Rewriting Eqs. (4.177), (4.178), and (4.179), we have

$$\frac{P(AB)}{P(B)} > P(A); \quad \text{positive correlation between A and B} \tag{4.180}$$

$$\frac{P(AB)}{P(B)} < P(A); \quad \text{negative correlation between A and B} \tag{4.181}$$

$$\frac{P(AB)}{P(B)} = P(A); \quad \text{correlation between A and B is 0} \qquad (4.182)$$

Equations (4.180), (4.181), and (4.182) suggest when the conditional probability will be larger or smaller than the marginal probability. Expressing the left hand side in terms of the conditional probability, Eq. (4.180) shows that

$$P(A|B) > P(A) \qquad (4.183)$$

Expressing the left hand side in terms of the conditional probability, Eq. (4.181) shows that

$$P(A|B) < P(A) \qquad (4.184)$$

Equation (4.182) suggests that the events A and B are independent. Such results were seen in Chap. 2.

4.7   Characteristic Functions, Laplace Transforms, and Mellin Transforms

The joint characteristic function, $\phi(\omega_x, \omega_y)$ of two random variables X and Y is given by (Papoulis and Pillai 2002; Shankar 2017)

$$\phi(\omega_x, \omega_y) = E\left(e^{j\omega_x X + j\omega_y Y}\right) = \int\int f(x, y) e^{j\omega_x X + j\omega_y Y} dxdy \qquad (4.185)$$

If the two variables are independent, Eq. (4.185) becomes

$$\phi(\omega_x, \omega_y) = \int\int f(x)f(y) e^{j\omega_x X + j\omega_y Y} dxdy = \phi(\omega_x)\phi(\omega_y) \qquad (4.186)$$

Therefore, for the case of two independent random variables, the joint characteristic function is the product of marginal characteristic functions.

While the characteristic functions play a role in estimating the moments (shown in Chap. 3 in connection), they play an important role in obtaining the densities of sums and differences of independent random variables. Invoking the convolution property, we can write the expression for the density of the sum (Z) of two independent random variables X and Y as

$$f_Z(z) = f_X(x) * f_Y(y) \qquad (4.187)$$

Taking the Fourier transform, we have

$$\phi_Z(\omega) = \phi_X(\omega)\phi_Y(\omega) \tag{4.188}$$

In other words, the characteristic function of the sum of two independent random variables is the product of the marginal characteristic functions. The characteristic function in Eq. (4.188) can be inverted to obtain the density of the sum, $f_Z(z)$.

Example 4.29 If X and Y are two independent identically distributed variables, each $N(0, \sigma^2)$, obtain the pdf of the sum $Z = X + Y$ using the characteristic functions.

Solution The characteristic functions of X and Y are

$$\phi_X(\omega) = \phi_Y(\omega) = \exp\left(-\frac{\omega^2\sigma^2}{2}\right)$$

The characteristic function of $Z = X + Y$ is

$$\phi_Z(\omega) = \phi_X(\omega)\phi_Y(\omega) = \exp\left(-\frac{\omega^2[2\sigma^2]}{2}\right)$$

By examining the characteristic functions of Z and X, it is seen that the density of Z is also Gaussian with a variance of $2\sigma^2$. In other words, the density of the sum of two independent identically distributed Gaussian variables is also Gaussian with a variance equal to the sum of the variances of X and Y. Therefore, Z is $N(0, 2\sigma^2)$. This approach can easily be extended to show that the density of the sum of two independent Gaussian variables will be another Gaussian density with a mean equal to the sum of the means and variance equal to the sum of the variances.

Example 4.30 If X and Y are two independent Poisson variables with parameters λ_x and λ_y respectively, show that $Z = X + Y$ is also Poisson distributed.

Solution The characteristic functions of X and Y are

$$\phi_X(\omega) = \exp\left(-\lambda_x[1 - j\omega]\right)$$
$$\phi_Y(\omega) = \exp\left(-\lambda_y[1 - j\omega]\right)$$

The characteristic function of Z is

$$\phi_Z(\omega) = \phi_X(\omega)\phi_Y(\omega) = \exp\left(-\lambda_x[1 - j\omega]\right)\exp\left(-\lambda_y[1 - j\omega]\right)$$
$$= \exp\left(-[\lambda_x + \lambda_y][1 - j\omega]\right)$$

The characteristic function of Z demonstrates the fact that Z is also Poisson distributed with a parameter equal to $\lambda_x + \lambda_y$.

When the random variable only takes positive, we may also use Laplace transforms. We may invoke the convolution property of Laplace transforms to obtain the pdf of the sum of two independent variables taking only positive values. If the Laplace transforms of X and Y are $L_X(s)$ and $L_Y(s)$, the Laplace transform of $Z = X + Y$ will be $L_Z(s)$ given as

$$L_Z(s) = L_X(s)L_Y(s) \tag{4.189}$$

The inverse Laplace can now be used to obtain the density of the sum (Shankar 2017).

A few properties of the Laplace transforms that are relevant and useful in finding the density of the sum of two random variables are given in Table 4.1 and Table 4.2. Laplace transforms of some commonly used densities are given in Table 4.3. All of these results were generated in Matlab.

Table 4.1 Properties of Laplace transforms	$g(t)$	$G(s) = \int_0^\infty g(t)e^{-st}dt$
	$b\,h(t) + a\,y(t)$	$b\,H(s) + a\,Y(s)$
	$y'(t)$	$sY(s) - y(0)$
	$y''(t)$	$s^2Y(s) - s\,y(0) - y'(0)$
	$y'''(t)$	$s^3Y(s) - s\,y'(0) - s^2y(0) - y''(0)$
	$\int_0^t h(\tau)y(t-\tau)d\tau$	$H(s)Y(s)$

Table 4.2 Laplace transforms of some known functions	$y(t)$	$Y(s) = \int_0^\infty y(t)e^{-st}dt$
	1	$\frac{1}{s}$
	$e^{a\,t}$	$-\frac{1}{a-s}$
	t^n	$\frac{\Gamma(n+1)}{s^{n+1}}$
	$\sin(a\,t)$	$\frac{a}{a^2+s^2}$
	$\cos(a\,t)$	$\frac{s}{a^2+s^2}$
	$e^{b\,t}\cos(a\,t)$	$-\frac{b-s}{a^2+(b-s)^2}$
	$e^{b\,t}\sin(a\,t)$	$\frac{a}{a^2+(b-s)^2}$
	$t^n\,e^{a\,t}$	$\frac{\Gamma(n+1)}{(s-a)^{n+1}}$
	$t^2\cos(a\,t)$	$\frac{8s^3}{(a^2+s^2)^3} - \frac{6s}{(a^2+s^2)^2}$
	$s > 0;\ a, b, c > s;\ n = 0, 1, 2, 3, ..$	

Table 4.3 Laplace transforms of some common densities (x > 0)	Probability Density Function	Laplace Transform
	$fX(x) = \frac{x^{m-1}e^{-\frac{mx}{a}}\left(\frac{m}{a}\right)^m}{\Gamma(m)}$	$\frac{m^m}{(m+a\,s)^m}$
	$fY(y) = \frac{x^{m-1}e^{-\frac{mx}{b}}\left(\frac{m}{b}\right)^m}{\Gamma(m)}$	$\frac{m^m}{(m+b\,s)^m}$
	$\int_0^z fX(x)fY(z-x)dx$	$\frac{m^{2m}}{(m+b\,s)^m(m+a\,s)^m}$
	$fW(w) = \frac{w^{n-1}e^{-\frac{nw}{b}}\left(\frac{n}{b}\right)^n}{\Gamma(n)}$	$\frac{n^n}{(n+b\,s)^n}$
	$\int_0^z fX(x)fW(z-x)dx$	$\frac{m^m n^n}{(n+b\,s)^n(m+a\,s)^m}$
	$fX(x) = \frac{xe^{-\frac{x^2}{2b^2}}}{b^2}$	$1 - \frac{\sqrt{2}bs\sqrt{\pi}e^{\frac{b^2s^2}{2}}\,\text{erfc}\left(\frac{\sqrt{2}bs}{2}\right)}{2}$

Example 4.31 Using Laplace transforms, obtain the density of the sum of two independent gamma variables (1) when they are identical (2) when they have different orders but have identical scaling factors, and (3) when they have identical orders, but different scaling factors.

Solution

Case (1)

$$f_X(x) = \frac{x^{m-1} e^{-\frac{x}{b}}}{b^m \Gamma(m)} \qquad \text{pdf of X}$$

$$f_Y(y) = \frac{y^{m-1} e^{-\frac{y}{b}}}{b^m \Gamma(m)} \qquad \text{pdf of Y}$$

$$L_X(s) = \frac{1}{(b\,s+1)^m} \qquad \text{Laplace transform of } f_X(x)$$

$$L_Y(s) = \frac{1}{(b\,s+1)^m} \qquad \text{Laplace transform of } f_Y(y)$$

$$L_Z(s) = \frac{1}{(b\,s+1)^{2m}} \qquad \text{Laplace transform of } f_Z(z) = L_X(s).L_Y(s)$$

$$f_Z(z) = \frac{z^{2m-1} e^{-\frac{z}{b}}}{b^{2m} \Gamma(2\,m)} \qquad pdf\,of\,Z = X + Y \;\Rightarrow G\,[2m, b]$$

Case (2)

Different orders: a and c

$$f_X(x) = \frac{x^{a-1} e^{-\frac{x}{b}}}{b^a \Gamma(a)} \qquad \text{pdf of X}$$

$$f_Y(y) = \frac{y^{c-1} e^{-\frac{y}{b}}}{b^c \Gamma(c)} \qquad \text{pdf of Y}$$

$$L_X(s) = \frac{1}{(b\,s+1)^a} \qquad \text{Laplace transform of } f_X(x)$$

$$L_Y(s) = \frac{1}{(b\,s+1)^c} \qquad \text{Laplace transform of } f_Y(y)$$

$$L_Z(s) = \frac{1}{(b\,s+1)^{a+c}} \qquad \text{Laplace transform of } f_Z(z) = L_X(s).L_Y(s)$$

$$f_Z(z) = \frac{z^{a+c-1} e^{-\frac{z}{b}}}{b^{a+c} \Gamma(a+c)} \qquad pdf\,of\,Z = X + Y \;\Rightarrow G\,[a+c, b]$$

Case (3)

Same order (m); different means

$$f_X(x) = \frac{x^{m-1}e^{-\frac{x}{a}}}{a^m \Gamma(m)}$$ 　　　pdf of X

$$f_Y(y) = \frac{y^{m-1}e^{-\frac{y}{b}}}{b^m \Gamma(m)}$$ 　　　pdf of Y

$$L_X(s) = \frac{1}{(a\,s+1)^m}$$ 　　　Laplace transform of $f_X(x)$

$$L_Y(s) = \frac{1}{(b\,s+1)^m}$$ 　　　Laplace transform of $f_Y(y)$

$$L_Z(s) = \frac{1}{((a\,s+1)(b\,s+1))^m}$$ 　　　Laplace transform of $f_z(z) = L_X(s).L_Y(s)$

$$f_Z(z) = \frac{z^{m-\frac{1}{2}}\sqrt{\pi}\,e^{-\frac{z\left(\frac{a}{2}+\frac{b}{2}\right)}{ab}}|a-b|^{\frac{1}{2}-m}I_{m-\frac{1}{2}}\left(\frac{z|a-b|}{2ab}\right)}{\sqrt{a}\sqrt{b}\Gamma(m)}$$

This pdf is the McKay density

While the Laplace or Fourier transforms allow us to obtain the density of the sum of two random variables, the Mellin transform may be used to obtain the density of the product or the ratio of two independent random variables (Epstein 1948; Erdelyi 1953; Springer and Thompson 1966; Mathai and Haubold 2008; Shankar 2017). The Mellin transform of a random variable X with the pdf of $f_X(x)$ is (Mathai and Haubold 2008; Shankar 2017)

$$M[f(x)] = M_X(s) = \int_0^\infty f(x)x^{s-1}dx \qquad (4.190)$$

The inverse Mellin transform gives the density as

$$f(x) = M^{-1}[M_X(s)] = \frac{1}{2\pi j}\int_{\lambda-j\infty}^{\lambda+j\infty} M_X(s)x^{-s}ds \qquad (4.191)$$

In Eq. (4.191), λ is a real number. While it is easy to obtain the Mellin transform of density using the integral representation in Eq. (4.190), inverse Mellin transformation is easily undertaken using the Table of Mellin transforms just as in the case of

Table 4.4 Properties of
Mellin transforms

$f(x)$	$M_f(s) = \int_0^\infty x^{s-1} f(x)\,dx$
$x^\nu f(x)$	$M_f(s + \nu)$
$f(bx)$	$b^{-s} M_f(s),\ b > 0$
$f(x^\rho)$	$\frac{1}{\rho} M_f\left(\frac{1}{\rho}\right),\ \rho > 0$
$\frac{d}{dx} f(x)$	$-(s-1) M_f(s-1)$
$\int_0^x f(w)\,dw$	$\frac{-1}{s} M_f(s+1)$
$A f(\alpha x) + B h(\beta x)$	$A\alpha^{-s} M_f(s) + B\beta^{-s} M_h(s)$

Laplace transforms. The properties of the Mellin transforms are given in Table 4.4.
These were generated in Matlab.

The convolution property of the Mellin transforms takes a form different from
that of the Laplace transforms. While convolution in the Laplace domain has the
integrand of the form $f_X(x) f_Y(z - x)$ with the limit of integration of 0 to z, the
"convolution integral" in the Mellin transform domain is (Mathai 1993)

$$f(w) = \int_0^\infty f_X(x) \frac{1}{x} f_Y\left(\frac{w}{x}\right) dx \qquad (4.192)$$

The convolution theorem in the Mellin domain is

$$M_W(s) = M_W[f(w)] = M_W\left[\int_0^\infty f_X(x) \frac{1}{x} f_Y\left(\frac{w}{x}\right) dx\right] = M_X(s) M_Y(s) \qquad (4.193)$$

Note that Eq. (4.193) gives the Mellin transform of the product of two independent random variables (each taking only positive values)

$$W = XY \qquad (4.194)$$

In other words, the pdf of W may be obtained by taking the inverse Mellin
transform of the product of the Mellin transforms of X and Y,

$$f_W(w) = M^{-1}[M_X(s) M_Y(s)] \qquad (4.195)$$

Mellin transforms are also useful in obtaining the density of the ratio of two
independent random variables U,

$$U = \frac{X}{Y} \qquad (4.196)$$

Using the concept of transformation of random variables,

$$f_U(u) = \int_0^\infty y f_X(uy) f_Y(y)\,dy \tag{4.197}$$

The Mellin convolution result now takes a different form

$$M_U(s) = M_U\left[f(u)\right] = M_W\left[\int_0^\infty y f_X(uy) f_Y(y)\,dy\right] = M_X(s)M_Y(2-s) \tag{4.198}$$

Thus, the density is obtained by taking the inverse Mellin transform of $M_U(s)$ as

$$f_U(u) = M^{-1}\left[M_X(s)M_Y(2-s)\right] \tag{4.199}$$

Mellin transforms of the some of the densities are given in Tables 4.5 and 4.6.

Example 4.32 Use Mellin transforms to verify the results of the following functions of two random variables:

1. Product of two independent and identically distributed gamma variables
2. Product of two independent and non-identical gamma variables

Table 4.5 Mellin transforms of some common densities

Probability Density Function	Mellin Transform
$\frac{x^{c-1}e^{-\frac{x}{b}}}{b^c\,\Gamma(c)}$	$\frac{b^{s-1}\Gamma(c+s-1)}{\Gamma(c)}$
$\frac{x^{m-1}e^{-\frac{mx}{b}}\left(\frac{m}{b}\right)^m}{\Gamma(m)}$	$\frac{b^{s-1}m^{1-s}\Gamma(m+s-1)}{\Gamma(m)}$
$\frac{e^{-\frac{x}{b}}}{b}$	$b^{s-1}\Gamma(s)$
e^{-x}	$\Gamma(s)$
$\frac{ax^{ac-1}e^{-\left(\frac{x}{b}\right)^a}}{b^{ac}\,\Gamma(c)}$	$\frac{b^{s-1}\Gamma\left(\frac{s+ac-1}{a}\right)}{\Gamma(c)}$
$\frac{xe^{-\frac{x^2}{2a^2}}}{a^2}$	$2^{\frac{s}{2}-1}a^{s-1}\Gamma\left(\frac{s}{2}+\frac{1}{2}\right)$

Table 4.6 Mellin transforms of densities containing modified Bessel functions

Probability Density Function	Mellin Transform
$\frac{2x^{c-1}K_0\left(\frac{2\sqrt{x}}{b}\right)}{b^{2c}\,\Gamma(c)^2}$	$\frac{b^{2\,s-2}\Gamma(c+s-1)^2}{\Gamma(c)^2}$
$\frac{2x^{\frac{a+b}{2}-1}K_{a-b}\left(\frac{2\sqrt{x}}{\sqrt{AB}}\right)}{\Gamma(a)\Gamma(b)(AB)^{\frac{a+b}{2}}}$	$\frac{A^{s-1}B^{s-1}\Gamma(a+s-1)\Gamma(b+s-1)}{\Gamma(a)\Gamma(b)}$
$\frac{zK_0\left(\left(\sqrt{\frac{1}{a^2}}\sqrt{\frac{z}{b^2}}\right)\right)}{a^2b^2}$	$2^{s-1}a^{s-1}b^{s-1}\Gamma\left(\frac{s}{2}+\frac{1}{2}\right)^2$
$\frac{2K_0\left(2\sqrt{\frac{z}{ab}}\right)}{ab}$	$a^{s-1}b^{s-1}\Gamma(s)^2$

Solution Case(1)

$$f_X(x) = \frac{x^{c-1}}{b^c \Gamma(c)} \exp\left(-\frac{x}{b}\right) U(x)$$

$$f_Y(y) = \frac{y^{c-1}}{b^c \Gamma(c)} \exp\left(-\frac{y}{b}\right) U(y)$$

If

$$Z = XY$$

$$f_Z(z) = \int_0^\infty \frac{1}{x} f_X(x) f_Y\left(\frac{z}{x}\right) dx = \frac{2}{b^{2c} \Gamma^2(c)} z^{c-1} K_0\left(\frac{2}{b}\sqrt{z}\right) U(z)$$

The Mellin transform of $f(x)$ is given by

$$M_X(s) = \int_0^\infty f(x) x^{s-1} dx = \int_0^\infty \frac{x^{c-1}}{b^c \Gamma(c)} \exp\left(-\frac{x}{b}\right) x^{s-1} dx = b^{s-1} \frac{\Gamma(c+s-1)}{\Gamma(c)}$$

X and Y being identical,

$$M_Y(s) = b^{s-1} \frac{\Gamma(c+s-1)}{\Gamma(c)}$$

The Mellin transform of $f(z)$ is

$$M_Z(s) = \int_0^\infty f_Z(z) z^{s-1} dz = \int_0^\infty \frac{2}{b^{2c} \Gamma^2(c)} z^{c-1} K_0\left(\frac{2}{b}\sqrt{z}\right) z^{s-1} dx = b^{2s-2} \frac{\Gamma^2(c+s-1)}{\Gamma^2(c)}$$

It can be seen that

$$M_Z(s) = M_X(s) M_Y(s)$$

This proves that Mellin transform of the density of the product of two independent gamma variables is equal to the product of the Mellin transforms of the individual densities as shown in Eq. (4.193).

Case (2)

$$f_X(x) = \frac{x^{a-1}}{A^a \Gamma(a)} \exp\left(-\frac{x}{A}\right) U(x)$$

$$f_Y(y) = \frac{y^{b-1}}{B^b \Gamma(b)} \exp\left(-\frac{y}{B}\right) U(y)$$

If

$$Z = XY$$

$$f_Z(z) = \int_0^\infty \frac{1}{x} f_X(x) f_Y\left(\frac{z}{x}\right) dx = \frac{2}{(\sqrt{AB})^{a+b} \Gamma(a)\Gamma(b)} z^{\frac{a+b}{2}-1} K_{a-b}\left(2\sqrt{\frac{z}{AB}}\right) U(z)$$

The Mellin transform of $f(x)$ is given by

$$M_X(s) = \int_0^\infty f(x) x^{s-1} dx = A^{s-1} \frac{\Gamma(s+a-1)}{\Gamma(a)}$$

X and Y being identical,

$$M_Y(s) = B^{s-1} \frac{\Gamma(s+b-1)}{\Gamma(b)}$$

The Mellin transform of $f(z)$ is

$$M_Z(s) = \int_0^\infty f_Z(z) z^{s-1} dz = A^{s-1} B^{s-1} b^{2s-2} \frac{\Gamma(a+s-1)\Gamma(b+s-1)}{\Gamma(a)\Gamma(b)}$$

It can be seen that

$$M_Z(s) = M_X(s) M_Y(s)$$

Example 4.33 For the case of two non-identical independent gamma variables X and Y, use Mellin transforms to verify the result in Eq. (4.198), namely, the ratio of two gamma variables.

Solution If W is the ratio of X to Y, we have

$$W = \frac{X}{Y}$$

$$f_W(w) = \int_0^\infty y f_X(wy) f_Y(y) dy = \int_0^\infty y \frac{(wy)^{a-1}}{A^a \Gamma(a)} \exp\left(-\frac{wy}{A}\right) \frac{y^{b-1}}{B^b \Gamma(b)} \exp\left(-\frac{y}{B}\right) dy$$

The density of W becomes

$$f_W(w) = \frac{\Gamma(a+b)}{\Gamma(a)\Gamma(b)} B^a A^b \frac{w^{a-1}}{(A+wB)^{a+b}} U(w)$$

The Mellin transform of $f(w)$ is

$$M_W(s) = \int_0^\infty f_W(w)w^{s-1}dw = B^{1-s}A^{s-1}\frac{\Gamma(a+s-1)\Gamma(1+b-s)}{\Gamma(a)\Gamma(b)}$$

Note that

$$M_Y(s) = B^{s-1}\frac{\Gamma(s+b-1)}{\Gamma(b)}$$

Therefore,

$$M_Y(2-s) = B^{1-s}\frac{\Gamma(b-s+1)}{\Gamma(b)}$$

It can now be seen that

$$M_W(s) = M_x(s)M_Y(2-s)$$

This proves the result in Eq. (4.198).

Additional example on the use of Laplace and Mellin transforms is given the next section where a discussion on multiple random variables is presented.

4.8 Multiple Random Variables

The concepts associated with a pair of random variables can be extended to the case of n random variables, $X_1, X_2,...,X_n$. The cumulative distribution function of n-random variables is

$$F_{X_1,X_2,...,X_n}(x_1,x_2,\ldots,x_n) = P(X_1 \le x_1, X_2 \le x_2, \ldots, X_n \le x_n) \qquad (4.200)$$

The joint density function

$$f_{X_1,X_2,...,X_n}(x_1,x_2,\ldots,x_n) = \frac{\partial^n}{\partial x_1 \partial x_2 \ldots \partial x_n} F_{X_1,X_2,...,X_n}(x_1,x_2,\ldots,x_n) \qquad (4.201)$$

The random variables are independent if the joint density can be expressed as the product of marginal density functions. The marginal density of any one of the variables is obtained by integrating the joint density w.r.t. the remaining variables.

While the field of multiple random variables is rather vast, we will limit ourselves to specific cases of interest that are pertinent to the undergraduate curriculum. We explore the examples of normalized sum of the random variables or the arithmetic mean, normalized product of the random variables, or the geometric mean and order statistics (maximum of a set of random variables, minimum of a set of random variables or the kth largest of a set of n random variables $k < n$). The analysis of the sums and products will lead us to the central limit theorem, while order statistics provides us with insight into the reliability of interconnected systems.

4.8.1 Order Statistics

Before we examine the general concepts of order statistics, let us examine two extreme cases of order statistics, namely, the maximum of a set of random variables and minimum of a set of random variables. For this analysis, we assume that the random variables are independent. If X_k, $k = 1,2,...,n$ represent n-independent random variables, each with a density of $f(x_k)$, $k = 1,2,..,n$.

If W is the maximum of the set, we have

$$W = \max \{X_1, X_2, .., X_n\} \tag{4.202}$$

Using the same logic invoked in obtaining the maximum of two random variables, the CDF of W will be

$$\begin{aligned} F_W(w) = P\{W \leq w\} &= P\{X_1 \leq w, X_2 \leq w, \cdots, X_n \leq w\} \\ &= P\{X_1 \leq w\}P\{X_2 \leq w\}\cdots P\{X_n \leq w\} \end{aligned} \tag{4.203}$$

$$F_W(w) = F_{X_1}(w)F_{X_2}(w)\cdots F_{X_n}(w) \tag{4.204}$$

If the random variables are also identical, Eq. (4.204) becomes

$$F_W(w) = [F_X(w)]^n \tag{4.205}$$

The pdf becomes

$$f_W(w) = n f_X(w)[F_X(w)]^{n-1} \tag{4.206}$$

If Z is the minimum of the set, we have

$$Z = \min \{X_1, X_2, .., X_n\} \tag{4.207}$$

While it is a simple task to extend the concepts of the maximum of two variables to maximum on n-variables, extending the concept of the minimum of two variables to the minimum of n-variables is less straightforward because it requires the use of total probability. Instead, we make use of the principle of the complementary events. If Z is the minimum of the set, the CDF is obtained by examining the probability associated with its complementary event, $Z > z$. This means that

$$P(Z < z) = 1 - P(Z > z) = 1 - P(X_1 > z, X_2 > z, \ldots, X_n > z) \qquad (4.208)$$

$$F_Z(z) = 1 - [P\{X_1 > z\}P\{X_2 > z\}\cdots P\{X_n > z\}] \qquad (4.209)$$

Treating the random variables as identically distributed, we have

$$F_Z(z) = 1 - [1 - F_X(z)]^n \qquad (4.210)$$

The pdf is obtained by differentiating the CDF in Eq. (4.210) as

$$f_Z(z) = nf_X(z)[1 - F_X(z)]^{n-1} \qquad (4.211)$$

We can now obtain the pdf of the kth largest variable. We assume that the variables are independent and identically distributed. Let Y_k be the kth largest variable. If $f_k(y)$ is the pdf of the variable Y_k, we can write

$$f_k(y)dy = P\{y \leq Y_k \leq y + dy\} \qquad (4.212)$$

Let us sort the variables in ascending order. The event $\{y \leq Y_k \leq y + dy\}$ occurs *iff*

- Exactly $(k-1)$ variables are less than y.
- One variable is the interval $\{y, y + dy\}$.
- There are $(n-k)$ variables greater than y.

If we identify the events above as A and B and C, respectively, they are

$$\begin{aligned} A &= (X \leq y) \\ B &= (y \leq X \leq y + dy) \\ C &= (X > y + dy) \end{aligned} \qquad (4.213)$$

It is easily seen that these are mutually exclusive events. The probability of each of these events is

$$\begin{aligned} \text{Prob}(A) &= F_X(y) \\ \text{Prob}(B) &= f_X(y)dy \\ \text{Prob}(C) &= 1 - F_X(y) \end{aligned} \qquad (4.214)$$

Note that the event A occurs $(k-1)$ times, event B occurs *just once* and the event C occurs $(n-k)$ times. This means that we can group these three events into a generalized Bernoulli trial with one occurring $(k-1)$ times, one (the chosen variable lies between y and $y + \Delta y$) occurring just once and the other one occurring $(n-k)$ times. Using the concept of generalized Bernoulli trial, Eq. (4.212) becomes

$$f_k(y)dy = \frac{n!}{1!(k-1)!(n-k)!}[F_X(y)]^{k-1}f_X(y)dy[1-F_X(y)]^{n-k} \qquad (4.215)$$

Equation (4.215) simplifies to

$$f_k(y) = \frac{n!}{(k-1)!(n-k)!}[F_X(y)]^{k-1}[1-F_X(y)]^{n-k}f_X(y) \qquad (4.216)$$

When k equals n, we get the pdf of the largest variable (maximum), and Eq. (4.216) becomes Eq. (4.206). If $k = 1$, we get the pdf of the smallest of the random variable (minimum) given in Eq. (4.211).

Example 4.34 If X_k, $k = 1,2,3$ are independent and identically distributed variables taking only positive values, obtain the probability that $X_1 < X_2 < X_3$. Generalize the result to the case of n $(n > 3)$ independent and identically distributed random variables.

Solution Since all the variables are identical, we have

$$f_{X_i}(x_i) = f(x), x > 0$$
$$F_{X_i}(x_i) = F(x), x > 0$$

$$P(X_1 < X_2 < X_3) = \int_0^\infty f(x_3) \int_0^{x_3} f(x_2) \int_0^{x_2} f(x_1)dx_1 dx_2 dx_3$$

$$P(X_1 < X_2 < X_3) = \int_0^\infty f(x_3) \int_0^{x_3} f(x_2)F(x_2)dx_2 dx_3$$

$$\int_0^{x_3} f(x_2)F(x_2)dx_2 = \int_0^{x_3} F(x_2)dF(x_2) = \frac{1}{2}F^2(x_3)$$

$$P(X_1 < X_2 < X_3) = \int_0^\infty \frac{1}{2}f(x_3)F^2(x_3)dx_3 = \int_0^\infty \frac{1}{2}f(x)F^2(x)dx = \frac{1}{2}\int_0^\infty F^2(x)dF(x)$$

$$= \frac{1}{2}\frac{1}{3} = \frac{1}{3!}$$

In general,

$$P(X_1 < X_2 < X_3 < \ldots < X_n) = \frac{1}{n!} = P(X_1 > X_2 > X_3 > \ldots > X_n)$$

4.8.2   Central Limit Theorem for the Sum

For the case of two independent random variables, the density of the sum was shown to be the convolution of the two marginal densities. If we extend this concept to a number of variables, we can obtain an expression for the density of the sum of n independent random variables. We may also choose the arithmetic mean instead of the sum. If Y is the arithmetic mean (sum of the samples divided by the number of samples), we have

$$Y = \frac{1}{n}\sum_{k=1}^{n} X_k = \sum_{k=1}^{n} \frac{X_k}{n} = \sum_{k=1}^{n} R_k \qquad (4.217)$$

In Eq. (4.217), each of the R's constitute a scaled version of X's, each one scaled by $1/n$. The density function of Y may be written as

$$f_Y(y) = f_{R_1}(r_1) * f_{R_2}(r_2) * \ldots * f_{R_n}(r_n) \qquad (4.218)$$

In Eq. (4.218), * represents the convolution. Under certain conditions, the density of the sum may be obtained using the central limit theorem.

If the continuous random variables are independent and none of the variables has a variance of infinity, the density of the sum approaches a Gaussian density when n becomes large (Rohatgi and Saleh 2001; Papoulis and Pillai 2002; Rodriguez-Lopez and Carrasquillo 2006; Price and Zhang 2007; Yates and Goodman 2014). In other words, we can express

$$f_Y(y) = \frac{1}{\sqrt{2\pi\sigma^2}} \exp\left[-\frac{(y-\mu)^2}{2\sigma^2}\right] \qquad (4.219)$$

In Eq. (4.219),

$$\mu = \sum_{i=1}^{n} E(R_i)$$
$$\sigma^2 = \sum_{i=1}^{n} \text{var}(R_i) \qquad (4.220)$$

The number of variables necessary to reach the Gaussian approximation will depend on the forms of the densities of the random variables (Fante 2001; Kwak and Kim 2017). It is obvious that the central limit theorem is not applicable if any one of the random variables has a Cauchy density because the Cauchy variable has infinite variance (Papoulis and Pillai 2002; Devore 2004).

If the characteristic function is available, the characteristic function of Z will be

$$\phi_Y(\omega) = \phi_{R_1}(\omega)\phi_{R_2}(\omega)\dots\phi_{R_n}(\omega) \tag{4.221}$$

If the Laplace transforms are available, it is possible to get the Laplace transform of Y as the product of the marginal Laplace transforms. The density of Y can be obtained using inverse transformation (Fourier domain if characteristic function is used and Laplace domain if Laplace transform is used).

Using the transformation of random variables, the pdf of any one of the variables on the right hand side of (4.217) becomes

$$f_R(r) = \left.\frac{f_X(x)}{\left|\frac{dR}{dx}\right|}\right|_{x=nr} = nf_X(rn) \tag{4.222}$$

In Eq. (4.222), $f_X(x)$ is the probability density function (pdf) of any one of the X's in Eq. (4.222). We will use Laplace transforms to obtain the density of Y under the assumption that the densities only exist for positive values. The Laplace transform of the density of Y is (Shankar 2017)

$$L_Y(s) = [L_R(s)]^n. \tag{4.223}$$

In Eq. (4.223), we have

$$L_R(s) = nL[\, f_x(nr)] \tag{4.224}$$

Using the scaling property of Laplace transforms,

$$L[\, f_x(nr)] = \frac{1}{n}L_X\left(\frac{s}{n}\right) \tag{4.225}$$

These manipulations lead us to

$$L_R(s) = n\frac{1}{n}L_X\left(\frac{s}{n}\right) = L_X\left(\frac{s}{n}\right) \tag{4.226}$$

$$L_Y(s) = \left[L_X\left(\frac{s}{n}\right)\right]^n \tag{4.227}$$

We can take the inverse Laplace transform of Eq. (4.227) and obtain the density of Y.

For the case when X is $U[a,b]$, with a and b are positive, the Laplace transform of the pdf is

$$L_X(s) = \frac{\exp(-as) - \exp(-bs)}{s(b-a)} \tag{4.228}$$

This leads to the Laplace transform of Y as

$$L_Y(s) = \frac{\left[\exp\left(-\frac{a}{n}s\right) - \exp\left(-\frac{b}{n}s\right)\right]^n}{\left[s\left(\frac{b}{n} - \frac{a}{n}\right)\right]^n} \tag{4.229}$$

The inverse transform of Eq. (4.229) can be performed in Matlab (symbolic toolbox), and the pdf is represented as a sum of signum functions. Choosing $a = 0$ and $b = 1$, the density of Y has been obtained and the densities are plotted for several values of n and shown in Fig. 4.8. It can be seen that the densities of the arithmetic mean appears to be more Gaussian like as n increases (Shankar 2017).

We can also examine what happens when X is a gamma distributed,

$$f_X(x) = G(a,b) = \frac{x^{a-1}}{b^a \Gamma(a)} \exp\left(\frac{x}{b}\right), x > 0, a > 0, b > 0 \tag{4.230}$$

Using the Laplace transform relationships, the density of the arithmetic mean Y becomes (Shankar 2017)

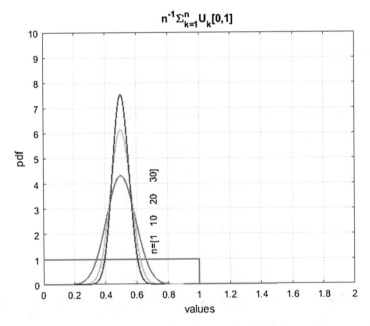

Fig. 4.8 The densities of the sample mean for $n = 10$, 20 and 30. For comparison, the density of U [0,1] corresponding to $n = 1$ is also shown

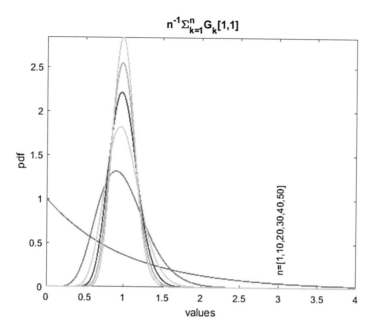

Fig. 4.9 The densities of the sample mean for $n = 10, 20, 30, 40,$ and 50. For comparison, the density of $G[1,1]$ corresponding to $n = 1$ is also shown

$$f_Y(y) = \frac{x^{an-1}}{\left(\frac{b}{n}\right)^{an}\Gamma(an)} \exp\left(\frac{n}{b}x\right) \equiv G\left(an, \frac{b}{n}\right) \qquad (4.231)$$

Equation (4.231) shows that the density function of the sample mean is also gamma density of order *an* and scaling factor (b/n). Note that when $G(a,b)$ becomes an exponential density when $a = 1$. Figure 4.9 shows the density of the arithmetic mean for $G(1,1)$ or an exponential density of mean $= 1$.

It is apparent that the pdf of the arithmetic mean approaches a Gaussian shape at a rate slower than the transition seen with the case of a uniform density. To see how close the densities match a Gaussian, the mean square error is estimated for each of these curves (density of the sample mean and the corresponding Gaussian fit). The mean square error (MSE) is defined as

$$\text{MSE} = \frac{1}{M} \sum_{k=1}^{M} [f(y_k) - f(g_k)]^2 \qquad (4.232)$$

In Eq. (4.232), $f(g)$ is the equivalent normal fit and M is the number of samples of the densities used.

The MSE values are given below:

$n = [10, 20, 30, 40, 50] \rightarrow \text{MSE} = [0.0041 \ 0.0028 \ 0.0022 \ 0.0019 \ 0.0017].$

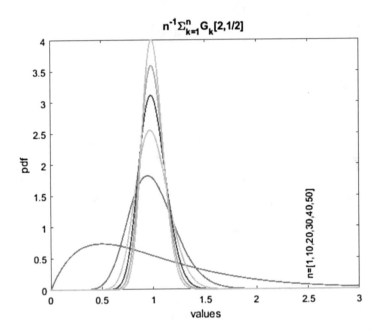

Fig. 4.10 The densities of the sample mean for $n = 10, 20, 30, 40,$ and 50. For comparison, the density of $G[2,1/2]$ corresponding to $n = 1$ is also shown

Figure 4.10 shows the densities for $G[2,1/2]$.

It is apparent that the pdf of the arithmetic mean approaches a Gaussian shape at a rate faster than the transition seen with the case of the exponential density. To see how close the densities match a Gaussian, the mean square error is estimated for each of these curves (density of the arithmetic mean and the corresponding Gaussian fit). The values are given below:

$n = [10, 20, 30, 40, 50] \rightarrow \text{MSE} = [0.0028 \ 0.0019 \ 0.0015 \ 0.0013 \ 0.0012]$.

Figure 4.11 shows the densities for $G[10,1/10]$ which is much more symmetric than $G[2,1/2]$.

It is apparent that the pdf of the arithmetic mean approaches a Gaussian shape at a rate faster than the transition seen with the case of the exponential and the gamma density of order 2. To see how close the densities match a Gaussian, the mean square error is estimated for each of these curves (density of the arithmetic mean and the corresponding Gaussian fit). The values are given below:

$n = [10, 20, 30, 40, 50] \rightarrow \text{MSE} = [0.0012 \ 0.0008 \ 0.0007 \ 0.0006 \ 0.0005]$.

Comparing the MSE values, it is clear that Gaussian fit is closer with the gamma density of order 10 compared to the exponential density. Note that the transition to Gaussian takes place faster with the symmetric density of $U[0,1]$.

While MSE provides a measure of match to a Gaussian, a more appropriate approach would require the use hypothesis testing described in Chap. 5.

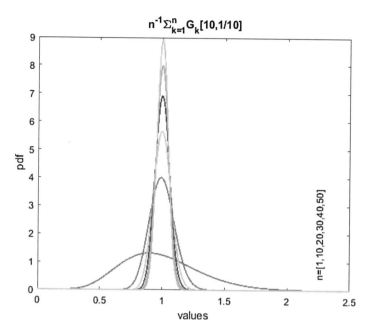

Fig. 4.11 The densities of the sample mean for $n = 10, 20, 30, 40$, and 50. For comparison, the density of $G[10,1/10]$ corresponding to $n = 1$ is also shown

While the central limit theorem is useful in engineering applications allowing the ease of obtaining the density of the sum of a number of independent variables, a similar version of the central limit theorem can be presented for the case of the product of several independent variables. To differentiate these two, the central limit theorem presented above is identified the central limit theorem for **sums**, while the one to be presented next will be identified as the central limit theorem for **products** (Papoulis and Pillai 2002).

4.8.3   Central Limit Theorem for the Products

Let us now examine the CLT for the products. In this case, the random variables X's are all identical and positive. Let us define the product W as

$$W = \prod_{k=1}^{n} X_k, \quad X_k > 0 \tag{4.233}$$

Let us take the logarithm and define a new random variable Z as

$$Z = \log_e(W) = \sum_{k=1}^{n} \log_e(X_k), \quad X_k > 0 \tag{4.234}$$

When n becomes large, the density of Z in Eq. (4.234) will be Gaussian. This statement constitutes the CLT for the products. It is clear that CLT for products follows the CLT for sums following the step of taking the logarithm. Just as we used the sample mean for the case of CLT for sums, we will use the geometric mean, and Eq. (4.234) becomes

$$Z = \left(\frac{1}{n}\right) \log_e(W) = \frac{1}{n}\sum_{k=1}^{n} \log_e(X_k), \quad X_k > 0 \tag{4.235}$$

Note that Z in Eq. (4.235) is a scaled version of Z in Eq. (4.234), and scaling will not impact the nature of the density of Z. The mean and variance of the ensuing Gaussian density will be different. For the remaining analysis, we will use Z in Eq. (4.235). It is clear that if Z is Gaussian, density of W will be lognormal as seen in Chap. 3. Let us see if we can obtain an analytical expression for the density of Z (Papoulis and Pillai 2002).

Let us assume that X's are gamma distributed. The Mellin transforms presented earlier offers an opportunity to obtain the densities of products of gamma, exponential, Rayleigh, and a few other types of variables (Springer and Thompson 1966; Mathai 1993). We will limit ourselves to the case of independent and identically distributed gamma variables.

Instead of deriving the density of the sum of the products of gamma random variables, we will work in reverse and start with the Meijer G function and prove that the densities expressed using Meijer G functions constitute the density of the product of gamma variables (Springer and Thompson 1966; Shankar 2017).

We will look at the following density function, $f(z)$ expressed as (Shankar 2017)

$$f(w) = \frac{1}{w\Gamma(a)^n} G_{0,n}^{n,0}\left(\frac{w}{b^n}\left|\begin{array}{c}-\\\underbrace{a, a, .., a}_{N-terms}\end{array}\right.\right) U(w) \tag{4.236}$$

In Eq. (4.236), $G_{0,n}^{n,0}(.)$ is the Meijer G function and $n = 1, 2, 3, \ldots$. Let choose $n = 1$ in Eq. (4.236) and write the new density as

$$f_1(w) = \frac{1}{w\Gamma(a)} G_{0,1}^{1,0}\left(\frac{w}{b}\left|\begin{array}{c}-\\a\end{array}\right.\right) U(w) \tag{4.237}$$

Let us obtain the Mellin transform of the density in Eq. (4.237),

$$M_1(s) = \int_0^\infty \frac{w^{s-1}}{w\Gamma(a)} G_{0,1}^{1,0}\left(\frac{w}{b}\left|\begin{array}{c}-\\a\end{array}\right.\right) dw \tag{4.238}$$

This integral can be evaluated in symbolic toolbox in Matlab, and we have

$$M_1(s) = b^{s-1} \frac{\Gamma(a+s-1)}{\Gamma(a)} \qquad (4.239)$$

Let us now put $n = 2$ in Eq. (4.236), and we have

$$f_2(w) = \frac{1}{w\Gamma(a)^2} G_{0,2}^{2,0}\left(\frac{w}{b^2} \middle| \begin{matrix} - \\ a, a \end{matrix}\right) U(w) \qquad (4.240)$$

The Mellin transform of the density in Eq. (4.240) is

$$M_2(s) = b^{2s-2} \frac{\Gamma^2(a+s-1)}{\Gamma^2(a)} = \left(b^{s-1} \frac{\Gamma(a+s-1)}{\Gamma(a)}\right)^2 \qquad (4.241)$$

The right hand side of Eq. (4.241) represents the square of $M_1(s)$ where

$$M_2(s) = M_1(s)M_1(s) \qquad (4.242)$$

In other words, Eq. (4.242) represents the Mellin transform of the product of two gamma densities that are independent and identically distributed and each with a marginal density given in Eq. (4.230).

Let us now put $n = 3$ in Eq. (4.236), and we have

$$f_3(w) = \frac{1}{w\Gamma(a)^3} G_{0,3}^{3,0}\left(\frac{w}{b^3} \middle| \begin{matrix} - \\ a, a, a \end{matrix}\right) U(w) \qquad (4.243)$$

Taking the Mellin transform of Eq. (4.243), we get

$$M_3(s) = \left(b^{s-1} \frac{\Gamma(a+s-1)}{\Gamma(a)}\right)^3 \qquad (4.244)$$

It can easily be seen that the density in Eq. (4.243) results from the product of three independent and identically distributed gamma variables.

Thus, the density expressed in Eq. (4.236) represents the density of the product of n independent and identically distributed random variables, and Mellin transforms offer a means to easily verify this property. Now that we have verified that the density in Eq. (4.236) represents the density of the products of n independent and identically distributed gamma variables, the density of Z in Eq. (4.235) may be obtained using transformation of variables as

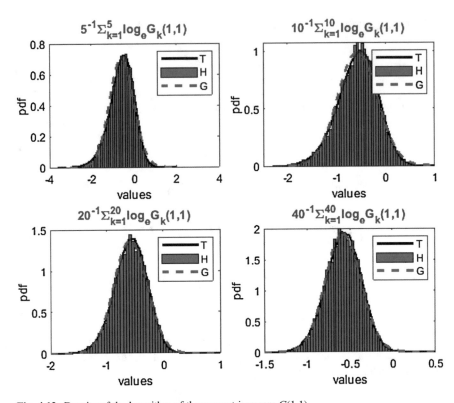

Fig. 4.12 Density of the logarithm of the geometric mean: $G(1,1)$

$$f_Z(z) = \frac{n}{[\Gamma(a)]^n} \, G_{0,n}^{n,0}\left(\frac{\exp{(zn)}}{b^n} \, \Bigg|_{\underbrace{a,a,..,a}_{n}}^{-} \right), \quad z > 0 \qquad (4.245)$$

Figure 4.12 shows the densities obtained by plotting Eq. (4.245) for $G(1,1)$.

For each value of n, an equivalent Gaussian fit is plotted along with a histogram of 10,000 samples of the data. The data set was used to estimate the mean and standard deviation of the normal density to obtain the Gaussian fit. It is seen that as n increases, the theoretical pdf plots and the Gaussian fits become closer.

Figure 4.13 shows the densities for $G(2,1/2)$. The densities appear to move faster to the normal because of the decreased levels of skewness of $G(2,1/2)$ compared to G $(1,1)$.

Thus, CLT for sum leads to a normal density for the sample mean, while the CLT for products results in a lognormal density for the products of positive random variables as the number of independent variables goes up.

The product model is used in the justification of the lognormal density for shadowing in wireless systems, while the additive model is used in the justification for the Rayleigh fading arising from multipath phenomenon in wireless systems.

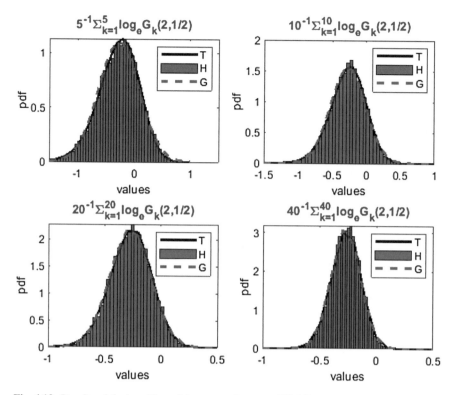

Fig. 4.13 Density of the logarithm of the geometric mean: $G(2,1/2)$

4.8.4   Densities of the Sum of the Squares of Normal Variables

While the central limit theorem offers a means to approximate the density of the sum of a number of independent random variables, there are a couple of additional densities of sum of random variables that are of interest to the engineering community.

If X_i, $i = 1,2,\ldots,n$ consist of n independent and identically distributed Gaussian variables of zero mean and unit variance, represented as $N(0,1)$, the density of the sum of the squares is identified as the chi squared density.

$$Z = \sum_{i=1}^{n} X_i^2 \tag{4.246}$$

$$f_Z(z) = \frac{z^{\left(\frac{n}{2}\right)-1}}{2^{\left(\frac{n}{2}\right)}\Gamma\left(\frac{n}{2}\right)} \exp\left(-\frac{z}{2}\right) U(z) \tag{4.247}$$

Equation (4.247) is identified as the chi-squared distribution with n degrees of freedom. Chi-squared (or chi square) distribution is used in hypothesis testing (Chap. 5). It can be easily seen that for $m = n/2$ where m is larger than 0, Eq. (4.247) is a gamma density. If m takes integer values that are even (starting with 2), (Eq. (4.247)) is identified as the Erlang density (Papoulis and Pillai 2002).

While Eq. (4.247) represents the density of the sum of squares of zero mean Gaussian variables of unit variance, the non-central chi square density represents the density of the sum of the squares of independent identically distributed Gaussian variables, each of which is $N(\delta, 1)$. The expression for the non-central chi square density is

$$f_Z(z) = \frac{1}{2}\left(\frac{z}{\lambda}\right)^{\left(\frac{n-2}{4}\right)} \exp\left(-\frac{z+\lambda}{2}\right) I_{\frac{n}{2}-1}\left(\sqrt{z\lambda}\right) U(z) \qquad (4.248)$$

In Eq. (4.248), I(.) is the modified Bessel function of the first kind (Gradshteyn and Ryzhik 2000) and

$$\lambda = n\delta \qquad (4.249)$$

We may explore some additional interesting results. For any two random variables X_1 and X_2,

$$E(X_1 \pm X_2) = E(X_1) \pm E(X_2) \qquad (4.250)$$

Equation (4.250) is true regardless of whether the variables are independent or not.

$$\text{var}(X_1 \pm X_2) = \text{var}(X_1) + \text{var}(X_2) \pm 2Cov(X_1 X_2) \qquad (4.251)$$

If X_1 and X_2 are independent,

$$\text{var}\left(X_1 \pm X_2\right) = \text{var}\left(X_1\right) + \text{var}\left(X_2\right) \qquad (4.252)$$

Equation (4.252)is also true if X_1 and X_2 are uncorrelated.
In general,

$$E\left(\sum_{k=1}^{n} X_k\right) = \sum_{k=1}^{n} E(X_k) \qquad (4.253)$$

$$\text{var}\left(\sum_{k=1}^{n} X_k\right) = \sum_{k=1}^{n} \text{var}(X_k) + 2\sum_{i<j} Cov\left(X_i X_j\right) \qquad (4.254)$$

4.9 Summary

We examined the properties of multiple random variables with emphasis on two random variables. Drawing on information from Chap. 2 on joint probabilities, it is possible to understand the concept of marginal, joint, and conditional densities. Engineering applications rely on modeling the outcomes using the concept of two or more variables and any data collected needs to be studied for analysis, processing of the data represented by two or more random variables becomes important. In this context, the concepts of transformation of a single variable introduced in Chap. 3 are restructured to operate on two random variables. Utilizing the notion of point conditioning of one variable, we can obtain the density of the function of two random variables. The density of the function of two random variables may also be obtained using graphical approaches which may require the use of Leibniz rule in one and two dimensions. The joint density of two functions of two random variables can be obtained relying on the Jacobian of the transformation, and this approach may even be used when interest exists only in finding the density a single function of the two random variables.

The density of the sum of two independent random variables can easily be obtained through the convolution principle.

The concepts of characteristic functions, Laplace transforms, and Mellin transforms are invoked to obtain the densities of the sums and products of independent variables. The transforms are also used for exploring the central limit theorem. A key application of multiple random variables is the study of order statistics allowing us to obtain the density of kth largest outcome in an experiment.

A number of important transformations and the equations necessary to obtain the densities are summarized. The Leibniz rule involving the differentiation of integrals is also given.

4.9.1 Density of a Function of Two or More Independent Variables

1. Density of $Z = X + Y$

$$f_Z(z) = \begin{cases} \int_0^z f_Y(z-x)f_X(x)dx, & x \geq 0, y \geq 0; z \geq 0 \\ 0, & x \geq 0, y \geq 0; z < 0 \end{cases} \tag{4.255}$$

$$f_Z(z) = \int_{-\infty}^{\infty} f_Y(z-x)f_X(x)dx, \quad -\infty \leq x \leq \infty, -\infty \leq y \leq \infty; -\infty \leq z \leq \infty \tag{4.256}$$

2. Density of $W = X - Y$

$$f_W(w) = \int_{-\infty}^{\infty} f_X(w+y)f_Y(y)\,dy, \quad -\infty \le x \le \infty, -\infty \le y \le \infty; -\infty \le w \le \infty$$

$$(4.257)$$

$$f_W(w) = \begin{cases} \int_0^{\infty} f_X(w+y)f_Y(y)\,dy, & x \ge 0, y \ge 0; w \ge 0 \\[2mm] \int_{-z}^{\infty} f_X(w+y)f_Y(y)\,dy, & x \ge 0, y \ge 0; w < 0 \end{cases}$$

$$(4.258)$$

3. Density of $W = \frac{X}{Y}$

$$f_W(w) = \int_{-\infty}^{\infty} |y| f_X(wy)f_Y(y)\,dy, \quad -\infty \le x \le \infty, -\infty \le y \le \infty; -\infty \le w \le \infty$$

$$(4.259)$$

$$\int_0^{\infty} y f_X(wy)f_Y(y)\,dy, \quad x \ge 0, y \ge 0; w \ge 0 \qquad (4.260)$$

4. Density of $V = XY$

$$f_V(v) = \int_{-\infty}^{\infty} \frac{1}{|y|} f_X\left(\frac{v}{y}\right) f_Y(y)\,dy, \quad -\infty \le x \le \infty, -\infty \le y \le \infty; -\infty \le v \le \infty$$

$$(4.261)$$

$$f_V(v) = \int_0^{\infty} \frac{1}{y} f_X\left(\frac{v}{y}\right) f_Y(y)\,dy, \quad x \ge 0, y \ge 0; v \ge 0 \qquad (4.262)$$

5. Density of $Z = \max(X, Y)$

$$f_Z(z) = F_X(z)f_Y(z) + F_Y(z)f_x(z) \qquad (4.263)$$

6. Density of $Z = \max(X_1, X_2, \cdots, X_n), \quad f_{X_i}(x_i) = f_X(x), \; i = 1, 2, \cdots, n$

All variables are independent.

$$f_Z(z) = n[F_X(z)]^{n-1} f_X(z) \qquad (4.264)$$

7. Density of $W = \min(X, Y)$

$$f_W(w) = f_x(w) + f_y(w) - F_x(w)f_y(w) - F_y(w)f_x(w) \qquad (4.265)$$

8. Density of $W = \min(X_1, X_2, \cdots, X_n)$, $f_{X_i}(x_i) = f_X(x)$, $i = 1, 2, \cdots, n$

All variables are independent.

$$f_W(w) = n[1 - F_X(w)]^{n-1} f_W(w) \qquad (4.266)$$

Special cases (Gaussian Variables)

9. Density of $Z = X^2 + Y^2$

$$f_Z(z) = \int_{-\sqrt{z}}^{\sqrt{z}} \frac{1}{2\sqrt{z - y^2}} \left[f_X\left(\sqrt{z - y^2}\right) + f_X\left(-\sqrt{z - y^2}\right) \right] f_Y(y)\, dy \qquad (4.267)$$

10. Density of $Z = \sqrt{X^2 + Y^2}$

$$f_Z(z) = \int_{-z}^{z} \frac{z}{\sqrt{z^2 - x^2}} \left[f_y\left(\sqrt{z^2 - x^2}\right) + f_y\left(-\sqrt{z^2 - x^2}\right) \right] f_X(x)\, dx \qquad (4.268)$$

11. Density of $Z = \sqrt{X^2 + Y^2}$ graphically

$$F_Z(z) = P\left[\sqrt{X^2 + Y^2} < z \right] = \iint\limits_{\text{shaded area}} f(x, y)\, dx\, dy \qquad (4.269)$$

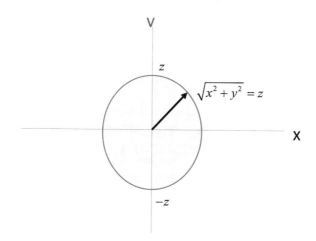

Note that $x^2 + y^2 \leq z^2$ represents the shaded area of radius z. The range of the shaded area is $-\sqrt{z^2 - y^2} < x < \sqrt{z^2 - y^2}$ and $-z < x < z$. This means that the CDF is

$$F_Z(z) = \iint\limits_{\text{shaded area}} f(x,y)dxdy = \int\limits_{y=-z}^{z} \int\limits_{x=-\sqrt{z^2-y^2}}^{\sqrt{z^2-y^2}} f(x,y)dxdy \qquad (4.270)$$

The density function is obtained by differentiating

$$f_Z(z) = \frac{d}{dz}F_Z(z) = \frac{d}{dz}\left[\int\limits_{y=-z}^{z} \int\limits_{x=-\sqrt{z^2-y^2}}^{\sqrt{z^2-y^2}} f(x,y)dxdy\right] \qquad (4.271)$$

Applying Leibniz theorem for the double integral,

$$f_Z(z) = \int\limits_{y=-z}^{z} \frac{z}{\sqrt{z^2 - y^2}}\left[f\left(\sqrt{z^2 - y^2}, y\right) + f\left(-\sqrt{z^2 - y^2}, y\right)\right]dy \qquad (4.272)$$

4.9.2   Density of the Difference of Two Independent Random Variables (Detailed Analysis)

Earlier we explored ways of obtaining the density of the difference of two independent and identically distributed exponential variables, two Gaussian variables, etc. We can generalize these results to include examples of cases where the random variables are non-identical.

Consider

$$Z = X - Y \qquad (4.273)$$

X and Y only take positive values,

$$f_X(x) = h(x)U(x) \qquad (4.274)$$

$$f_Y(y) = g(y)U(y) \qquad (4.275)$$

Rewriting Eq. (4.273) as the sum of two variables, we have

$$Z = X + V \tag{4.276}$$

In Eq. (4.276),

$$V = -Y \tag{4.277}$$

The pdf of V will be

$$f_V(v) = g(-v)U(-v) \tag{4.278}$$

Since the pdf of the sum of two independent random variables is the convolution of the marginal densities, we have

$$f_Z(z) = f_X(x) * f_V(v) = \int f_X(x) f_V(z - x) dx \tag{4.279}$$

In terms of the marginal densities, Eq. (4.279) becomes

$$f_Z(z) = \int_{-\infty}^{\infty} h(x)U(x)g\left[-(z-x)\right]U\left[-(z-x)\right]dx$$

$$= \int_{-\infty}^{\infty} h(x)U(x)g(x-z)U(x-z)dx \tag{4.280}$$

Depending on whether $X > Y$ or $X < Y$, Z can be positive or negative.
Case 1: $z > 0$ implying $X > Y$
In this case, $U(x - z)$ must be such that $x > z$. Equation (4.280) now becomes

$$f_Z(z) = \int_{z}^{\infty} h(x)g(x-z)dx. \tag{4.281}$$

Case 2: $z < 0$ implying $Y > X$
Since X is always positive, $x - z$ will always be positive (z is negative). Therefore, the limits of integration now are $x = 0$ to $x = \infty$. Equation (4.280) now becomes

$$f_Z(z) = \int_{0}^{\infty} h(x)g(x-z)dx \tag{4.282}$$

Example X and Y are i. i. d exponential variables, each with a parameter a. We have

$$f(x) = \frac{1}{a}\exp\left(-\frac{x}{a}\right)U(x) \tag{4.283}$$

$$f(y) = \frac{1}{a} \exp\left(-\frac{y}{a}\right) U(y) \tag{4.284}$$

Case 1: $z > 0$ Eq. (4.281) becomes

$$f_Z(z) = \int_z^\infty \frac{1}{a} \exp\left(-\frac{x}{a}\right) \frac{1}{a} \exp\left(\frac{z-x}{a}\right) dx = \frac{1}{a^2} \exp\left(\frac{z}{a}\right) \int_z^\infty \exp\left(-\frac{2x}{a}\right) dx$$

$$= \frac{1}{2a} \exp\left(-\frac{z}{a}\right) \tag{4.285}$$

Case 2: $z < 0$ Eq. (4.282) becomes

$$f_Z(z) = \int_0^\infty \frac{1}{a} \exp\left(-\frac{x}{a}\right) \frac{1}{a} \exp\left(\frac{z-x}{a}\right) dx = \frac{1}{a^2} \exp\left(\frac{z}{a}\right) \int_0^\infty \exp\left(-\frac{2x}{a}\right) dx$$

$$= \frac{1}{2a} \exp\left(\frac{z}{a}\right) \tag{4.286}$$

Combining both conditions, we have

$$f_Z(z) = \begin{cases} \frac{1}{2a} \exp\left(\frac{z}{a}\right), z < 0 \\ \frac{1}{2a} \exp\left(-\frac{z}{a}\right), z > 0 \end{cases} = \frac{1}{2a} \exp\left(-\frac{|z|}{a}\right) \tag{4.287}$$

Different approach:

$$F(z) = P(X - Y < z) \tag{4.288}$$

Choosing $Y = y$, we have

$$F(z|y) = P(X < z + y) = F_X(z + y) \tag{4.289}$$

Differentiating Eq. (4.289) w. r. z, we have

$$f(z|y) = f_X(z + y) \tag{4.290}$$

Therefore, the joint density of Z and Y is

$$f(z, y) = f(z|y) f_Y(y) = f_X(z + y) f_Y(y) \tag{4.291}$$

The pdf of Z is

$$f(z) = \int f(z, y) dy = \int f_X(z + y) f_Y(y) dy \tag{4.292}$$

Consider two exponentially distributed random variables with

$$f_X(x) = \frac{1}{a} \exp\left(-\frac{x}{a}\right) U(x)$$
$$f_Y(y) = \frac{1}{b} \exp\left(-\frac{y}{b}\right) U(y)$$

$$(4.293)$$

Substituting Eq. (4.293) in Eq. (4.292), we have

$$f(z) = \int_{-\infty}^{\infty} \frac{1}{ab} \exp\left(-\frac{z+y}{a}\right) U(z+y) \exp\left(-\frac{y}{b}\right) U(y) dy \qquad (4.294)$$

Case 1: $z > 0$. In this case, $(z + y)$ and y will be positive when $0 < y < \infty$. Therefore, the limits of integration in Eq. (4.294) will be 0 and ∞.

$$f(z) = \int_{0}^{\infty} f_X(z+y) f_Y(y) dy, \qquad z > 0 \qquad (4.295)$$

$$f(z) = \int_{0}^{\infty} \frac{1}{ab} \exp\left(-\frac{z+y}{a}\right) \exp\left(-\frac{y}{b}\right) dy = \frac{1}{a+b} \exp\left(-\frac{z}{a}\right) \qquad (4.296)$$

Case 2: When $z < 0$. This means that for $z + y$ to be positive

$$z + y > 0 \quad \Rightarrow \quad y > -z \qquad (4.297)$$

This means that the limits of y will be $-z$ to ∞. Note that since z is negative, $-z$ will be positive, and therefore the requirement that $y > 0$ for the term $U(y)$ is also met. Thus, the limits of integration in Eq. (4.294) will be $-z$ and ∞.

$$f(z) = \int_{-z}^{\infty} \frac{1}{ab} \exp\left(-\frac{z+y}{a}\right) \exp\left(-\frac{y}{b}\right) dy = \frac{1}{a+b} \exp\left(\frac{z}{b}\right) \qquad (4.298)$$

The density can therefore be written as

$$f(z) = \begin{cases} \dfrac{1}{a+b} \exp\left(\dfrac{z}{b}\right), & z < 0 \\[3mm] \dfrac{1}{a+b} \exp\left(-\dfrac{z}{a}\right), & z > 0 \end{cases} \qquad (4.299)$$

When $a = b$, Eq. (4.299) becomes Eq. (4.287). *An interesting set of expressions is obtained by combining both approaches leading to identical limits of integration*

(but integration over different variables). Using Eq. (4.282) for the case of $z < 0$ and using Eq. (4.295) for the case when $z > 0$, we can write

$$f(z) = \int_0^\infty f_Y(x - z) f_X(x) dx, \quad z < 0 \tag{4.300}$$

$$f(z) = \int_0^\infty f_X(z + y) f_Y(y) dy, \quad z > 0 \tag{4.301}$$

For the case of two non-identical exponential random variables with densities given in Eqs. (4.293), (4.300) and (4.301) become

$$f(z) = \int_0^\infty f_Y(x - z) f_X(x) dx = \int_0^\infty \frac{1}{ab} \exp\left(-\frac{x - z}{b}\right) \exp\left(-\frac{x}{a}\right) dx \quad z < 0 \tag{4.302}$$

$$f(z) = \int_0^\infty f_X(z + y) f_Y(y) dy = \int_0^\infty \frac{1}{ab} \exp\left(-\frac{z + y}{a}\right) \exp\left(-\frac{y}{b}\right) dx \quad z > 0 \tag{4.303}$$

Completing the integration we get,

$$f(z) = \begin{cases} \dfrac{1}{a + b} \exp\left(\dfrac{z}{b}\right), z < 0 \\[3mm] \dfrac{1}{a + b} \exp\left(-\dfrac{z}{a}\right), z > 0 \end{cases} \tag{4.304}$$

Let us now look at the joint density of the sum and difference of two independent and non-identical exponential variables with Z being the difference and let W be the sum,

$$W = X + Y \tag{4.305}$$

The Jacobian of the transformation of $Z = X - Y$ and $X + Y$ was obtained earlier in Example 4.28 and the joint density of Z and W becomes

$$f_{Z,W}(z, w) = \frac{f(x, y)}{2} = \frac{1}{2ab} \exp\left(-\frac{z + w}{2a}\right) \exp\left(-\frac{w - z}{2b}\right), 0 < |z| < w \tag{4.306}$$

$$< \infty$$

The marginal density of the sum is

$$f_W(w) = \int\limits_{-\infty}^{\infty} f(z, w)dz = \int\limits_{-w}^{w} \frac{1}{2ab} \exp\left(-\frac{z+w}{2a}\right) \exp\left(-\frac{w-z}{2b}\right)dz \qquad (4.307)$$

Equation (4.307) becomes

$$f(w) = \frac{1}{a-b}\left[\exp\left(-\frac{w}{a}\right) - \exp\left(-\frac{w}{b}\right)\right]U(w) \qquad (4.308)$$

The marginal density of the difference is

$$f_Z(z) = \int\limits_{-\infty}^{\infty} f(z, w)dw = \int\limits_{|z|}^{\infty} \frac{1}{2ab} \exp\left(-\frac{z+w}{2a}\right) \exp\left(-\frac{w-z}{2b}\right)dw \qquad (4.309)$$

Equation (4.309) becomes

$$f_Z(z) = \frac{1}{a+b}\left[\exp\left(-\frac{z+|z|}{2a}\right) \exp\left(\frac{z-|z|}{2b}\right)\right], \quad -\infty < z < \infty \qquad (4.310)$$

Notice that Eq. (4.310) is identical to Eq. (4.304)

We will now look at the case of two independent and identically distributed gamma variables each with densities given by

$$\begin{aligned} f_X(x) &= \frac{x^{a-1}}{b^a\Gamma(a)} \exp\left(-\frac{x}{b}\right)U(x) \\ f_Y(y) &= \frac{y^{a-1}}{b^a\Gamma(a)} \exp\left(-\frac{y}{b}\right)U(y) \end{aligned} \qquad (4.311)$$

We have

$$f(z) = \int\limits_0^{\infty} f_Y(x-z) f_X(x)dx = \int\limits_0^{\infty} \frac{(x-z)^{a-1} x^{a-1}}{b^{2a}\Gamma^2(a)} \exp\left(-\frac{x-z}{b}\right) \exp\left(-\frac{x}{b}\right)dx \quad z < 0 \qquad (4.312)$$

$$f(z) = \int\limits_0^{\infty} f_X(z+y) f_Y(y)dy = \int\limits_0^{\infty} \frac{(y+z)^{a-1} y^{a-1}}{b^{2a}\Gamma^2(a)} \exp\left(-\frac{y+z}{b}\right) \exp\left(-\frac{y}{b}\right)dy \quad z > 0 \qquad (4.313)$$

Equation (4.312) can be solved using 3.383.3 (Gradshteyn and Ryzhik 2000), and Eq. (4.313) can be solved using 3.383.8 (Gradshteyn and Ryzhik 2000).

$$\int_0^\infty x^{\mu-1}(x-u)^{\mu-1}e^{-\beta x}dx = \frac{1}{\sqrt{\pi}}\left(\frac{u}{\beta}\right)^{\mu-\frac{1}{2}}\Gamma(\mu)\exp\left(-\frac{\beta u}{2}\right)K_{\mu-\frac{1}{2}}\left(\frac{\beta u}{2}\right) \qquad (4.314)$$

$$\int_0^\infty x^{\nu-1}(x+\beta)^{\nu-1}e^{-\mu x}dx = \frac{1}{\sqrt{\pi}}\left(\frac{\beta}{\mu}\right)^{\nu-\frac{1}{2}}\Gamma(\nu)\exp\left(\frac{\beta\mu}{2}\right)K_{\frac{1}{2}-\nu}\left(\frac{\beta\mu}{2}\right) \qquad (4.315)$$

Noting the symmetry, we have the density of $Z = X - Y$ as

$$f_Z(z) = \frac{1}{\sqrt{\pi}\Gamma(a)b^{a+\frac{1}{2}}}\left|\frac{z}{2}\right|^{a-\frac{1}{2}}K_{a-\frac{1}{2}}\left(\frac{|z|}{b}\right) \qquad (4.316)$$

4.9.3 Leibniz Rule: Differentiation of a Definite Integral

Case (1)

Single integral

$$\frac{\partial}{\partial z}\int_{a(z)}^{b(z)} f(x,z)\,dx = \int_{a(z)}^{b(z)}\frac{\partial f(x,z)}{\partial z}\,dx + \frac{\partial b(z)}{\partial z}f(b(z),z) - \frac{\partial a(z)}{\partial z}f(a(z),z)$$

$$(4.317)$$

Case (2)
Double integral

$$\frac{\partial}{\partial z}\int_{\theta(z)}^{\gamma(z)}\int_{\alpha(z)}^{\beta(z)} f(x,y)\,dx\,dy = \int_{\theta(z)}^{\gamma(z)}\left[\frac{\partial\beta(z)}{\partial z}f(\beta(z),y) - \frac{\partial\alpha(z)}{\partial z}f(\alpha(z),y)\right]dy +$$

$$\frac{\partial\gamma(z)}{\partial z}\int_{\alpha(z)}^{\beta(z)} f(x,\gamma(z))\,dx - \frac{\partial\theta(z)}{\partial z}\int_{\alpha(z)}^{\beta(z)} f(x,\theta(z))\,dx$$

$$(4.318)$$

Exercises

4.1 X and Y are independent and identically distributed uniform random variables in [0,5] each. Obtain the following probabilities.

$P(X + Y > 3)$ $P(X - Y < 3)$ $P(|X - Y| < 3)$ $P(X/Y < 3)$
$P(XY > 1)$

4.2 Verify the results in Problem 1 using random number generation.

4.3 X and Y are uniform and independent identically distributed random variables in $[0,a]$ each. Obtain the density of $Z = X + Y$.

4.4 Verify the results of problem 3 through random number simulation. Choose $a = 2$.

4.5 X and Y are uniform and independent and identically distributed random variables in $[0,b]$ each. Obtain the pdf of $Z = X - Y$ and then, $W = |Z|$.

4.6 Verify the results of problem 5 using random number simulation. Choose $b = 3$;

4.7 X and Y are independent and identical variables, each uniform in $[0,5]$. Obtain the pdf of $Z = XY$.

4.8 Verify the solution in Problem 4.7 through random number simulations.

4.9 If X and Y are independent and identically distributed random variables, each uniform in $[0,b]$, obtain the pdf of $W = X/Y$.

4.10 Verify the results in Problem 4.9 through random number simulations.

4.11 Obtain the CDF and pdf of max (X,Y) and min (X,Y) CDF and pdf of maximum and minimum when two random variables have different ranges. X is $U[0,a]$ and Y is $[0,b]$ with $b > a$.

4.12 The joint density of X and Y is

$$f(x, y) = k \sin (x + y), \ \ 0 < x < \frac{\pi}{2}, 0 < y < \frac{\pi}{2}$$

Obtain the value of k and determine $f(x|y)$ and $f(y|x)$.

4.13 The joint density of X and Y is

$$f(x, y) = \frac{k}{(a^2 + x^2)(b^2 + y^2)}, \ \ -\infty \le x \le \infty, \ -\infty \le y \le \infty, a > 0, b > 0$$

What is the value of k? Obtain the marginal pdfs and CDFs. Obtain the expression for the joint CDF.

4.14 The joint density of X and Y is

$$f(x, y) = \frac{1}{4\pi}, \ \sqrt{x^2 + y^2} \le 2$$

What are the marginal density functions of X and Y?

4.15 The joint density of X and Y is

$$f(x, y) = \frac{1}{12\pi}, \ \ \sqrt{\frac{x^2}{9} + \frac{y^2}{16}} \le 1.$$

Obtain the marginal densities of X and Y as well as the conditional densities of X and Y, with conditioning imposed by the other variable.

4.16 X and Y are independent random variables with densities

$$f(y) = 2\exp(-2y)U(y)$$
$$f(x) = 2[\exp(-x) - \exp(-2x)]U(x)$$

Find the joint pdf given that $X > Y$.

4.17 X is exponentially distributed with mean a and Y is exponentially distributed with mean b. X and Y are also independent.

(a) what is the probability that $Y > X$
(b) What is the probability that $X > Y + 2$?

4.18 A random variable obeys Poisson statistics; however, it appears that the mean is a random variable uniform in [3,6].

$$P(X = k | A = a) = \frac{a^k}{k!}\exp(-a)$$

What is the conditional density of A given that the Poisson event has occurred with $k = m$?

4.19 The lifetime of a driver circuit at the front end of device is modeled as Y and the lifetime of the device is modeled as X. It is obvious that the device will cease to operate when the driver circuit fails, while the converse is not true. The device may fail even when the driver circuit is working. The joint density of these two life times is

$$f(x, y) = K\exp\left(-\frac{y}{10}\right), 0 < x < y < \infty$$

Obtain the value of K so that $f(x,y)$ is a valid density. Obtain the mean life times of the driver circuit and the device. If the device fails, what is the expected lifetime of the driver circuit?

4.20 X and Y have the joint pdf

$$f(x, y) = y\exp[-y(1+x)]U(X)U(y)$$

Obtain the marginal and conditional densities.

4.21 X and Y are independent random variables with densities of

$$f(x) = \frac{1}{\sqrt{2\pi}}\exp\left(-\frac{x^2}{2}\right)$$

$$f(y) = y \exp\left(-\frac{y^2}{2}\right) U(y)$$

Obtain the density of $Z = XY$?

4.22 Verify the results of Problem 4.21 through random number simulation.

4.23 If X is $N(0,4)$ and Y is $N(0,9)$, obtain the pdf of

$$W = \sqrt{X^2 + Y^2}$$

4.24 Verify the results of problem 4.23 through random number simulation.

4.25 It is usual practice in signal processing (wireless, imaging in medicine, target detection in radar, sonar or IR) to collect two sets of data and combine the signals to improve the performance. The data sets are identically distributed and independent. For the case of a pair of gamma random variables, obtain the improvement in performance after the processing. The improvement is defined as the ratio of the mean to the standard deviation. Three types of signal processing algorithms are employed. If X and Y are the two independent and identically distributed pair and the three algorithms are

$$Z = \frac{X + Y}{2}$$
$$V = \sqrt{XY}$$
$$W = \max(X, Y)$$

Note that Z is the arithmetic mean, V is the geometric mean and W is the maximum of the two.

4.26 Verify the results of Problem 4.25 using random number simulation. Choose $a = 1.5$ and $b = 2$;

4.27 Obtain the pdf of $Z = X - Y$ using characteristic functions. X and Y are independent and identically distributed random variables having exponential density.

4.28 If X and Y are independent identically distributed Gaussian variables, $N(0,b^2)$, obtain the density of $W = X^2 + Y^2$ using the concept of Laplace transforms.

4.29 If X and Y are independent and identically distributed gamma variables, obtain the joint pdf of

$$Z = X + Y$$
$$W = \frac{X}{Y}$$

4.30 X and Y are exponentially distributed and independent with densities

$$f(x) = \frac{1}{a} \exp\left(\frac{-x}{a}\right) U(x)$$

$$f(y) = \frac{1}{b} \exp\left(\frac{-y}{b}\right) U(y)$$

Obtain $P\left[\frac{X}{X+Y} > c\right]$

4.31 The joint pdf of X and Y is

$$f(x,y) = \begin{cases} \frac{1}{6}, & 0 < x < 3, 0 < y < 2 \\ 0, \text{elsewhere} \end{cases}$$

(a) What is the $P\left[X + Y < \frac{3}{2}\right]$?
(b) What is the $P[XY > 1]$

4.32 If X and Y are independent exponentially distributed random variables with means a and b respectively, what is the probability that $\max\{X,Y\} > a$?

4.33 X, Y and Z are independent random variables with densities

$$f(x) = \exp(-x)U(x)$$

$$f(y) = y\exp(-y)U(y)$$

$$f(z) = \frac{z^2}{2} \exp(-z)U(z)$$

(a) What is the probability that $Z > Y > X$?
(b) What is the probability that Z is the largest of the three variables?

4.34 Verify the results from Problem 4.33 using random number simulation.

4.35 Prove that the results of Problem 4.33 can be explained in terms of mutually exclusive events (what are the mutually exclusive events in this case?)

4.36 The joint density of X and Y is

$$f(x,y) = k(x^2 + y^2), 0 < x < 2, 0 < y < 2$$

Obtain E(X|Y) first and then, obtain E(X). Verify the result directly.

4.37 In Problem 4.23, the joint density of the independent Gaussian variables is expressed as

$$f(x,y) = \frac{1}{2\pi ab} \exp\left(-\frac{x^2}{2a^2} - \frac{y^2}{2b^2}\right)$$

The cross-sectional appearance in this case will be an ellipse. Obtain an expression for the density of the phase,

$$W = \tan^{-1}\left(\frac{Y}{X}\right)$$

4.38 Verify the results of problem 4.37 through random number simulation by choosing appropriate values of a and b. Choose $a = 3$ and $b = [2,3,4,5]$; Plot also the histogram of the phase in rose plot

4.39 If X and Y are independent identically distributed exponential variables, obtain the density of $Z = XY$.

4.40 Verify the result in Problem 4.39. Chose $a = 3.5$

4.41 Using the result of Problem 4.39 or otherwise, obtain the density of $W = X_1.X_2$. X_3 where X's are independent and identically distributed exponential variables.

4.42 Verify the results of Problem 4.41 using random number simulation.

The following three problems rely on the fact that the sum of two independent Gaussian variables is also Gaussian, the sum of two i. i. d gamma variables is another gamma variable and the sum of two i.i.d Poisson variables is another Poisson variable. For these exercises, generate two sets of 1000 samples of data belonging to hypothesis H_0 (target absent) and two sets 900 samples of data belonging to hypothesis H_1 (target present). Use one set each for the respective hypothesis. In each case, obtain the density plots of the data, corresponding ROC curves, areas under the ROC curves (all from random numbers). Obtain the ROC, area under the ROC curve and optimum operating point, using theoretical densities. Compare the area under the ROC curve (theory vs. simulation)

4.43 (compare to 3-139)

```
N1=1000; N2=900;
m1=-1/2;sig1=2*sqrt(2);
m2=3;sig2=2*sqrt(2);
x=random('normal',m1,sig1,2,N1);
y=random('normal',m2,sig2,2,N2);
```

4.44 (compare to 3-144)

```
N1=1000; N2=900;
a1=2;b1=4;
a2=3;b2=6;
x=random('gamma',a1,b1,2,N1);
y=random('gamma',a2,b2,2,N2);
```

4.45 (3-149 discrete case)

```
N1=1000; N2=900;
lam1=5;
```

```
lam2=8;
x=random('poisson',lam1,2,N1);
y=random('poisson',lam2,2,N2);
```

For the following set of exercises, generate two sets of 1000 samples of data belonging to hypothesis H_0 (target absent) and two sets 900 samples of data belonging to hypothesis H_1 (target present). Add the cohorts in each set. In each case, obtain the density plots of the data, corresponding ROC curves, areas under the ROC curves (all from random numbers). Obtain the optimal operation point. These should show improvement over the problems from Chap. 3.

```
4.46 ((3-145)
N1=1000;N2=900;
b1=3;
s=3;sig=6;
x=random('rayleigh',b1,2,N1);
y=random('rician',s,sig,2,N2);
```

```
4.47 (3-146)
N1=1000;N2=900;
s1=1.2;sig1=1;
s2=2.5;sig2=2;
x=random('rician',s1,sig1,2,N1);
y=random('rician',s2,sig2,2,N2);
```

```
4.48 (3-147)
N1=1000;N2=900;
m1=1.2;omega1=3;
s2=2.5;sig2=2;
x=random('nakagami',m1,omega1,2,N1);
y=random('rician',s2,sig2,2,N2);
```

```
4.49 (3-148)
N1=1000;N2=900;
m1=1.2;omega1=2;
a2=3;b2=5;
x=random('nakagami',m1,omega1,2,N1);
y=random('weibull',a2,b2,2,N2);
```

4.50

In this case, undertake the analysis for one set first before undertaking it for two sets

```
N1=1000;N2=900;
a1=1.2;b1=6;
m2=2;omega2=3;
```

```
x=random('weibull',a1,b1,1,N1);
y=random('nakagami',m2,omega2,1,N2);
```

For two sets

```
x=random('weibull',a1,b1,2,N1);
y=random('nakagami',m2,omega2,2,N2);
```

```
4.51-4.60
```

For the pair of independent uniform random variables X and Y, obtain the densities of the sum and difference, Z=X+Y and W=X-Y. Plot the densities.

```
4.51 X->U[0,2], Y->U[0,3]
4.52 X->U[1,2], Y->U[1,3]
4.53 X->U[0,1], Y->U[-3,0]
4.54 X->U[-1,2], Y->U[-2,1]
4.55 X->U[-1,1], Y->U[0,3]
4.56 X->U[-1,0], Y->U[1,2]
4.57 X->U[1,2], Y->U[-1,0]
4.58 X->U[-2,2], Y->U[0,1]
4.59 X->U[-2,-1], Y->U[1,2]
4.60 X->U[0,2], Y->U[-2,0]
```

References

Abramowitz M, Segun IA (1972) Handbook of mathematical functions with formulas, graphs, and mathematical tables. Dover Publications, New York

Devore JL (2004) Probability and statistics for engineering and the sciences. Thomson, Belmont

Epstein B (1948) Some applications of Mellin transform in statistics. Ann Math Stat 19(3):370–379

Erdelyi A (1953) Table of integral transforms. McGraw-Hill, Inc., New York

Fante RL (2001) Central limit theorem: use with caution. IEEE Trans Aerosp Electron Syst 37 (2):739–740

Gradshteyn IS, Ryzhik IM (2000) Table of integrals, series, and products, 6th edn. Academic, New York

Kwak SG, Kim JH (2017) Central limit theorem: the cornerstone of modern statistics. Korean J Anesthesiol 70(2):144–156. https://doi.org/10.4097/kjae.2017.70.2.144

Mathai AM (1993) A handbook of generalized special functions for statistical and physical sciences. Oxford University Press, Oxford

Mathai AM, Haubold HJ (2008) Special functions for applied scientists. Springer, New York

Nakagami M (1960) The m-distribution—a general formula of intensity distribution of rapid fading. In: Hoffman WC (ed) Statistical methods in radio wave propagation. Pergamon, Elmsford

Papoulis A, Pillai U (2002) Probability, random variables, and stochastic processes. McGraw-Hill, New York

Price BA, Zhang X (2007) The power of doing: a learning exercise that brings the central limit theorem to life. Decis Sci J Innov Educ 5(2):405–411

Rodríguez-López M, Carrasquillo A Jr (2006) Improving conceptions in analytical chemistry: the central limit theorem. J Chem Educ 83(11):1645–1648

Rohatgi VK, Saleh AKME (2001) An introduction to probability and statistics. Wiley, New York

Shankar PM (2015) A composite shadowed fading model based on the McKay distribution and Meijer G functions. Wirel Pers Commun 81(3):1017–1030

Shankar PM (2016) Performance of cognitive radio in N*Nakagami cascaded channels. Wirel Pers Commun 88(3):657–667

Shankar PM (2017) Fading and shadowing in wireless systems, 2nd edn. Springer, Cham

Springer M, Thompson W (1966) The distribution of products of independent random variables. SIAM J Appl Math 14(3):511–526

Yates RD, Goodman DJ (2014) Probability and stochastic processes: a friendly introduction for electrical and computer engineers. Wiley, Hoboken

Chapter 5
Applications to Data Analytics and Modeling

5.1 Introduction

The applications of statistics are vast and spread across multiple disciplines in business, social sciences, sciences, engineering, economics, medicine, etc. A perusal of relevant literature shows the use of densities and distributions in communication and computer systems (performance analysis), image analysis and interpretation, pattern recognition, machine intelligence, etc. with overlap in medical diagnostics. It can also be seen that the process of application involves a number of steps, often taking two paths after a few initial steps. Researchers in communication systems rely on error rates and outage probabilities to quantify the performance of the systems. On the other hand, researchers exploring cognitive radio, pattern recognition, medical diagnostics, machine vision, and related areas are generally interested in receiver operating characteristics, positive predictive values, etc. Both groups of researchers share the common starting point of modeling the statistics and establishing a best statistical fit for the data (Helstrom 1968; Van Trees 1968; Metz 2006). Most of these technical problems involve the determination of whether the data came from one the two possible scenarios. These could be the determination of the presence or absence of target based on the data collected, the received signal being a "1" or a "0" in binary communication systems, the decision to be made as to whether the subject of a medical test suffers from an illness or not, etc. The aim is to identify the categories to which the data sets belong, quantify the ability of the processing to predict the presence or absence of any specific feature (presence of a target, confirmation of an illness, etc.) and examine if we can improve the prediction capability.

Another important topic of interest in data analytics is bootstrapping (Efron and Tibshirani 1986; Shankar 2020a). Bootstrapping makes it possible to extract

Electronic supplementary material: The online version of this chapter (https://doi.org/10.1007/978-3-030-56259-5_5) contains supplementary material, which is available to authorized users.

P. M. Shankar, *Probability, Random Variables, and Data Analytics with Engineering Applications*, https://doi.org/10.1007/978-3-030-56259-5_5

statistical information on the data collected even though we may only have conducted a single set of experiments. Bootstrapping also allows us to compare the performance of two sensors in a machine system or the efficacies of two competing drugs or treatments to cure a specific illness. We will study the process of bootstrapping and examine how it can be implemented and used in data analytics.

Two additional areas of exploration are possible with the topics that have been covered so far in previous chapters. These are the diversity in wireless systems to mitigate the effects of signal strength fluctuations (known as fading) and study of reliability of engineering systems. Diversity can be analyzed in terms of the performance of error rates and outage probabilities (Brennan 1959; Shankar 2017). To facilitate the discussion of diversity, a cluster-based approach is presented to model the statistics of fading channels (Shankar 2015, 2017). Similar approaches are also in use to improve the performance of medical diagnostic systems and machine vision systems.

The studies on reliability allow management of interconnected systems making it possible to plan for outages and undertake repairs in a timely fashion.

We start the discussion with decision theory applied to machine vision problems characterized in terms of the area under the ROC curve (AUC) and positive predictive values (PPV), topics presented in Chaps. 2 and 3. These are also relevant in medical diagnostics. The various steps involved in this process will take us through sections from previous chapters. We will add and fill the missing pieces as needed.

5.2   Receiver Operating Characteristics (ROC) Curves

In Chaps. 2 and 3, we explored data analytics with two sets of data acquired during the test phase. These correspond to the case when there is a target present and to the case when a target is absent (Helstrom 1968; Shankar 2017). These tests might represent the study undertaken to examine the effectiveness of a radio receiver to detect the presence of a target at a certain distance, sonar system to detect the presence of an unidentified object, channel characteristics to transmit binary data, clinical tests to determine the efficacy of a new drug, imaging modality, etc. While in most of these cases, it is possible to have both sets of data to be of equal size, the clinical case illustrates an example where the data collected on subjects with no illness is likely to be larger than the number of subjects with an illness. If N_0 is the number of samples of data corresponding to the case of the absence of the target and N_1 is the number of samples of data corresponding to the case of the presence of the target, we assume that $N_0 \geq N_1$. It must be stated that such an assumption is not necessary. It merely reflects the facts in medical data analytics.

A simple approach to finding a means to interpret the data was presented in Chap. 3 using the concept of the receiver operating characteristics (ROC) curves. We will now explore the ROC further and examine ways of undertaking a statistical analysis to improve the performance of the machine vision or clinical diagnostic system that produced the data.

5.2.1 Optimal Operating Point (OOP) and Positive Predictive Value (PPV)

An example is provided with a sample data set consisting of 40 (N_0) values collected when a target is absent and 30 (N_1) values collected when a target is present in the field of radar, sonar, IR, or any other modality used to detect the presence of a target. The data sets and analysis are displayed in Table 5.1. Two data sets are first put in a single column with corresponding values of the "gold standard" or "labels," 0's representing "no target" and "1" representing the presence of the target. These two columns are then sorted in descending order of the values. Details are provided in Chap. 3.

To estimate the probability of false alarm and probability of detection, it is necessary to choose a threshold. The highest value of the observed quantity is chosen as the first threshold and the number of 1's above the threshold are counted as the number of correct decisions (N_C) and the number of 0's are counted as the "false" decisions (N_F). The threshold values are then chosen from the sorted values and the process of counting continued until the last value of the threshold is reached, with a value of '0' appended to the data values. The probability of detection is obtained as N_C/N_1 and the probability of false alarm is N_F/N_0. Details of the procedure are presented in Chap. 3.

A plot of the ROC curve appears in Fig. 5.1. An important question pertains to the basis for the appropriate interpretation of the ROC curve. To answer this question, we need a strategy. A simple strategy presented in Chap. 3 is based on the realization that in our experiments, we need to have the maximum value of the probability of detection and the minimum value of the probability of false alarm. This is known as the Neyman-Pearson strategy (Helstrom 1968; van Trees 1968). It is implemented by determining the minimum distance to the top left hand corner of the ROC plot [$P_F = 0$; $P_D = 1$] as

$$d = \sqrt{P_F^2 + P_M^2}. \tag{5.1}$$

$$d_{\min} = \sqrt{P_{F_{\min}}^2 + (1 - P_{D_{\max}})^2} \tag{5.2}$$

Equation (5.2) is the same as Eq. (3.230). It should be noted that $PF_{\min}$ and $PD_{\max}$ are based on the shortest distance to the top left corner.

The point on the ROC curve where the distance to the top left corner is the minimum gives us the **optimal operating point** (OOP) in terms of the lowest false alarm and the highest probability of detection. The optimum threshold is obtained by inverting either the optimum P_F or P_D. This point is also shown in the ROC plot. This was seen in Chap. 3.

If a specific threshold is chosen to perform data analysis to determine whether a target is present or not, the next issue of interest is the percentage of truth that is associated with the decision. In other words, if a decision is made to conclude that a target exists on the basis of the threshold, how much trust can be placed on that decision? Issues similar to this were addressed in Chap. 2 where a posteriori

Table 5.1 Data set consisting of 40 samples of target absent and 30 samples of target present

No target	Target	Labels	Values	Sorted values	Labels	NC	NF	PD	PF
1.141545	4.766717	0	1.141545	11.9558	1	0	0	0	0
1.976186	2.308291	0	1.976186	11.9419	1	1	0	0.0333	0
3.816114	1.960845	0	3.816114	9.4362	1	2	0	0.0667	0
1.142871	5.160745	0	1.142871	8.2583	1	3	0	0.1	0
3.048602	1.779052	0	3.048602	7.8343	1	4	0	0.1333	0
0.883718	1.50689	0	0.883718	7.211	1	5	0	0.1667	0
4.547587	7.211023	0	4.547587	6.7219	1	6	0	0.2	0
2.711339	5.319057	0	2.711339	6.4333	1	7	0	0.2333	0
3.855104	4.174724	0	3.855104	6.4002	1	8	0	0.2667	0
2.237817	1.708625	0	2.237817	6.258	1	9	0	0.3	0
3.827987	3.050167	0	3.827987	6.0182	0	10	0	0.3333	0
3.505546	7.834319	0	3.505546	5.9558	1	10	1	0.3333	0.025
1.276531	2.314778	0	1.276531	5.3191	1	11	1	0.3667	0.025
6.018166	8.25828	0	6.018166	5.2092	1	12	1	0.4	0.025
3.272193	5.097232	0	3.272193	5.1607	1	13	1	0.4333	0.025
3.490266	4.735725	0	3.490266	5.1073	1	14	1	0.4667	0.025
0.613291	5.955787	0	0.613291	5.0972	1	15	1	0.5	0.025
2.88149	11.94186	0	2.88149	5.0482	1	16	1	0.5333	0.025
1.935876	5.048158	0	1.935876	4.7667	1	17	1	0.5667	0.025
2.597112	6.433276	0	2.597112	4.7357	1	18	1	0.6	0.025
4.642246	9.436206	0	4.642246	4.6422	0	19	1	0.6333	0.025
3.193532	6.257952	0	3.193532	4.5476	0	19	2	0.6333	0.05
1.874784	6.721949	0	1.874784	4.1747	1	19	3	0.6333	0.075
0.832295	5.20917	0	0.832295	**3.8551**	0	20	3	**0.6667**	**0.075**
2.336328	6.400249	0	2.336328	3.828	0	20	4	0.6667	0.1
2.450363	2.973267	0	2.450363	3.8161	0	20	5	0.6667	0.125
0.212409	3.449114	0	0.212409	3.5055	0	20	6	0.6667	0.15
0.303238	5.107348	0	0.303238	3.4903	0	20	7	0.6667	0.175
1.116619	11.95581	0	1.116619	3.4491	1	20	8	0.6667	0.2
1.504661	2.455447	0	1.504661	3.3913	0	21	8	0.7	0.2
1.439038		0	1.439038	3.2722	0	21	9	0.7	0.225
3.258827		0	3.258827	3.2588	0	21	10	0.7	0.25
1.938363		0	1.938363	3.1935	0	21	11	0.7	0.275
1.168983		0	1.168983	3.0502	1	21	12	0.7	0.3
0.925057		0	0.925057	3.0486	0	22	12	0.7333	0.3
2.362685		0	2.362685	2.9733	1	22	13	0.7333	0.325
0.791781		0	0.791781	2.8815	0	23	13	0.7667	0.325
3.391318		0	3.391318	2.7113	0	23	14	0.7667	0.35
0.939172		0	0.939172	2.5971	0	23	15	0.7667	0.375
1.838244		0	1.838244	2.4554	1	23	16	0.7667	0.4
		1	4.766717	2.4504	0	24	16	0.8	0.4
		1	2.308291	2.3627	0	24	17	0.8	0.425

(continued)

Table 5.1 (continued)

No target	Target	Labels	Values	Sorted values	Labels	NC	NF	PD	PF
		1	1.960845	2.3363	0	24	18	0.8	0.45
		1	5.160745	2.3148	1	24	19	0.8	0.475
		1	1.779052	2.3083	1	25	19	0.8333	0.475
		1	1.50689	2.2378	0	26	19	0.8667	0.475
		1	7.211023	1.9762	0	26	20	0.8667	0.5
		1	5.319057	1.9608	1	26	21	0.8667	0.525
		1	4.174724	1.9384	0	27	21	0.9	0.525
		1	1.708625	1.9359	0	27	22	0.9	0.55
		1	3.050167	1.8748	0	27	23	0.9	0.575
		1	7.834319	1.8382	0	27	24	0.9	0.6
		1	2.314778	1.7791	1	27	25	0.9	0.625
		1	8.25828	1.7086	1	28	25	0.9333	0.625
		1	5.097232	1.5069	1	29	25	0.9667	0.625
		1	4.735725	1.5047	0	30	25	1	0.625
		1	5.955787	1.439	0	30	26	1	0.65
		1	11.94186	1.2765	0	30	27	1	0.675
		1	5.048158	1.169	0	30	28	1	0.7
		1	6.433276	1.1429	0	30	29	1	0.725
		1	9.436206	1.1415	0	30	30	1	0.75
		1	6.257952	1.1166	0	30	31	1	0.775
		1	6.721949	0.9392	0	30	32	1	0.8
		1	5.20917	0.9251	0	30	33	1	0.825
		1	6.400249	0.8837	0	30	34	1	0.85
		1	2.973267	0.8323	0	30	35	1	0.875
		1	3.449114	0.7918	0	30	36	1	0.9
		1	5.107348	0.6133	0	30	37	1	0.925
		1	11.95581	0.3032	0	30	38	1	0.95
		1	2.455447	0.2124	0	30	39	1	0.975
				0.0000		30	40	1	1

probabilities were calculated. The calculations of PPV were presented in Chap. 3 and the expression for PPV was given as (the counts are obtained on the basis of the optimum threshold)

$$PPV = \frac{N_C}{N_C + N_F} \tag{5.3}$$

The optimal operating point is highlighted in Table 5.1. The analysis is now repeated using Matlab and the result is shown in Fig. 5.1. Histograms of the two sets are obtained and shown in Fig. 5.2 with the optimal threshold and the corresponding probabilities of false alarm and miss. These plots are obtained using *ksdensity(.)* in Matlab.

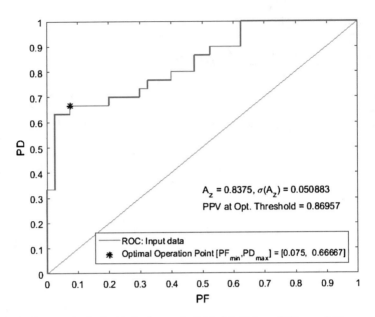

Fig. 5.1 ROC curve obtained with the data set in Table 5.1. Values of PF_{min} and PD_{max} are based on the shortest distance

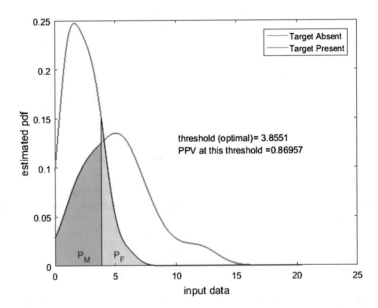

Fig. 5.2 Density fits for the target absent and target present cohorts of data from Table 5.1

5.2.2 Statistical Variation of the Area under the ROC Curve

If a single set of measurements is undertaken with a limited set of samples, it is essential to determine the confidence levels of the area under the ROC curve (McClish 1989; Metz 2006). This is possible through the bootstrapping which will be described later in this chapter. Another approach is to use the formula available in medical literature (Hanley and McNeil 1982). The standard deviation of the area under the ROC curve (the confidence interval is taken to be $\pm$ one standard deviation) is given as

$$\sigma(A_z) = \sqrt{\frac{A_z(1 - A_z) + (N_1 - 1)(A_1 - A_z^2) + (N_0 - 1)(A_2 - A_z^2)}{N_1 N_0}}. \tag{5.4}$$

In Eq. (5.4),

$$A_1 = \frac{A_z}{2 - A_z} \tag{5.5}$$

$$A_2 = \frac{2A_z^2}{1 + A_z} \tag{5.6}$$

For the example shown,

$$\sigma(A_z) = 0.0509. \tag{5.7}$$

The value of $\sigma(A_z)$ is displayed in Fig. 5.1. Note that A_z is the AUC (area under the curve).

5.2.3 Improvement in Performance and Statistical Modeling

Once a single set of measurements is available, it is possible to improve the performance of the receiver through signal processing (Brennan 1959). For example, if we can create additional data sets that have the same statistical characteristics as the data collected, it will be possible to enhance the performance through signal processing algorithms that combine multiple sets of data. To achieve this goal, it requires that we establish the statistics of the data sets (target absent and target present). Thus, there is a need to see if the data fits any of the known densities and if not, whether it is possible to model the statistics and propose a new statistical model for the data which will lead to a new probability density. While the latter step is beyond the scope of the book, the former is possible based on distributions, densities, and moments of random variables discussed in Chaps. 3 and 4. The different steps involved in obtaining the densities and confirming their validities along with the process of combining the data using different algorithms are now described.

5.2.3.1 Statistics of the Data (Probability Density)

Based on past observations, it is always possible to expect that the statistics of the data might follow a specific density. It is also possible that a particular density can be associated with the measurements based on the theoretical modeling undertaken to justify the measurements. This means that we need to test whether the estimated density of the data fits the expected or known theoretical density. The parameter estimation described below may be skipped because parameters are easily estimated in Matlab and other computational tools.

Parameter Estimation: Method of Moments (MoM) and Maximum Likelihood Estimation (MLE)

One of the first steps in this process is the estimation of the parameters of the distribution assuming that the data set fits the specific probability density. For example, the exponential and Rayleigh are single parameter distributions. The Gaussian, gamma, Weibull, Rician, etc. are two parameter distributions. The simplest means of obtaining the parameters is through method of moments (MoM). Since the first moment of the exponential density is its parameter, the density is completely defined by the first moment of the data collected. The Rayleigh density is also characterized by the first moment. The Gaussian density has two parameters requiring estimation of the two lowest order moments, namely, the first and second. The same applies to gamma, Weibull, Nakagami, and Rician densities as well.

An example of a set of data assumed to follow the exponential density is given in Table 5.2. From Chap. 3, it is seen that for the exponential density with a parameter a, the mean a is the first moment.

The mean of the data is 2.8936, very close to the parameter of the exponential density of 3. Accuracy of estimation can certainly be improved by increasing the number of samples.

Now, let us consider the case of a data set assumed to follow the gamma density. Based on the moments of the gamma density $G(a,b)$ seen in Chap. 3,

Table 5.2 A sample set of 50 exponential random numbers ($a = 3$)

3.862	2.089	0.858	0.125	0.520	3.150	3.135	3.757	7.737	6.123
1.158	0.123	4.098	1.809	4.108	4.880	0.556	0.834	8.759	1.693
1.269	3.233	2.044	5.928	0.616	4.146	1.607	0.848	1.900	2.269
5.449	1.607	1.074	5.706	4.238	1.453	1.795	2.899	0.749	13.293
6.386	4.491	0.347	4.070	0.220	2.244	0.259	1.698	0.205	3.262

Table 5.3 A sample set of 50 gamma random variables $G(a,b)$ with $a = 2.5$ and $b = 4$

14.616	1.189	39.539	13.488	10.896	1.415	7.409	7.668	6.155	13.163
9.269	19.339	23.373	6.746	8.802	12.103	7.050	2.952	2.867	7.567
10.579	15.645	19.585	2.303	1.906	8.464	3.603	2.628	4.568	7.002
11.883	20.176	6.190	10.435	4.635	16.155	22.525	11.181	12.346	8.972
23.158	7.554	16.876	2.947	3.485	5.104	3.172	7.922	3.611	21.124

Table 5.4 A sample set of 100 gamma random variables $G(a,b)$ with $a = 2.5$ and $b = 4$

7.490	21.158	6.245	29.536	8.667	1.106	13.075	2.766	6.194	32.213
9.791	3.790	3.781	15.989	15.214	9.146	3.959	9.671	20.617	6.937
15.543	18.158	10.897	5.636	10.896	11.322	9.499	7.156	14.156	3.886
9.487	2.479	9.786	11.401	15.375	14.761	10.568	6.132	7.361	15.469
10.299	7.781	8.178	11.751	5.204	11.591	4.263	13.199	7.099	4.465
7.746	7.218	9.259	9.062	17.751	14.497	1.199	14.957	2.575	5.778
5.896	11.185	5.023	10.738	8.523	15.101	5.465	4.821	4.671	5.725
4.465	4.198	10.108	16.042	16.171	7.829	7.256	24.862	10.801	4.342
5.114	6.007	6.652	10.270	8.674	1.212	8.544	9.290	7.078	2.019
7.583	8.740	26.876	2.981	0.723	3.132	28.852	9.353	1.729	7.991

$$M_1 = E(X) = ab \tag{5.8}$$

$$M_2 = E(X^2) = ab^2 + a^2b^2 \tag{5.9}$$

Equations (5.8) and (5.9) can be solved to obtain the values of a and b as

$$a = \frac{M_1^2}{M_2 - M_1^2} \tag{5.10}$$

$$b = \frac{M_2 - M_1^2}{M_1} \tag{5.11}$$

Table 5.3 contains 50 samples of gamma density, $G(2.5, 4)$. Using the method of moments, it can be seen that the mean is 10.22 and the second moment is 160.33. This leads to $b = 6$ and $a = 1.67$. The accuracy does not appear to be good. Table 5.4 contains 100 samples and one can see that the parameter estimates (mean now is 9.57 and the second moment is 130.33 resulting in $a = 2.36$ and $b = 4.04$) are closer to the actual values.

From these examples, it is seen that the method of moments may lead to estimates that are less accurate. Accuracy of parameter estimation could be improved if more samples are available. That is not an option in most of the situations because data collection modalities employed often limits the size of the data.

A better means for estimating the parameters is the method of maximum likelihood estimation (MLE). If we treat each sample to be obtained from an independent iteration of the pdf, the sample set can be described in terms of a joint pdf. If $X_1, X_2,...,$ X_n constitute the n- samples, the joint density is

$$f(x_1, x_2, \cdots, x_n; \theta) = f(x_1; \theta)f(x_2; \theta) \ldots f(x_n; \theta) = \prod_{i=1}^{n} f(x_i; \theta). \qquad (5.12)$$

The parameter of the density is θ (for the exponential density, θ will be the mean), and it is assumed that the observations are independent allowing us to express the joint density as the product of the marginal densities (Chap. 4). The product term on the right hand side of Eq. (5.12) is identified as the likelihood function (LF) expressed as (Rohatgi and Saleh 2001; Papoulis and Pillai 2002)

$$L(\theta; x_1, x_2, \cdots, x_n) = \prod_{i=1}^{n} f(x_i; \theta). \qquad (5.13)$$

While eq. (5.13) is expressed as a function of a single scalar parameter θ (as in the case of exponential or Rayleigh densities), θ can even be treated as a vector (gamma and normal densities with two parameters, generalized gamma density with three or four parameters, etc.) without losing the meaning of the likelihood function. Since logarithm is a monotonic function, we can study the properties of LF by taking logarithm of LF allowing us to operate with a summation instead of a product. The log-likelihood function (LLF) becomes

$$\log\left[L(\theta; x_1, x_2, \cdots, x_n)\right] = \sum_{i=1}^{n} \log\left[f(x_i; \theta)\right] \qquad (5.14)$$

Maximum likelihood estimation consists of choosing an estimator that maximizes LF or the log-likelihood function. If $\widehat{\theta}$ represents the estimate of θ and if $\widehat{\theta}$ is a vector of size k, we have

$$\widehat{\theta} = \theta_j, j = 1, 2, \cdots, k. \qquad (5.15)$$

Since we are seeking estimators that maximize the log-likelihood ratio, this means that

$$\frac{\partial}{\partial \theta_j}\left\{\log\left[L(\theta; x_1, x_2, \cdots, x_n)\right]\right\} = \frac{\partial}{\partial \theta_j}\sum_{i=1}^{n} \log\left[f(x_i; \theta)\right]_{j=1,2,\cdots,k} = 0 \qquad (5.16)$$

A non-trivial solution to Eq. (5.16) leads to the maximum likelihood estimate $\widehat{\theta}$.

Let us examine the MLE approach first by considering the example of the exponential density expressed as

$$f(x_i; \theta) = \frac{1}{\theta} \exp\left(-\frac{x_i}{\theta}\right) \qquad (5.17)$$

The likelihood function is

$$L(\theta; x_1, x_2, \cdots, x_n) = \prod_{i=1}^{n} f(x_i; \theta) = \left(\frac{1}{\theta}\right)^n \prod_{i=1}^{n} \exp\left(-\frac{x_i}{\theta}\right) \tag{5.18}$$

Taking the logarithm, we have

$$\log_e[L(\theta; x_1, x_2, \cdots, x_n)] = -n \log_e(\theta) - \frac{1}{\theta} \sum_{i=1}^{n} x_i \tag{5.19}$$

Taking the derivative w. r. t. θ and setting it to 0, Eq. (5.19) results in

$$\frac{n}{\theta} = \frac{1}{\theta^2} \sum_{i=1}^{n} x_i \tag{5.20}$$

Solving for θ, the MLE of θ becomes

$$\hat{\theta} = \frac{1}{n} \sum_{i=1}^{n} x_i. \tag{5.21}$$

Equation (5.21) shows the MLE of θ is the sample mean. It is clear that for the exponential density, MoM and MLE, lead to the same estimate.

Let us consider the case of the normal pdf represented by $N(\mu, \sigma^2)$. This means that we have

$$f(x_i; \mu, \sigma^2) = \frac{1}{\sqrt{2\pi\sigma^2}} \exp\left[-\frac{(x_i - \mu)^2}{2\sigma^2}\right], \quad i = 1, 2, \cdots, n \tag{5.22}$$

The likelihood function is

$$L(\theta; x_1, x_2, \cdots, x_n) = \prod_{i=1}^{n} f(x_i; \theta) = \left(\frac{1}{\sqrt{2\pi\sigma^2}}\right)^n \prod_{i=1}^{n} \exp\left[-\frac{(x_i - \mu)^2}{2\sigma^2}\right] \tag{5.23}$$

Taking the logarithms, we have

$$\log\left(L(\theta; x_1, x_2, \cdots, x_n)\right) = -\frac{n}{2}\left[\log(2\pi) + \log(\sigma^2)\right] - \frac{1}{2\sigma^2} \sum_{i=1}^{n} (x_i - \mu)^2 \tag{5.24}$$

Taking the derivative of eq. (5.24) w. r. t. μ and setting it to zero,

$$\frac{1}{\sigma^2} \sum_{i=1}^{n} (x_i - \mu) = 0 \tag{5.25}$$

Simplifying, Eq. (5.25) leads to

$$n\widehat{\mu} = \sum_{i=1}^{n} x_i \quad \Rightarrow \quad \widehat{\mu} = \frac{1}{n}\sum_{i=1}^{n} x_i \tag{5.26}$$

Taking the derivative of Eq. (5.24) w. r. t. σ^2 and setting it to zero, we have

$$-\frac{n}{2\sigma^2} + \frac{1}{2(\sigma^2)^2}\sum_{i=1}^{n}(x_i - \mu)^2 = 0 \tag{5.27}$$

Simplifying, we get the estimate of σ^2 as

$$\widehat{\sigma^2} = \frac{1}{n}\sum_{i=1}^{n}(x_i - \widehat{\mu})^2 \tag{5.28}$$

Note that $\widehat{\sigma^2}$ is a biased estimator and the unbiased estimator (see Appendix A) is

$$\widehat{\sigma^2}\Big|_{unbiased} = \frac{1}{n-1}\sum_{i=1}^{n}(x_i - \widehat{\mu})^2 \tag{5.29}$$

Let us now consider the case of the gamma density. For the gamma density we have,

$$f(x_i; a, b) = \frac{x_i^{a-1}}{b^a\Gamma(a)}\exp\left(-\frac{x_i}{b}\right), i = 1, 2, \cdots, n \tag{5.30}$$

The likelihood function is

$$L(\theta; x_1, x_2, \cdots, x_n) = \prod_{i=1}^{n} f(x_i; \theta) = \left(\frac{1}{b^a\Gamma(a)}\right)^n \prod_{i=1}^{n} x_i^{a-1}\left[\exp\left(-\frac{x_i}{b}\right)\right] \tag{5.31}$$

Taking the logarithm, we have

$$\log_e[L(\theta; x_1, x_2, \cdots, x_n)] = -n\log_e(b^a\Gamma(a)) + (a-1)\sum_{i=1}^{n}\log(x_i) - \frac{1}{b}$$

$$\times \sum_{i=1}^{n} x_i \tag{5.32}$$

Since we have two parameters, we need to take the derivatives separately and equate them to zero. Taking the derivative w.r.t. b, we have

$$\frac{\partial}{\partial b} \log_e[L(\theta; x_1, x_2, \cdots, x_n)] = -n\frac{a}{b} + \frac{1}{b^2} \sum_{i=1}^{n} x_i \qquad (5.33)$$

Taking the derivative w.r.t. a, we have

$$\frac{\partial}{\partial a}[\log_e[L(\theta; x_1, x_2, \cdots, x_n)]] = -n \log(b) - n\frac{d}{da}[\log_e \Gamma(a)]$$

$$+ \sum_{i=1}^{n} \log(x_i) \qquad (5.34)$$

Equating Eq. (5.33) to zero, we have

$$ab = \frac{1}{n} \sum_{i=1}^{n} x_i = \overline{X} \qquad (5.35)$$

Equating Eq. (5.34) to zero and substituting for b from Eq. (5.35), we have

$$-n\left[\log(\overline{X}) - \log(a)\right] - n\psi(a) + \sum_{i=1}^{n} \log(x_i) = 0 \qquad (5.36)$$

In Eq. (5.36), $\psi(a)$ is the digamma function (Abramowitz and Segun 1972; Gradshteyn and Ryzhik 2000):

$$\psi(a) = \frac{d}{da}[\log_e \Gamma(a)] \qquad (5.37)$$

Solving for a, we have

$$-\left[\log(\overline{X}) - \log(a)\right] - \psi(a) + \frac{1}{n} \sum_{i=1}^{n} \log(x_i) = 0 \qquad (5.38)$$

Simplifying further, we can express the transcendental equation for the MLE estimate of a as

$$\log(\hat{a}) - \psi(\hat{a}) - \log(\overline{X}) + \frac{1}{n} \sum_{i=1}^{n} \log(x_i) = 0 \qquad (5.39)$$

Once the MLE of a is obtained, MLE of b is obtained from Eq. (5.35) as

$$\hat{b} = \frac{\overline{X}}{\hat{a}} \qquad (5.40)$$

While Eqs. (5.39) and (5.40) show that it is possible to get estimates of the parameters using the first moment and first moment of the logarithm thereby eliminating the need for more noisy second (or higher order moments), Eq. (5.39) is of transcendental type requiring numerical or graphical techniques for obtaining the solution.

Applying the MLE approach directly in Matlab (see the discussion on chi square testing), with 50 samples used earlier, we get $a = 1.9624$ and $b = 5.213$, while with 100 samples, we get $a = 2.4347$ and $b = 3.9316$. These estimates are better than those obtained directly using the method of moments. But, estimation required the use of numerical techniques for solving Eq. (5.39).

Matlab provides a means to estimate the parameters of several known densities (use *fitdist*(.) in Matlab). It is also possible to implement the maximum likelihood estimation for other densities that are not part of the Matlab menu of densities through the "custom pdf" option.

5.2.3.2 Hypothesis Testing: Chi-Square (χ^2) Tests

Once the parameters of the density associated with the data collected are estimated, the next step is to check whether the estimated density of the data matches the expected density through hypothesis testing. One of the most common hypotheses tests used to examine whether a sample came from the population with a specified probability density is the **chi-square test** (Rohatgi and Saleh 2001, Papoulis and Pillai 2002, Shankar 2019b). It can be applied to any univariate distribution provided its cumulative distribution is available (analytically or numerically). The test relies on sorting the data, putting them into bins and counting the populations in these bins. Therefore, the test is influenced by the number of bins used. It also requires a reasonable sample size to populate the bins. The procedure for carrying out the test can be summarized in a few steps given below.

Step #1 The data set is sorted and values are arranged in ascending order.

Step #2 Estimate the parameters of the underlying assumed distribution from the data. This would require the estimation of one, two, three, or more parameters as discussed in the previous sections. Note that steps #1 and #2 are interchangeable.

Step #3 The sorted data set (Step #1) is divided into k bins.

It is expected that the number of bins must be such that each bin is populated and no bin exists with a frequency of 0 even though it has been suggested that the lower limit is 5 (Cochran 1954; Watson 1958; White et al. 2009; Shankar 2019a, 2019b). Generally, number of bins between 5 and 10 are sufficient. Number of bins below 5 is too low and number of bins above 20 may be too high. The choice of the bin size is also dependent on the sparseness the data, and this aspect will be discussed when the limitations of chi-square tests are described. Let the observed frequency (counts in each bin) be denoted by O_i. The total number of samples is N. O_i is an integer lying between 1 and N and

$$\sum_{i=1}^{k} O_i = N \qquad (5.41)$$

Step #4 Use the estimated parameters from step #2 and recreate the theoretical (model) cumulative distribution function and estimate the probabilities for the random variable to stay within the corresponding bins. This step therefore recreates the theoretical probabilities for the data to exist in the ranges defined by the bin edges. If E_i is the number of observations expected using the theoretical model,

$$E_i = N p_i, \quad i = 1, 2, \cdots, k \qquad (5.42)$$

In Eq. (5.42), p_i is the probability that the random variable lies in the ith bin and is given by

$$p_i = F_X(x_{i+1}) - F_X(x_i), \quad i = 1, 2, \cdots, k \qquad (5.43)$$

The test statistic χ^2 is formed as (Rohatgi and Saleh 2001; Papoulis and Pillai 2002)

$$\chi^2 = \sum_{i=1}^{k} \frac{(O_i - E_i)^2}{E_i} \qquad (5.44)$$

This value of χ^2 may also be identified as the test statistic, χ^2_{test}.

Step #5 The test statistic, χ^2_{test}, follows a chi-squared distribution (Rohatgi and Saleh 2001). The proof of this result is beyond the scope of this book. The chi-squared density is specified in terms of its degree of freedom (Rohatgi and Saleh 2001, Papoulis and Pillai 2002). The number of degrees of freedom is $(k-c)$, with c to be defined.

If no parameters are estimated from the data, $c = 1$. In this case, χ^2_{test} follows a chi-squared distribution with $(k-1)$ degrees of freedom. Although there are k bins, once $(k-1)$ bins are filled, the remaining one is fixed. Therefore, the number of freely determined bins is $(k-1)$. The number of degrees of freedom is $(k-1)$ if no parameters are estimated from the data. If the number of parameters estimated from the data is q (note that q is an integer), the test statistic χ^2_{test} follows a chi-squared distribution with $(k - q - 1)$ degrees of freedom or $c = q + 1$. Thus, larger number of parameters estimated from the data leads to a lower number of degrees of freedom of the chi-squared distribution associated with the test.

The hypothesis that the data set follows a specified density is rejected if the test statistic exceeds a threshold chosen on the basis of a significance level α. The threshold value (also known as the critical value) is

$$\chi_{thr} = \chi^2_{1-\alpha,k-c} \tag{5.45}$$

The chi-squared density with ν degrees of freedom is (Rohatgi and Saleh 2001, Papoulis and Pillai 2002)

$$f(\lambda;\nu) = \frac{1}{2^{\frac{\nu}{2}}\Gamma\left(\frac{\nu}{2}\right)} \lambda^{\frac{\nu-2}{2}} e^{-\frac{\lambda}{2}} U(\lambda) \tag{5.46}$$

For the case of a Rayleigh or exponential random variable with a single parameter, $c = 2$. For a Gaussian distribution with two parameters $c = 3$ etc. In most cases, a value of $\alpha = 0.05$ (level of significance of 5%) is sufficient.

The hypothesis (referred to as the null hypothesis) that the data belongs to a population specified by the distribution is rejected if the χ^2 from eq. (5.44) exceeds the threshold (Fig. 5.3). This relationship can be expressed as

$$\begin{aligned}
\chi^2 > \chi^2_{1-\alpha,k-c} &\Rightarrow \text{Reject } H_0 \text{ at a significance level of } \alpha \\
\chi^2 < \chi^2_{1-\alpha,k-c} &\Rightarrow \text{Do not reject } H_0 \text{ at a significance level of } \alpha
\end{aligned} \tag{5.47}$$

In other words, we cannot reject the null hypothesis that the data came from the specific density if χ^2 from eq. (5.44) is less than the threshold.

$\chi^2_{1-\alpha,k-c}$ is the chi square critical value with $(k - c)$ degrees of freedom and a significance level α expressed as

$$\chi^2_{thr} = F^{-1}(1 - \alpha | k - c) \tag{5.48}$$

$$\text{Prob}\left(\chi^2 > \chi^2_{1-\alpha,k-c}\right) = \alpha = 0.05 \tag{5.49}$$

Depending on the test statistic χ^2_{test} and the degrees of freedom, we may calculate the p-value of the test as

$$p\text{-value} = 1 - F_{\chi^2}\left(\chi^2_{\text{test}}, k - c\right) \tag{5.50}$$

In Eq. (5.50), $F(.)$ is the CDF of the chi square density. The p-value may be evaluated directly in Matlab using the command $chi2cdf(\chi^2_{\text{test}},$ Deg. Of Freedom, 'upper').

The chi-squared density plotted in Fig. 5.3 shows the threshold and the area (probability) $\alpha = 0.05$.

Let us undertake a chi square test of the data in Table 5.5 to determine the fit to gamma and exponential densities.

The test is undertaken with seven bins. The parameters of the gamma and exponential densities are estimated (MLE). The gamma parameters are $[a, b] = [3.4369, 1.47319]$ and the parameter of the exponential density $a = 5.06323$. The histogram of the data is given in Fig. 5.4.

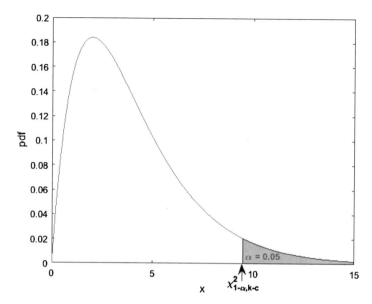

Fig. 5.3 The χ^2 density function. Shaded area $= \alpha$

Table 5.5 Data set used to demonstrate the implementation of chi square test

Input unsorted data				
6.5889	2.8042	3.8164	2.8874	1.7348
6.2764	3.4547	2.2632	4.8881	3.3568
5.7135	3.3896	2.9765	9.1957	5.7611
5.3399	5.7077	6.2818	5.7068	1.6729
6.1216	4.1935	1.944	4.8768	0.9561
2.5663	3.8185	4.4903	13.5919	3.0162
6.3338	6.253	5.8799	5.3804	7.0976
3.222	7.4027	7.2966	0.8844	1.4173
4.1875	6.5021	4.8783	10.7859	7.6758
4.9922	4.2004	4.857	3.8925	6.3321
7.2075	4.6937	3.5838	5.7283	8.8146
5.1427	6.913	6.9682	1.8229	0.2315
5.1357	3.2421	2.8491	3.1159	7.628
3.546	5.0351	3.53	9.0868	13.1373
5.8516	4.3793	5.3213	4.4847	1.0822
9.053	6.2066	3.7034	4.9232	8.8616
7.9149	4.6885	4.4253	2.8418	5.4464
3.7641	1.035	2.5628	5.6141	2.7637
9.0439	5.6855	2.4353	11.317	2.3767
4.849	6.5785	4.6542	8.9624	3.2207

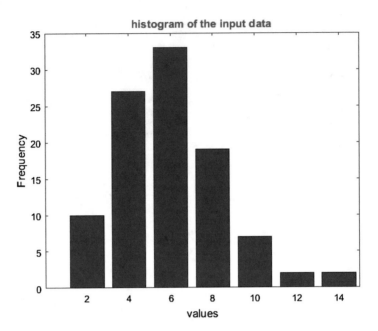

Fig. 5.4 Histogram of the data

Table 5.6 Bin counts and the corresponding probabilities (created in Matlab)

χ^2 test results

$N = 100,\ E_j = p_j N,\ j = 1,2,\ldots k$

Bin (j)	O(j)	$p_{gamma}(j)$	$p_{exp}(j)$	$E_{gamma}(j)$	$E_{exp}(j)$
1	10	0.097	0.326	9.68	32.63
2	27	0.31	0.22	30.98	21.98
3	33	0.286	0.148	28.57	14.81
4	19	0.17	0.1	17.03	9.98
5	7	0.082	0.067	8.19	6.72
6	2	0.035	0.045	3.46	4.53
7	2	0.021	0.093	2.1	9.35

$$\chi^2_{gamma} = \sum_{j=1}^{j=k} \frac{(O_j - E_j)^2}{E_j} = 2.2356$$

$$\chi^2_{exp} = \sum_{j=1}^{j=k} \frac{(O_i - E_j)^2}{E_j} = 54.5421$$

The bin counts and the corresponding probabilities are shown in Table 5.6. The calculated χ^2 test statistic values are also given in Table 5.6.

Table 5.7 provides an interpretation of the test statistic obtained. The two columns represent the degrees of freedom and $\chi^2_{threshold}$. It is the table of the critical values (level of significance of 5%) which we may use to make a decision on rejecting the hypothesis. With seven bins, the number of degrees of freedom associated with the

Table 5.7 Degrees of freedom and threshold (critical) values

Degrees of freedom and $\chi^2_{\text{threshold}}$		Deg. of Freedom	$\chi^2_{\text{threshold}}$
		2	5.9915
		3	7.8147
Gamma-pdf ⇒		4	9.4377
Exp-pdf ⇒		5	11.0705
		6	12.5916
		7	14.0671
Deg. of freedom = no. bins −1− no. parameters		3	15.5073
Deg. of freedom (gamma) = no. bins −1−2		9	16.919
Deg. of freedom (exp) = no. bins −1−1		10	13.307
No. bins = 7		11	19.6751
Do not REJECT gamma: $\chi^2_{\text{gamma}} < \chi^2_{\text{thr}}$		12	21.0261
$\chi^2_{\text{gamma}} = 2.2356$		13	22.362
REJECT exponential: $\chi^2_{\text{exp}} > \chi^2_{\text{thr}}$		14	23.6343
$\chi^2_{\text{exp}} = 54.5421$		15	24.9953

hypothesis that the data set fits gamma density is $7 - 1 - 2 = 4$, while the number of degrees of freedom associated with the hypothesis that the data set fits exponential density is $7 - 1 - 1 = 5$. From the test statistic values, it is clear that we <u>cannot reject</u> the hypothesis that the data sets a gamma distribution and we may <u>reject</u> the hypothesis that the data set fits an exponential distribution.

Chi-square testing can also be undertaken in Matlab for a number of densities. If the density being investigated is not part of the Matlab menu of densities, it is possible to create a script in Matlab to undertake chi square testing.

5.2.3.3   Simultaneous Testing of Multiple Hypothesis: A Comprehensive Approach

The example shown with the aid of entries in Table 5.5 dealt with two hypotheses. In engineering applications, it may be necessary to explore whether the data fits one of several densities, and some of these densities may be a single parameter ones, while the others may be two or three parameter ones (Barton 1956; Baggerly 2001). The interest is to determine the density that is the best fit or the closest fit to the data when several densities produce chi-square test values below the threshold. In other words, multiple null hypotheses exist that could not be rejected. Another interesting case is one where all the densities produce chi-square test values exceeding the threshold leading to the rejection of all densities. This means that no densities fit and we may either choose a density based on a different criterion or reduced the level of significance from 5% to 1%.

A parameter that may provide a quantitative measure of the fit is the reduced chi-square test statistics given as the ratio of the test statistic to the degrees of freedom, χ^2_{Re} (Hooper et al. 2008; Andrae et al. 2010)

$$\chi^2_{Re} = \frac{\chi^2_{stat}}{DoF} \tag{5.51}$$

This means that the density with the lowest value of the reduced chi-square test statistic (χ^2_{Re}) may be the closest fit. We may also choose the density with the highest p-value as the closest fit.

Another issue pertained to the choice of the number of bins and the minimum population in the bins required. This issue is mitigated by having a constant or fixed population in bins instead of the traditional fixed bin width. This means that the bin width will be adjusted such that every bin will have the same number of samples (if the number of entries is not an integer multiple of the number of bins, the last bin is assigned a larger number of entries). For example, if the number of samples is 100 and the number of bins is chosen as 7, bins 1–6 will have population of 14 each; bin #7 will have 16 entries.

We will use the data in Table 5.8 in this demo. Figure 5.5 shows the traditional results of fixed bin width one group of densities. It is seen that Rayleigh, gamma, and Nakagami all result is high p-values (much higher than 0.05), and none of these three could be rejected at a level of significance of 5%. Exponential density certainly is rejected. Fig. 5.6 shows the results with the constant population. While the p-values

Table 5.8 Data set used in the demo

1.0875	3.2982	1.9952	0.7042	10.532
10.0614	0.5532	4.6879	1.2521	1.6252
11.4336	8.9444	5.0528	9.3585	4.192
7.3546	7.9666	1.1607	7.1031	9.4288
3.9672	7.4142	5.5894	8.688	2.4782
2.7201	0.5781	1.7064	0.0946	7.828
0.5502	0.6816	2.1614	3.1137	2.6564
2.9793	8.4962	5.0535	3.5661	2.5696
5.097	12.1552	9.7406	4.5094	1.7923
6.99	13.7233	10.7134	9.7685	0.991
0.7713	1.2	2.3225	7.5445	4.9557
2.3447	10.4994	0.5977	2.24	2.0531
5.5552	1.9862	14.1403	9.489	10.9764
6.4288	6.5407	0.8011	4.0202	4.035
9.3621	10.4272	7.841	5.1621	4.6378
9.0806	4.1889	1.2814	6.9819	5.9836
0.6903	0.3517	7.1243	10.2132	0.9922
4.7312	9.7355	6.037	5.429	3.2503
13.7327	12.2296	3.595	5.6011	10.3967
3.8835	11.7988	6.4795	6.3879	1.7977

χ^2 test (fixed bin width): No. of samples = 100, No. of bins = 7

bin$_j$	O$_j$	gamma		Nakagami		Exponential		Rayleigh	
		P$_j$	E$_j$	P$_j$	E$_j$	P$_j$	E$_j$	P$_j$	E$_j$
j = 1	10	0.0968	9.675	0.103	10.301	0.3263	32.632	0.1173	11.728
2	27	0.3098	30.983	0.2752	27.515	0.2198	21.984	0.2756	27.557
3	33	0.2857	28.566	0.2935	29.346	0.1481	14.81	0.2818	28.175
4	19	0.1703	17.025	0.1966	19.655	0.0998	9.977	0.1895	18.951
5	7	0.0819	8.185	0.0916	9.159	0.0672	6.721	0.0917	9.167
6	2	0.0346	3.464	0.0309	3.089	0.0453	4.528	0.033	3.301
7	2	0.021	2.101	0.0093	0.935	0.0935	9.348	0.0112	1.121

Density	Parameters		DoF	$\chi^2_{0.05}$	χ^2_T	χ^2_R	pvalue	Reject (Yes/No)
gamma	[3.44	1.47]	4	9.49	2.24	0.56	0.69252	No
Rayleigh	[4	---]	5	11.07	2.81	0.56	0.72987	No
Nakagami	[1.08	32.07]	4	9.49	2.6	0.65	0.62628	No
Exponential	[5.06	---]	5	11.07	54.54	10.91	1.6215e-10	Yes

Fig. 5.5 The reject (yes/no) choice at a level of significance of 5%

χ^2 test (fixed population count): No. of samples = 100, No. of bins = 7

bin$_j$	O$_j$	gamma		Nakagami		Exponential		Rayleigh	
		P$_j$	E$_j$	P$_j$	E$_j$	P$_j$	E$_j$	P$_j$	E$_j$
j = 1	14	0.1724	17.243	0.1686	16.86	0.3972	39.719	0.1852	18.521
2	14	0.1322	13.219	0.1158	11.58	0.0908	9.082	0.1159	11.593
3	14	0.1709	17.09	0.1601	16.012	0.0947	9.471	0.1559	15.59
4	14	0.1065	10.651	0.108	10.795	0.0546	5.462	0.1036	10.365
5	14	0.0963	9.628	0.1037	10.365	0.0496	4.957	0.0991	9.911
6	14	0.1298	12.981	0.1483	14.829	0.0722	7.221	0.1423	14.232
7	16	0.1919	19.189	0.1956	19.558	0.2409	24.087	0.1979	19.788

Density	Parameters		DoF	$\chi^2_{0.05}$	χ^2_T	χ^2_R	pvalue	Reject (Yes/No)
Nakagami	[1.08	32.07]	4	9.49	4.16	1.04	0.38436	No
Rayleigh	[4	---]	5	11.07	5.46	1.09	0.36264	No
gamma	[3.44	1.47]	4	9.49	4.86	1.22	0.3016	No
Exponential	[5.06	---]	5	11.07	60.4	12.08	1.0038e-11	Yes

Fig. 5.6 The reject (yes/no) choice at a level of significance of 5%

$$\chi^2 \text{ test: No. of samples} = 100$$

Parameters

gamma [3.44 1.47]   Weibull [5.71 2.08]
Nakagami [1.08 32.07]   exponential 5.06
Rician [0.62 3.98]   Rayleigh   4

Density	DoF	$\chi^2_{0.05}$	χ^2_T	χ^2_R	pvalue	Reject (at 5%)
Fixed width — No. of bins = 7						
gamma	4	9.49	2.24	0.56	0.6925	No
Rayleigh	5	11.07	2.81	0.56	0.7299	No
Nakagami	4	9.49	2.6	0.65	0.6263	No
Rician	4	9.49	2.81	0.7	0.5908	No
Weibull	4	9.49	2.83	0.71	0.5871	No
exponential	5	11.07	54.54	10.91	< 0.0001	Yes
Fixed population — No. of bins = 7						
Nakagami	4	9.49	4.16	1.04	0.3844	No
Rayleigh	5	11.07	5.46	1.09	0.3626	No
Weibull	4	9.49	4.47	1.12	0.3457	No
gamma	4	9.49	4.86	1.22	0.3016	No
Rician	4	9.49	5.46	1.37	0.2431	No
exponential	5	11.07	60.4	12.08	< 0.0001	Yes

Fig. 5.7 The reject (yes/no) choice at a level of significance of 5%

are different, the trends match those seen with fixed bin width seen in Figs. 5.5 and 5.7 shows the results by including additional densities for testing.

5.2.3.4 Cautionary Note on Chi-Square Testing (Variability)

While the use of the reduced chi square test statistic is one way to identify the closest fit, another way to obtain the closest fit when multiple hypotheses are tested is to rely on the p-value. The density with the highest p-value is chosen. But, one needs to be careful not to rely too much on these choices because when the data collected is random, the test statistic and the p-value are also random. Additionally, number of samples and number of bins impact the test. To demonstrate the dependence of the number of samples on the test, the chi-square test is undertaken with 100 samples (Table 5.9). Next, 60 samples are picked randomly from this set and the chi square tests are repeated. This is done a number of times, each time picking 60 out of 100 samples using *randperm(.)* command in Matlab. Figure 5.8 shows the results with all 100 samples. Figure 5.9 shows the density fits to the data using the parameters estimated using MLE. Figures 5.10, 5.11, 5.12, 5.13, 5.14, 5.15, 5.16, and 5.17 show the test results for iterations with 60 randomly chosen samples. Results show that the densities that appear as the best fits change as the process of random sampling to pick 60 samples is repeated.

Table 5.9 Data set used to examine the effects of randomness of samples

Sorted data				
0.914	4.721	7.563	10.782	15.674
1.54	4.765	7.573	10.845	16.298
1.813	4.809	7.906	11.042	16.337
2.755	4.938	8.273	11.607	16.896
3.022	4.99	8.301	11.631	16.976
3.323	5.591	8.491	11.717	17.051
3.324	5.641	8.549	11.871	17.144
3.455	5.84	8.803	12.015	18.41
3.515	5.889	8.86	12.068	19.307
3.625	6.841	8.865	12.893	19.975
3.763	7.02	9.13	12.913	20.194
3.81	7.022	9.183	13.449	20.729
3.908	7.091	9.299	13.789	21.185
4.085	7.131	9.418	13.86	21.779
4.419	7.159	9.794	14.211	22.245
4.504	7.214	9.817	14.49	24.569
4.521	7.309	10.499	14.497	31.179
4.571	7.321	10.603	14.797	33.669
4.604	7.448	10.606	14.799	34.542
4.666	7.515	10.642	15.022	41.63

χ^2 **test: No. of samples = 100**

Parameters

gamma	$\begin{bmatrix} 2.48 & 4.37 \end{bmatrix}$	Weibull	$\begin{bmatrix} 12.15 & 1.6 \end{bmatrix}$
Nakagami	$\begin{bmatrix} 0.74 & 170.77 \end{bmatrix}$	exponential	10.83
Rician	$\begin{bmatrix} 0.52 & 9.23 \end{bmatrix}$	Rayleigh	9.24

	Density	DoF	$\chi^2_{0.05}$	χ^2_T	χ^2_R	pvalue	Reject (at 5%)
Fixed width **No. of bins = 7**	gamma	4	9.49	5.44	1.36	0.2451	No
	Weibull	4	9.49	7.5	1.88	0.1115	No
	Nakagami	4	9.49	13.17	3.29	0.0105	Yes
	exponential	5	11.07	19.13	3.83	0.0018	Yes
	Rayleigh	5	11.07	41.6	8.32	< 0.0001	Yes
	Rician	4	9.49	41.54	10.39	< 0.0001	Yes
Fixed population **No. of bins = 7**	gamma	4	9.49	1.1	0.28	0.8945	No
	Weibull	4	9.49	2.89	0.72	0.5756	No
	Nakagami	4	9.49	4.59	1.15	0.3322	No
	Rayleigh	5	11.07	9.78	1.96	0.0817	No
	Rician	4	9.49	9.79	2.45	0.0441	Yes
	exponential	5	11.07	21.81	4.36	0.0006	Yes

Fig. 5.8 Summary of chi-square test

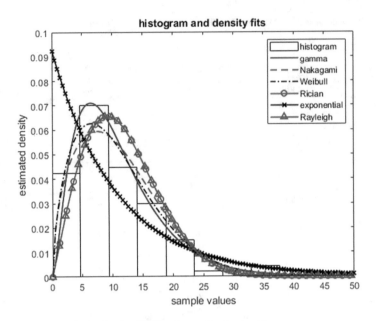

Fig. 5.9 Density fits obtained using MLE estimates for the densities

χ^2 test: No. of samples = 60

Parameters				
gamma	[2.68 3.8]		Weibull	[11.44 1.67]
Nakagami	[0.8 145.74]		exponential	10.17
Rician	[0.37 8.53]		Rayleigh	8.54

	Density	DoF	$\chi^2_{0.05}$	χ^2_T	χ^2_R	pvalue	Reject (at 5%)
Fixed width No. of bins = 7	gamma	4	9.49	6.01	1.5	0.1984	No
	Weibull	4	9.49	9.31	2.33	0.0538	No
	exponential	5	11.07	13.6	2.72	0.0184	Yes
	Nakagami	4	9.49	15.44	3.86	0.0039	Yes
	Rayleigh	5	11.07	34.57	6.91	< 0.0001	Yes
	Rician	4	9.49	34.54	8.64	< 0.0001	Yes
Fixed population No. of bins = 7	gamma	4	9.49	2.96	0.74	0.5638	No
	Rayleigh	5	11.07	6.58	1.32	0.2534	No
	Weibull	4	9.49	5.32	1.33	0.2561	No
	Nakagami	4	9.49	6.16	1.54	0.1878	No
	Rician	4	9.49	6.59	1.65	0.1593	No
	exponential	5	11.07	20.99	4.2	0.0008	Yes

Fig. 5.10 Summary of chi-square test using 60 of the 100 samples

χ^2 test: No. of samples = 60

Parameters						
gamma	[2.39	4.56	]	Weibull	[12.23	1.57]
Nakagami	[0.72	175.49	]	exponential	10.93	
Rician	[0.49	9.36	]	Rayleigh	9.37	

	Density	DoF	$\chi^2_{0.05}$	χ^2_T	χ^2_R	pvalue	Reject (at 5%)
Fixed width No. of bins = 7	gamma	4	9.49	2.16	0.54	0.7056	No
	Weibull	4	9.49	4.12	1.03	0.3895	No
	Nakagami	4	9.49	8.41	2.1	0.0778	No
	exponential	5	11.07	12.22	2.44	0.0319	Yes
	Rayleigh	5	11.07	32.25	6.45	< 0.0001	Yes
	Rician	4	9.49	32.2	8.05	< 0.0001	Yes
Fixed population No. of bins = 7	gamma	4	9.49	0.59	0.15	0.9642	No
	Weibull	4	9.49	1.49	0.37	0.8278	No
	Nakagami	4	9.49	2.57	0.64	0.6324	No
	Rayleigh	5	11.07	6.46	1.29	0.2637	No
	Rician	4	9.49	6.47	1.62	0.1667	No
	exponential	5	11.07	11.75	2.35	0.0384	Yes

Fig. 5.11 Summary of chi-square test using 60 of the 100 samples

χ^2 test: No. of samples = 60

Parameters						
gamma	[2.53	4.28	]	Weibull	[12.14	1.61]
Nakagami	[0.76	169.39	]	exponential	10.82	
Rician	[0.49	9.2	]	Rayleigh	9.2	

	Density	DoF	$\chi^2_{0.05}$	χ^2_T	χ^2_R	pvalue	Reject (at 5%)
Fixed width No. of bins = 7	gamma	4	9.49	5.58	1.4	0.233	No
	Weibull	4	9.49	7.35	1.84	0.1183	No
	exponential	5	11.07	14.74	2.95	0.0115	Yes
	Nakagami	4	9.49	13.07	3.27	0.0109	Yes
	Rayleigh	5	11.07	42.07	8.41	< 0.0001	Yes
	Rician	4	9.49	42	10.5	< 0.0001	Yes
Fixed population No. of bins = 7	gamma	4	9.49	2.94	0.74	0.5675	No
	Weibull	4	9.49	3.16	0.79	0.5313	No
	Nakagami	4	9.49	3.68	0.92	0.4517	No
	Rayleigh	5	11.07	6.55	1.31	0.2559	No
	Rician	4	9.49	6.56	1.64	0.161	No
	exponential	5	11.07	16.12	3.22	0.0065	Yes

Fig. 5.12 Summary of chi-square test using 60 of the 100 samples

χ^2 **test: No. of samples = 60**

Parameters

gamma	[2.3	5.02	]	Weibull	[12.9	1.53	]
Nakagami	[0.69	200.26]		exponential	11.55		
Rician	[0.54	10	]	Rayleigh	10.01		

	Density	DoF	$\chi^2_{0.05}$	χ^2_T	χ^2_R	pvalue	Reject (at 5%)
Fixed width No. of bins = 7	gamma	4	9.49	7.23	1.81	0.1241	No
	Weibull	4	9.49	8.51	2.13	0.0747	No
	Nakagami	4	9.49	11.32	2.83	0.0232	Yes
	exponential	5	11.07	16.06	3.21	0.0067	Yes
	Rayleigh	5	11.07	27.96	5.59	< 0.0001	Yes
	Rician	4	9.49	27.93	6.98	< 0.0001	Yes
Fixed population No. of bins = 7	gamma	4	9.49	8.54	2.13	0.0738	No
	Weibull	4	9.49	11.29	2.82	0.0235	Yes
	Rayleigh	5	11.07	14.85	2.97	0.011	Yes
	Nakagami	4	9.49	13.15	3.29	0.0106	Yes
	Rician	4	9.49	14.86	3.72	0.005	Yes
	exponential	5	11.07	26.53	5.31	< 0.0001	Yes

Fig. 5.13 Summary of chi-square test using 60 of the 100 samples

χ^2 **test: No. of samples = 60**

Parameters

gamma	[2.54	4.44	]	Weibull	[12.66	1.59	]
Nakagami	[0.75	186.07]		exponential	11.27		
Rician	[0.53	9.64	]	Rayleigh	9.65		

	Density	DoF	$\chi^2_{0.05}$	χ^2_T	χ^2_R	pvalue	Reject (at 5%)
Fixed width No. of bins = 7	gamma	4	9.49	5.6	1.4	0.2311	No
	Weibull	4	9.49	6.85	1.71	0.1438	No
	exponential	5	11.07	12.61	2.52	0.0273	Yes
	Nakagami	4	9.49	11.18	2.8	0.0246	Yes
	Rayleigh	5	11.07	30.47	6.09	< 0.0001	Yes
	Rician	4	9.49	30.43	7.61	< 0.0001	Yes
Fixed population No. of bins = 7	gamma	4	9.49	2.07	0.52	0.722	No
	Weibull	4	9.49	2.71	0.68	0.6073	No
	Nakagami	4	9.49	3.65	0.91	0.4554	No
	Rayleigh	5	11.07	7.09	1.42	0.2139	No
	Rician	4	9.49	7.1	1.78	0.1307	No
	exponential	5	11.07	14.99	3	0.0104	Yes

Fig. 5.14 Summary of chi-square test using 60 of the 100 samples

χ^2 test: No. of samples = 60

Parameters	gamma [2.78 3.49] Weibull [10.92 1.73]	
	Nakagami [0.84 129.75] exponential 9.69	
	Rician [0.4 8.05] Rayleigh 8.05	

	Density	DoF	$\chi^2_{0.05}$	χ^2_T	χ^2_R	pvalue	Reject (at 5%)
Fixed width No. of bins = 7	gamma	4	9.49	2.68	0.67	0.6133	No
	Weibull	4	9.49	5.25	1.31	0.2626	No
	Nakagami	4	9.49	9.3	2.33	0.054	No
	exponential	5	11.07	13.72	2.74	0.0175	Yes
	Rayleigh	5	11.07	19.9	3.98	0.0013	Yes
	Rician	4	9.49	19.88	4.97	0.0005	Yes
Fixed population No. of bins = 7	gamma	4	9.49	0.44	0.11	0.979	No
	Weibull	4	9.49	1.27	0.32	0.8672	No
	Nakagami	4	9.49	1.58	0.4	0.8115	No
	Rayleigh	5	11.07	2.22	0.44	0.8184	No
	Rician	4	9.49	2.22	0.56	0.6953	No
	exponential	5	11.07	17.06	3.41	0.0044	Yes

Fig. 5.15 Summary of chi-square test using 60 of the 100 samples

χ^2 test: No. of samples = 60

Parameters	gamma [2.6 4.28] Weibull [12.54 1.69]	
	Nakagami [0.8 173.58] exponential 11.15	
	Rician [0.53 9.31] Rayleigh 9.32	

	Density	DoF	$\chi^2_{0.05}$	χ^2_T	χ^2_R	pvalue	Reject (at 5%)
Fixed width No. of bins = 7	Weibull	4	9.49	5.22	1.31	0.2654	No
	gamma	4	9.49	5.5	1.38	0.2397	No
	Nakagami	4	9.49	7.11	1.78	0.13	No
	exponential	5	11.07	10.28	2.06	0.0677	No
	Rayleigh	5	11.07	17.11	3.42	0.0043	Yes
	Rician	4	9.49	17.1	4.28	0.0018	Yes
Fixed population No. of bins = 7	Weibull	4	9.49	2.91	0.73	0.5728	No
	gamma	4	9.49	3.2	0.8	0.5249	No
	Nakagami	4	9.49	3.23	0.81	0.5208	No
	Rayleigh	5	11.07	5.82	1.16	0.3237	No
	Rician	4	9.49	5.83	1.46	0.2123	No
	exponential	5	11.07	15.49	3.1	0.0085	Yes

Fig. 5.16 Summary of chi-square test using 60 of the 100 samples

χ^2 **test: No. of samples = 60**

Parameters

gamma	[2.19	5.34	]	Weibull	[13.05	1.53	]
Nakagami	[0.69	202.33	]	exponential	11.69		
Rician	[0.44	10.05	]	Rayleigh	10.06		

Density	DoF	$\chi^2_{0.05}$	χ^2_T	χ^2_R	pvalue	Reject (at 5%)
Fixed width — No. of bins = 7						
gamma	4	9.49	2.17	0.54	0.704	No
Weibull	4	9.49	2.68	0.67	0.6125	No
Nakagami	4	9.49	4.51	1.13	0.3416	No
exponential	5	11.07	8.76	1.75	0.1189	No
Rayleigh	5	11.07	19.87	3.97	0.0013	Yes
Rician	4	9.49	19.86	4.97	0.0005	Yes
Fixed population — No. of bins = 7						
Weibull	4	9.49	1.58	0.4	0.8119	No
gamma	4	9.49	1.67	0.42	0.7965	No
Nakagami	4	9.49	2.17	0.54	0.7037	No
Rayleigh	5	11.07	9.06	1.81	0.1068	No
exponential	5	11.07	9.91	1.98	0.0778	No
Rician	4	9.49	9.06	2.27	0.0596	No

Fig. 5.17 Summary of chi-square test using 60 of the 100 samples

5.2.3.5 Additional Data Creation and Analysis

Having seen the PPV and area under the ROC curve (AUC), and the ability to test whether the data sets fit certain densities, the question one should ask is whether it is possible to improve the performance so that the area under the ROC curve and PPV will go up. This aspect is important since we often may only have a single opportunity to conduct the experiment and we are interested in maximizing the performance based on the data collected once. In this context, it is appropriate to introduce another quantitative measure of the performance of the receiver. The performance index κ, known as contrast-to-noise ratio in image analysis (Timischl 2015), is defined as

$$\kappa = \frac{\left| \text{mean}\left(\text{data}_{\text{target Absent}}\right) - \text{mean}\left(\text{data}_{\text{target Present}}\right) \right|}{\sqrt{\text{var}\left(\text{data}_{\text{target Present}}\right) + \text{var}\left(\text{data}_{\text{target Absent}}\right)}}. \tag{5.52}$$

To understand the notion of observing improvement, consider the case of data collected in an experiment, 100 samples when a target is present and 100 samples when no target is present. The data sets are displayed in the Tables 5.10 and 5.11.

The densities of these two data sets and the performance index values are shown in Fig. 5.18. The results also display the outcomes of the chi square tests conducted to establish the densities matching to the data sets.

Table 5.10 Data used to demonstrate performance improvement (Target absent)

Target absent									
4.881	2.888	1.951	2.719	1.81	3.238	2.036	1.803	1.174	1.016
0.453	3.395	2.48	4.298	2.087	2.96	2.977	1.82	2.177	3.694
3.055	4.389	1.382	2.554	1.402	0.631	0.339	2.749	2.227	4.061
1.948	2.546	1.786	2.592	2.522	2.458	0.826	1.386	0.548	2.742
0.988	1.752	2.339	0.853	3.027	2.056	7.033	3.282	1.621	0.861
0.523	2.684	4.589	1.549	5.277	2.358	1.472	2.458	2.616	2.302
3.986	5.382	6.246	1.666	0.673	2.166	1.161	2.119	0.55	1.617
1.46	1.588	2.235	5.465	2.841	2.473	4.156	5.516	3.054	3.621
1.857	2.448	3.349	0.892	1.571	3.798	3.865	3.201	3.81	2.809
2.324	4.154	1.495	1.103	1.898	3.255	2.405	5.435	2.029	2.174

Table 5.11 Data used to demonstrate performance improvement (Target present)

Target present									
7.741	5.138	1.522	7.482	4.956	4.918	11.878	6.705	5.352	4.541
7.804	1.981	20.771	7.622	12.78	6.348	10.079	9.397	1.671	5.711
10.361	2.781	4.039	4.383	11.755	7.506	4.023	13.256	3.848	7.428
9.078	3.726	8.222	2.487	1.721	12.229	4.886	10.853	8.826	8.203
7.78	6.506	1.448	6.948	11.882	8.341	2.447	5.446	4.611	6.864
6.54	13.66	6.955	3.72	11.125	3.822	2.433	6.053	6.854	4.231
3.15	3.766	10.267	2.564	4.627	7.546	6.882	7.09	5.951	4.626
1.933	9.66	11.562	6.353	7.077	8.702	4.65	5.43	5.662	8.895
4.645	1.617	0.53	1.225	1.635	5.819	7.947	4.936	4.567	6.058
9.726	4.551	4.49	3.568	5.239	4.432	3.254	5.075	5.08	3.09

Three different algorithms are explored to see how the performance can be improved. Using the statistics (density), two sets of data are generated (target absent and target present), and the sets are combined to get the arithmetic mean, geometric mean, and the maximum for the two cases. The performance indices are evaluated again. The results are displayed in Fig. 5.19. It is seen that there is an improvement in performance index following the implementation of the signal processing algorithms.

5.2.3.6   ROC Analysis and Performance Enhancement

Now that we have examined the methods of parameter estimation, hypothesis testing and simple signal processing approaches to improve performance, it is time to revisit the ROC analysis in an expanded way (Swets 1979, 1988; Bradley 1997; Kester and Buntinx 2000; Hand and Till 2001; Obuchowski et al. 2004; Obuchowski 2005; Liu 2012). We are given two data sets collected in a machine vision experiment with 70 samples of target absent in the field of view and 60 samples of target present in the field of view. The analysis is demonstrated in 4 separate parts.

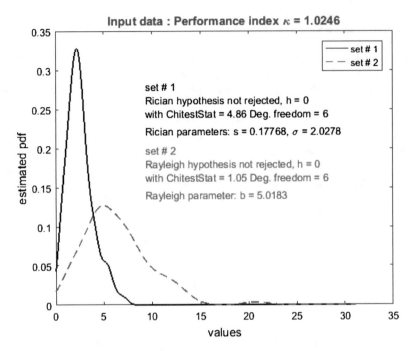

Fig. 5.18 Histograms of the two data sets

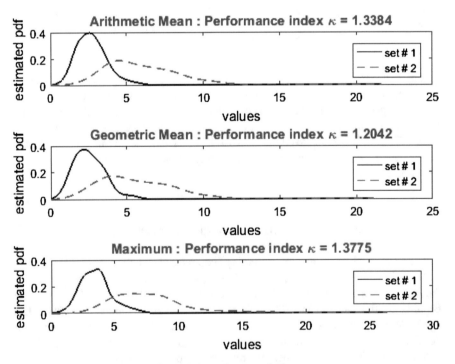

Fig. 5.19 Densities after processing and the performance indices

Part 1 ROC Analysis, Theoretical ROC Based on Bigamma Models, Etc.

ROC analysis is carried out. We examine whether the optimal threshold leads to improved performance, i.e., lower error rates. It should be noted that we are not minimizing the error rate; we are choosing an operation point to be as close to the ideal one as seen Eq. (5.2). The data values will be positive because the machine vision sensor collects either the power or amplitude implying that the values theoretically span the range of 0 to ∞. Therefore, we may also check whether a bigamma model is a good fit to the data in the following way.

Assuming that the data set (target absent) fits a gamma density, we estimate the parameters. Similarly, assuming that the data set (target present) fits a gamma density, we estimate the parameters. Using the method described in Chap. 3, we now superimpose a bigamma theoretical ROC plot and estimate the area under the ROC curve (Rota and Antolini 2014; Shankar 2020c). The standard deviation of the AUC is determined using the Eq. (5.4). Results are displayed in several figures. The data sets are displayed in Table 5.12. Figure 5.20 displays the histograms and the continuous density fits obtained using the *ksdensity*(.). Figure 5.21 displays the ROC curve and the theoretical ROC based on a bigamma fit along with AUC values. We see that the fit is excellent because the AUC values are very close. The standard deviation of the AUC (data based ROC) is also shown. Figure 5.22 displays the confusion and transition matrices based on the intersection threshold, while Fig. 5.23 displays the confusion and transition matrices based on the optimal threshold. It is seen that the error rate is less with the optimal threshold. Figure 5.24 displays the densities and the optimal threshold (Shankar 2019c, 2020b).

Part 2 Hypothesis Testing and Theoretical ROC

Chi-square tests are undertaken to determine the best fit. The best fits are used to obtain the theoretical ROC and corresponding AUC. In addition, 5000 pairs of random variables (fitting the two hypotheses) are created and 5000 samples of AUC are obtained to estimate the mean and standard deviation of the theoretical AUC. Figure 5.25 displays the summary of the chi square tests and the choice of the best fit based on the highest *p*-value. Figure 5.26 displays the histograms and the best fits. Figure 5.27 shows the ROC curve, bigamma fit, and the ROC obtained using the best fits. In addition, it also displays the mean and standard deviation of AUC obtained from 5000 pairs of data sets matching the best fits.

Part 3 ROC Analysis, Diversity, and Performance Improvement

In this segment, we examine the improvement in performance when we have two sets of data belonging to either cohort. In other words, we create a pair of 70 samples matching the density fit (target absent) and create a pair of 60 samples matching the density fit (target absent). We apply signal processing techniques of AM, GM, and

Table 5.12 Data sets used in the comprehensive ROC analysis

Target absent							Target present					
0.0417	0.4975	2.4658	0.1856	0.31	1.6165	1.6053	3.8512	3.2526	0.4546	3.2042	3.4836	2.6542
0.6093	0.4879	1.4913	0.3173	0.8336	0.6355	0.7442	1.118	0.2712	1.1744	10.8196	3.5073	0.6482
0.9142	0.3872	0.168	0.16	0.5997	0.8393	4.0264	3.2501	0.4149	4.9029	1.0372	2.3228	12.0799
0.0715	0.728	0.2706	0.0655	1.8017	0.7809	0.38	5.7053	3.4297	1.4847	3.6267	7.7312	2.6015
0.4109	1.2556	0.6694	0.3387	0.9671	0.7718	0.893	4.0862	4.2458	3.562	2.6836	5.3086	1.4573
1.9195	3.1118	0.4225	1.6583	0.8788	0.1435	0.3317	5.7655	5.5355	3.963	2.2996	2.7605	5.8009
2.1212	3.0926	0.6476	0.4509	0.3279	0.4179	0.9568	1.9516	4.3466	1.5988	4.9497	2.6621	1.6743
1.9869	1.1641	0.189	1.4747	1.2443	0.543	0.1737	6.6566	5.4364	0.4496	5.6089	0.9655	2.1666
0.3102	1.2432	0.6533	1.2117	2.9444	1.0255	0.4258	3.1963	5.6629	2.0495	1.8079	1.7515	2.1469
0.8157	0.0786	0.3236	0.1874	0.2765	3.0399	1.6026	6.064	2.9357	2.7209	9.3232	4.0503	1.6281

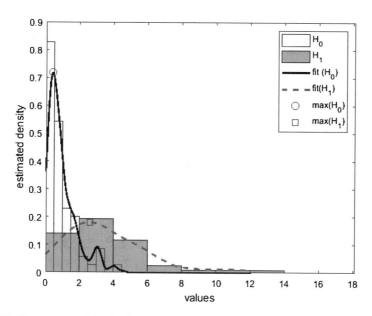

Fig. 5.20   Histograms and density fits

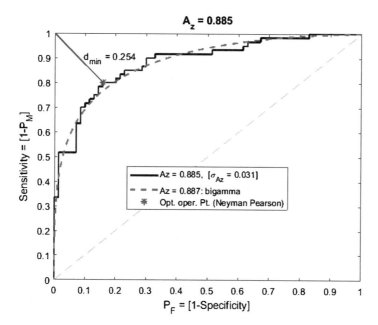

Fig. 5.21   ROC curve and the theoretical ROC (bigamma fit) along with AUC values

Threshold (v_T) = 1.8079 (intersection)

	Target Not Detected (D_n)	Target Detected (D_p)	Total Samples
Target Absent	61	False Alarm →9 ← N_F	70 ← N_0
Target Present	16 ← Miss	44 ← N_C	60 ← N_1
Total Decisions	77	53	130 ← N

$$\text{ERROR RATE} = \frac{N_F + (N_1 - N_C)}{N} = \frac{25}{130} \quad \text{PPV} = \frac{N_C}{N_F + N_C} = \frac{44}{53}$$

$$T_X = \begin{bmatrix} P(D_n|H_0) & P(D_n|H_1) \\ P(D_p|H_0) & P(D_p|H_1) \end{bmatrix}$$

$$T_X = \begin{bmatrix} 1 - P_F & P_M \\ P_F & 1 - P_M \end{bmatrix} = \begin{bmatrix} 61 & 16 \\ 9 & 44 \end{bmatrix} \begin{bmatrix} \frac{1}{70} & 0 \\ 0 & \frac{1}{60} \end{bmatrix} = \begin{bmatrix} \frac{61}{70} & \frac{16}{60} \\ \frac{9}{70} & \frac{44}{60} \end{bmatrix}$$

$$\text{a priori prob.} \rightarrow \begin{bmatrix} P(H_0) \\ P(H_1) \end{bmatrix} = \frac{1}{N} \begin{bmatrix} N_0 \\ N_1 \end{bmatrix} = \frac{1}{130} \begin{bmatrix} 70 \\ 60 \end{bmatrix}$$

$$\begin{bmatrix} P(D_n) \\ P(D_p) \end{bmatrix} = T_X \begin{bmatrix} P(H_0) \\ P(H_1) \end{bmatrix} = \frac{1}{130} \begin{bmatrix} 77 \\ 53 \end{bmatrix}$$

Fig. 5.22 Confusion and transition matrices based on intersection threshold

Threshold (v_T) = 1.6165 (optimal: Neyman Pearson)

	Target Not Detected (D_n)	Target Detected (D_p)	Total Samples
Target Absent	59	False Alarm →11 ← N_F	70 ← N_0
Target Present	12 ← Miss	48 ← N_C	60 ← N_1
Total Decisions	71	59	130 ← N

$$\text{ERROR RATE} = \frac{N_F + (N_1 - N_C)}{N} = \frac{23}{130} \quad \text{PPV} = \frac{N_C}{N_F + N_C} = \frac{48}{59}$$

$$T_X = \begin{bmatrix} P(D_n|H_0) & P(D_n|H_1) \\ P(D_p|H_0) & P(D_p|H_1) \end{bmatrix} = \begin{bmatrix} F_V(v_T|H_0) & F_V(v_T|H_1) \\ S_V(v_T|H_0) & S_V(v_T|H_1) \end{bmatrix}$$

$$T_X = \begin{bmatrix} 1 - P_F & P_M \\ P_F & 1 - P_M \end{bmatrix} = \begin{bmatrix} 59 & 12 \\ 11 & 48 \end{bmatrix} \begin{bmatrix} \frac{1}{70} & 0 \\ 0 & \frac{1}{60} \end{bmatrix} = \begin{bmatrix} \frac{59}{70} & \frac{12}{60} \\ \frac{11}{70} & \frac{48}{60} \end{bmatrix}$$

$$\text{a priori prob.} \rightarrow \begin{bmatrix} P(H_0) \\ P(H_1) \end{bmatrix} = \frac{1}{N} \begin{bmatrix} N_0 \\ N_1 \end{bmatrix} = \frac{1}{130} \begin{bmatrix} 70 \\ 60 \end{bmatrix}$$

$$\begin{bmatrix} P(D_n) \\ P(D_p) \end{bmatrix} = T_X \begin{bmatrix} P(H_0) \\ P(H_1) \end{bmatrix} = \frac{1}{130} \begin{bmatrix} 71 \\ 59 \end{bmatrix}$$

Fig. 5.23 Confusion and transition matrices based on optimal threshold

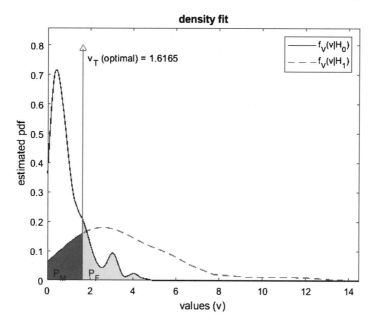

Fig. 5.24 Densities, probabilities of false alarm and miss (optimal threshold)

Hypothesis testing-Target Absent (70)

*Nakagami hypothesis not rejected, h = 0 with χ^2_{statR} = 1.165, DoF = 2, p-val = 0.3115

Nakagami parameters: m = 0.449, Ω = 1.606

Hypothesis testing-Target Present (60)

*gamma hypothesis not rejected, h = 0 with χ^2_{statR} = 1.223, DoF = 3, p-val = 0.2993

gamma parameters: a = 2.085, b = 1.697

summary of χ^2 tests (5% significance level)

	h	DoF	χ^2_{stat}	p-val	χ^2_{statR}	h	DoF	χ^2_{stat}	p-val	χ^2_{statR}
Weibull	0	3	4.92	0.1775	1.64	0	3	4.131	0.248	1.377
Nakagami	0	2	2.33	0.3115	1.165	0	2	2.66	0.2651	1.33
Gamma	0	3	4.779	0.189	1.593	0	3	3.669	0.2993	1.223
Rician	1	2	34.87	0	17.435	1	2	6.25	0.0439	3.125
Rayleigh	1	3	34.86	0	11.62	0	3	6.249	0.1002	2.083

Target Absent ⇑   **Target Present** ⇑

* **corresponds to the highest p-value**

$$\chi^2_{statR} = \frac{\chi^2_{stat}}{DoF}$$

Fig. 5.25 Results of the chi-square testing on target absent and target present data along with the names of the best fits based on the highest p-value

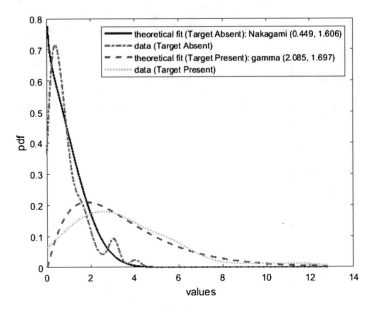

Fig. 5.26 Histograms and best fits

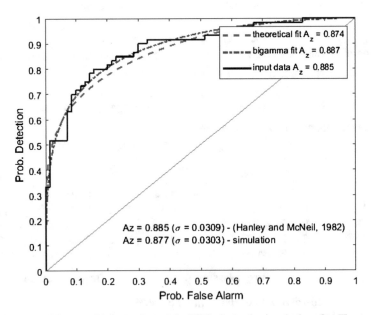

Fig. 5.27 The ROC curve, bigamma fit, and the ROC obtained using the best fits. The mean and standard deviation of AUC obtained from 5000 pairs of data sets matching the best fits

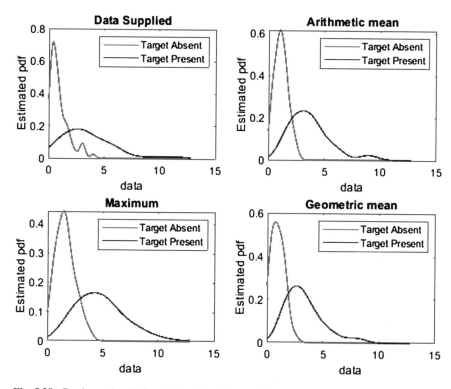

Fig. 5.28 Continous density fits of the original data and the processed sets

Max to each pair, creating a set of 70 samples of target absent and 60 samples of target present. This set is used for the estimation of AUC, PPV (optimal threshold) and performance index to demonstrate that improvement in AUC, PPV, and performance index is achieved (Shankar 2019c, 2020b). Figure 5.28 shows the densities of the processed sets along with that of the original data (top left corner). Figure 5.29 shows the ROC curves and the AUC values clearly indicating improvement. Figures 5.30, 5.31, 5.32 and 5.33 provide the confusion and transition matrices. Figure 5.34 provides the summary of the analysis clearly showing the improvement in PPV, AUC and performance index.

Part 4 ROC and Diversity Analysis

While Part 3 clearly demonstrated the potential of signal processing techniques (diversity), the results need to be tested because we only conducted the analysis once. This means that we do not know the variability of the performance metrics (AUC, PPV, error rate, and performance index) arising from the simulation (Shankar 2019c). In light of this, a simulation is carried out 500 times and the mean and standard deviation of the metrics are obtained to show the variability. Figure 5.35

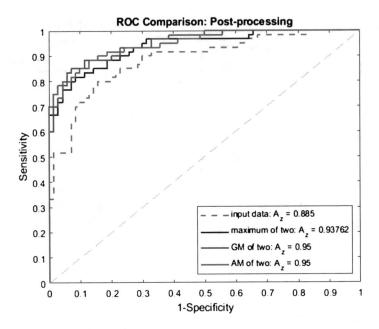

Fig. 5.29 ROC curves and the AUC values

Input Data

	Target Not Detected (D_n)	Target Detected (D_p)	Total Samples
Target Absent	59	**False Alarm** $\rightarrow 11 \leftarrow$ **N_F**	70 $\leftarrow$ **N_0**
Target Present	12 $\leftarrow$ **Miss**	48 $\leftarrow$ **N_C**	60 $\leftarrow$ **N_1**
Total Decisions	71	59	130 $\leftarrow$ **N**

$$\text{ERROR RATE} = \frac{N_F + (N_1 - N_C)}{N} = \frac{23}{130} \quad \text{PPV} = \frac{N_C}{N_F + N_C} = \frac{48}{59}$$

$$T_X = \begin{bmatrix} P(D_n|H_0) & P(D_n|H_1) \\ P(D_p|H_0) & P(D_p|H_1) \end{bmatrix}$$

$$T_X = \begin{bmatrix} 1 - P_F & P_M \\ P_F & 1 - P_M \end{bmatrix} = \begin{bmatrix} 59 & 12 \\ 11 & 48 \end{bmatrix} \begin{bmatrix} \frac{1}{70} & 0 \\ 0 & \frac{1}{60} \end{bmatrix} = \begin{bmatrix} \frac{59}{70} & \frac{12}{60} \\ \frac{11}{70} & \frac{48}{60} \end{bmatrix}$$

$$\text{a priori prob.} \rightarrow \begin{bmatrix} P(H_0) \\ P(H_1) \end{bmatrix} = \frac{1}{N} \begin{bmatrix} N_0 \\ N_1 \end{bmatrix} = \frac{1}{130} \begin{bmatrix} 70 \\ 60 \end{bmatrix}$$

$$\begin{bmatrix} P(D_n) \\ P(D_p) \end{bmatrix} = T_X \begin{bmatrix} P(H_0) \\ P(H_1) \end{bmatrix} = \frac{1}{130} \begin{bmatrix} 71 \\ 59 \end{bmatrix}$$

Fig. 5.30 Confusion and transition matrices

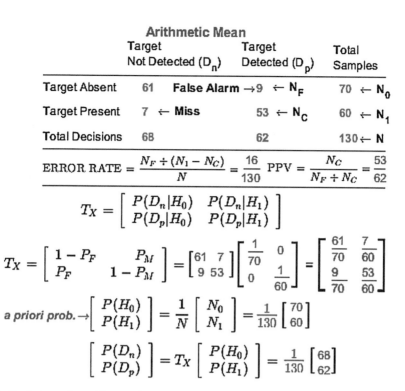

Arithmetic Mean

	Target Not Detected (D_n)	Target Detected (D_p)	Total Samples
Target Absent	61	False Alarm →9 ← N_F	70 ← N_0
Target Present	7 ← Miss	53 ← N_C	60 ← N_1
Total Decisions	68	62	130 ← N

$$\text{ERROR RATE} = \frac{N_F + (N_1 - N_C)}{N} = \frac{16}{130} \qquad \text{PPV} = \frac{N_C}{N_F + N_C} = \frac{53}{62}$$

$$T_X = \begin{bmatrix} P(D_n|H_0) & P(D_n|H_1) \\ P(D_p|H_0) & P(D_p|H_1) \end{bmatrix}$$

$$T_X = \begin{bmatrix} 1 - P_F & P_M \\ P_F & 1 - P_M \end{bmatrix} = \begin{bmatrix} 61 & 7 \\ 9 & 53 \end{bmatrix} \begin{bmatrix} \frac{1}{70} & 0 \\ 0 & \frac{1}{60} \end{bmatrix} = \begin{bmatrix} \frac{61}{70} & \frac{7}{60} \\ \frac{9}{70} & \frac{53}{60} \end{bmatrix}$$

$$\textit{a priori prob.} \rightarrow \begin{bmatrix} P(H_0) \\ P(H_1) \end{bmatrix} = \frac{1}{N} \begin{bmatrix} N_0 \\ N_1 \end{bmatrix} = \frac{1}{130} \begin{bmatrix} 70 \\ 60 \end{bmatrix}$$

$$\begin{bmatrix} P(D_n) \\ P(D_p) \end{bmatrix} = T_X \begin{bmatrix} P(H_0) \\ P(H_1) \end{bmatrix} = \frac{1}{130} \begin{bmatrix} 68 \\ 62 \end{bmatrix}$$

Fig. 5.31 Confusion and transition matrices

Maximum

	Target Not Detected (D_n)	Target Detected (D_p)	Total Samples
Target Absent	62	False Alarm →8 ← N_F	70 ← N_0
Target Present	10 ← Miss	50 ← N_C	60 ← N_1
Total Decisions	72	58	130 ← N

$$\text{ERROR RATE} = \frac{N_F + (N_1 - N_C)}{N} = \frac{18}{130} \qquad \text{PPV} = \frac{N_C}{N_F + N_C} = \frac{50}{58}$$

$$T_X = \begin{bmatrix} P(D_n|H_0) & P(D_n|H_1) \\ P(D_p|H_0) & P(D_p|H_1) \end{bmatrix}$$

$$T_X = \begin{bmatrix} 1 - P_F & P_M \\ P_F & 1 - P_M \end{bmatrix} = \begin{bmatrix} 62 & 10 \\ 8 & 50 \end{bmatrix} \begin{bmatrix} \frac{1}{70} & 0 \\ 0 & \frac{1}{60} \end{bmatrix} = \begin{bmatrix} \frac{62}{70} & \frac{10}{60} \\ \frac{8}{70} & \frac{50}{60} \end{bmatrix}$$

$$\textit{a priori prob.} \rightarrow \begin{bmatrix} P(H_0) \\ P(H_1) \end{bmatrix} = \frac{1}{N} \begin{bmatrix} N_0 \\ N_1 \end{bmatrix} = \frac{1}{130} \begin{bmatrix} 70 \\ 60 \end{bmatrix}$$

$$\begin{bmatrix} P(D_n) \\ P(D_p) \end{bmatrix} = T_X \begin{bmatrix} P(H_0) \\ P(H_1) \end{bmatrix} = \frac{1}{130} \begin{bmatrix} 72 \\ 58 \end{bmatrix}$$

Fig. 5.32 Confusion and transition matrices

Confusion Matrix (vertical label)

Transition Matrix (vertical label)

Geometric Mean

	Target Not Detected (D_n)	Target Detected (D_p)	Total Samples
Target Absent	62 False Alarm $\rightarrow$ 8 $\leftarrow N_F$		70 $\leftarrow N_0$
Target Present	7 $\leftarrow$ Miss	53 $\leftarrow N_C$	60 $\leftarrow N_1$
Total Decisions	69	61	130 $\leftarrow$ N

$$\text{ERROR RATE} = \frac{N_F + (N_1 - N_C)}{N} = \frac{15}{130} \quad \text{PPV} = \frac{N_C}{N_F + N_C} = \frac{53}{61}$$

$$T_X = \begin{bmatrix} P(D_n|H_0) & P(D_n|H_1) \\ P(D_p|H_0) & P(D_p|H_1) \end{bmatrix}$$

$$T_X = \begin{bmatrix} 1 - P_F & P_M \\ P_F & 1 - P_M \end{bmatrix} = \begin{bmatrix} 62 & 7 \\ 8 & 53 \end{bmatrix} \begin{bmatrix} \frac{1}{70} & 0 \\ 0 & \frac{1}{60} \end{bmatrix} = \begin{bmatrix} \frac{62}{70} & \frac{7}{60} \\ \frac{8}{70} & \frac{53}{60} \end{bmatrix}$$

$$\text{a priori prob.} \rightarrow \begin{bmatrix} P(H_0) \\ P(H_1) \end{bmatrix} = \frac{1}{N} \begin{bmatrix} N_0 \\ N_1 \end{bmatrix} = \frac{1}{130} \begin{bmatrix} 70 \\ 60 \end{bmatrix}$$

$$\begin{bmatrix} P(D_n) \\ P(D_p) \end{bmatrix} = T_X \begin{bmatrix} P(H_0) \\ P(H_1) \end{bmatrix} = \frac{1}{130} \begin{bmatrix} 69 \\ 61 \end{bmatrix}$$

Fig. 5.33 Confusion and transition matrices

Summary

Hypothesis testing-Target Absent (70)

Nakagami hypothesis not rejected, h = 0 with χ^2_{statR} = 1.17, DoF = 2, p-val = 0.311

Nakagami parameters: m = 0.449, Ω = 1.606

Hypothesis testing-Target Present (60)

gamma hypothesis not rejected, h = 0 with χ^2_{statR} = 1.2233, DoF = 3, p-val = 0.2994

gamma parameters: a = 2.085, b = 1.697

Performance measures

Positive Predictive Value at optimal threshold = 0.81356

Performance Index = 1.0095

Area under the ROC curve   Az = 0.885

Post-Processing Results

Positive Predictive Value at optimal threshold

0.85484 (Arithmetic Mean), 0.86885 (Geometric Mean), 0.86207 (Maximum)

Performance Index

1.3167 (Arithmetic Mean), 1.3271 (Geometric Mean), 1.2147 (Maximum)

Area under the ROC curve

0.95 (Arithmetic Mean), 0.95 (Geometric Mean), 0.93762 (Maximum)

Fig. 5.34 Summary of the analysis showing improvement in PPV, AUC, and performance index

Goodness of fit test (level of significance: 5%)

H_0: Nakagami pdf (m = 0.45, Ω = 1.61) not rejected with χ^2_{stat} = 2.33, DoF = 2, p-val = 0.311

H_1: gamma pdf (a = 2.08, b = 1.7) not rejected: χ^2_{stat} = 3.67, DoF = 2, p-val = 0.2993

Performance Metrics

Positive Predictive Value at optimal threshold: PPV = 0.8136

Area under the ROC curve: AUC = 0.885

Performance Index: K = 1.0095

Error Rate: ER = 0.1769

Metric	Arithmetic Mean	Maximum	Geometric Mean
PPV	0.8966 (0.038)	0.8783 (0.0411)	0.8803 (0.0395)
AUC	0.9531 (0.0174)	0.94 (0.0201)	0.9466 (0.0184)
K	1.4139 (0.1443)	1.294 (0.14)	1.3643 (0.1401)
ER	0.1003 (0.026)	0.1178 (0.028)	0.1127 (0.027)

Left margin labels: Data Analytics — Input — Diversity

Fig. 5.35 Summary of the diversity analysis

summarizes the results clearly showing the benefits of diversity as indicated by reduced error, higher values of PPV, AUC, and performance index. Figure 5.36 represent the theoretical ROC curves (assuming bigamma fits for the original data) of the processed data. In this case, the theoretical densities of the processed outputs were obtained numerically for the plots. The appropriate densities used to obtain the ROC curves in Fig. 5.36 are tabulated in Fig. 5.37.

5.3 Bayesian Decision Theory

The analysis described so far has been undertaken in terms of the simple Neyman-Person strategy. For the PPV calculations, it relied on the sample sizes to obtain the a priori probabilities. Such a strategy cannot be pursued in medical diagnostics where a priori probabilities depend on the disease prevalence in the general population. The costs of decisions are also not considered in the Neyman-Pearson strategy. A formal approach to decision strategy can be developed on the basis of what is termed as the Bayes' decision strategy. The strategy relies on two simple principles (Helstrom 1968; van Trees 1968):

- We choose from two hypotheses (H_0 and H_1) to make a final decision.
- Every decision has a cost associated with it regardless of whether it is a correct or incorrect.

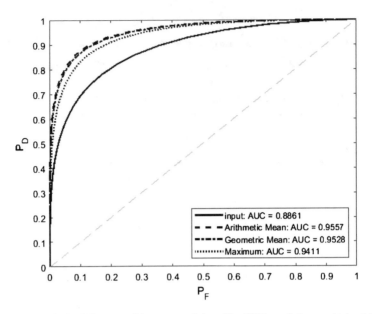

Fig. 5.36 Theoretical ROC curves of the processed data. The CDFs needed were obtained through numerical integration

Densities of the input and outputs assuming input to be gamma

pdf of input $\qquad f(x) = \dfrac{x^{a-1} e^{-\frac{x}{a}}}{b^a \Gamma(a)}$

pdf of AM $\qquad f(w) = \dfrac{w^{2a-1} e^{-\frac{2w}{b}} \left(\frac{2}{b}\right)^{2a}}{\Gamma(2a)}$

pdf of GM $\qquad f(w) = \dfrac{4 w^{2a-1} K_0\left(\frac{2w}{b}\right)}{b^{2a} \Gamma(a)^2}$

pdf of Max $\qquad f(w) = \dfrac{2 w^{a-1} e^{-\frac{w}{b}} \left(\Gamma(a) - \Gamma\left(a, \frac{w}{b}\right)\right)}{b^a \Gamma(a)^2}$

Fig. 5.37 Appropriat densities used to obtain theoretical ROC curves in Fig. 5.36

A criterion needs to be developed to make the "best choice" or "best course of action" by incorporating the cost into the decision-making process. To understand the second item above, we may start assigning costs to all the decisions. There are four decisions being made and we assign a cost to each of them. They are

$$\begin{bmatrix} C_{00} \\ C_{01} \\ C_{10} \\ C_{11} \end{bmatrix} = \begin{bmatrix} \text{Cost of choosing } H_0 \text{ when } H_0 \text{ is actually true} \\ \text{Cost of choosing } H_0 \text{ when } H_1 \text{ is actually true} \\ \text{Cost of choosing } H_1 \text{ when } H_0 \text{ is actually true} \\ \text{Cost of choosing } H_1 \text{ when } H_1 \text{ is actually true} \end{bmatrix} \quad (5.53)$$

Instead of representing them as a column vector, we represent them in matrix form. The cost matrix C is

$$C = \begin{bmatrix} C_{00} & C_{01} \\ C_{10} & C_{11} \end{bmatrix} \quad (5.54)$$

The terms in Eq. (5.54) consist of

$$C_{ij} = \text{cost}\big(\text{choose } H_i | H_j \text{ true}\big) \quad (5.55)$$

A note about the costs is important at this point. If a cancer is missed and subject informed that everything is normal, there is a substantial harm to the subject in terms of future costs and consequences from the cancer spreading. If a healthy person is informed that cancer is present, the cost of such a wrong conclusion is also extensive (both in terms of the expenses for further tests, unnecessary surgeries and mental anguish). We also should not ignore the fact that there is also a cost associated with correct decision.

We can now express the conditional costs associated with the decision threshold Z and the corresponding hypothesis noting that the choice of the threshold is critical to the process of decision making. The conditional cost (given H_0 is true) associated with choosing a threshold Z is

$$C(Z|H_0) = C_{00}P(\text{decide } H_0|H_0 \text{ true}) + C_{10}P(\text{decide } H_1|H_0 \text{ true}) \quad (5.56)$$

The conditional cost (given H_1 is true) is

$$C(Z|H_1) = C_{01}P(\text{decide } H_0|H_1 \text{ true}) + C_{11}P(\text{decide } H_1|H_1 \text{ true}) \quad (5.57)$$

The overall cost of the decision-making strategy is (Bayes' total probability rule)

$$C(Z) = C(Z|H_0)p(H_0) + C(Z|H_1)p(H_1) \quad (5.58)$$

Using the conditional costs, the overall cost (also known as the Bayes' risk) is

$$\begin{aligned} C(Z) = &[C_{00}P(\text{decide } H_0|H_0 \text{ true}) + C_{10}P(\text{decide } H_1|H_0 \text{ true})]p(H_0) \\ &+ [C_{01}P(\text{decide } H_0|H_1 \text{ true}) + C_{11}P(\text{decide } H_1|H_1 \text{ true})]p(H_1) \end{aligned} \quad (5.59)$$

In simple terms, the threshold Z needs to be chosen so that we minimize the Bayes' risk in Eq. (5.59).

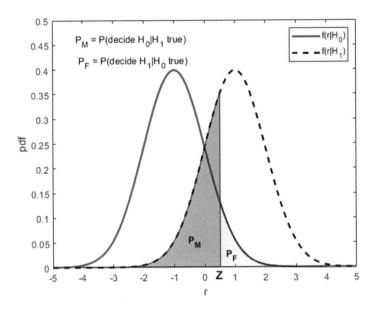

Fig. 5.38 Two densities and errors

If R represents the random variable (measured data), the densities corresponding to the two hypotheses are shown in Fig. 5.38. From the figure, we can express the following relationships (Chap. 3)

$$\begin{aligned}
P(\text{decide } H_1 | H_0 \text{ true}) &= P_F \\
P(\text{decide } H_0 | H_0 \text{ true}) &= 1 - P_F \\
P(\text{decide } H_0 | H_1 \text{ true}) &= P_M \\
P(\text{decide } H_1 | H_1 \text{ true}) &= 1 - P_M
\end{aligned}$$
(5.60)

The probabilities of miss (P_M) and false alarm (P_F) are

$$P_M = \int_{-\infty}^{Z} f(r | H_1) dr$$

$$1 - P_M = \int_{Z}^{\infty} f(r | H_1) dr$$

(5.61)

$$P_F = \int_{Z}^{\infty} f(r | H_0) dr$$

$$1 - P_F = \int_{-\infty}^{Z} f(r | H_0) dr$$

We represent the a priori probabilities as

$$
\begin{aligned}
p(H_0) &= p \\
p(H_1) &= q
\end{aligned}
\tag{5.62}
$$

Note that the probability of false alarm is also called the error of the first kind (type I) and the probability of miss is also called the error of the second kind (type II). The relative cost of the error of the first kind is $C_{10} - C_{00}$, and the relative cost of the error of the second kind is $C_{01} - C_{11}$.

Values in Eq. (5.60) are obtained on the basis of an arbitrarily chosen threshold Z as seen in Eq. (5.61). Using Eq. (5.60), we can express the Bayes' risk in eq. (5.59) as

$$
C(Z) = [C_{00}(1 - P_F) + C_{10}P_F]p(H_0) + [C_{01}P_M + C_{11}(1 - P_M)]p(H_1)
\tag{5.63}
$$

Using integral representation, Eq. (5.63) becomes

$$
\begin{aligned}
C(Z) = {}& \left[C_{00} \int_{-\infty}^{Z} f(r|H_0)dr + C_{10}\left(1 - \int_{-\infty}^{Z} f(r|H_0)dr\right) \right] p(H_0) \\
&+ \left[C_{01} \int_{-\infty}^{Z} f(r|H_1)dr + C_{11}\left(1 - \int_{-\infty}^{Z} f(r|H_1)dr\right) \right] p(H_1)
\end{aligned}
\tag{5.64}
$$

Simplifying Eq. (5.63), we get

$$
C(Z) = C_{10}p + C_{11}q + \int_{-\infty}^{Z} \left([q(C_{01} - C_{11})f(r|H_1)] - [p(C_{10} - C_{00})f(r|H_0)] \right) dr
\tag{5.65}
$$

The first two terms in Eq. (5.65) are always positive and fixed because they do not depend on the threshold. If we make the realistic assumption that the wrong decisions are far more costly than correct decisions, we have

$$
\begin{aligned}
C_{10} &> C_{00} \\
C_{01} &> C_{11}
\end{aligned}
\tag{5.66}
$$

Let us identify the two terns in eq. (5.65) as

$$
\begin{aligned}
A &= [q(C_{01} - C_{11})f(r|H_1)] \\
B &= [p(C_{10} - C_{00})f(r|H_0)]
\end{aligned}
\tag{5.67}
$$

By virtue of Eq. (5.66), quantities A and B in Eq. (5.67) are positive. Additionally, if A is less than B, the result of integration will be negative minimizing the cost. In this case, we conclude that the decision should be made in favor of H_0 (since $B > A$). If $A > B$, result of integration will also be minimum, and we conclude that the decision should be made in favor of H_1. Since the range of integration is identical, we can use the integrand alone to translate our decision strategy (minimize the cost) described here as

$$A \underset{H_0}{\overset{H_1}{\gtrless}} B \tag{5.68}$$

Rewriting Eq. (5.68) using the densities and costs, we have

$$[q(C_{01} - C_{11})f(r|H_1)] \underset{H_0}{\overset{H_1}{\gtrless}} [p(C_{10} - C_{00})f(r|H_0)] \tag{5.69}$$

Rearranging, we have

$$\frac{f(r|H_1)}{f(r|H_0)} \underset{H_0}{\overset{H_1}{\gtrless}} \left(\frac{p}{q}\right)\left(\frac{C_{10} - C_{00}}{C_{01} - C_{11}}\right) \tag{5.70}$$

The left-hand side is the likelihood ratio and the right-hand side offers a means to obtain the threshold. If we simplify further using Eq. (5.66), Eq. (5.70) becomes

$$\frac{f(r|H_1)}{f(r|H_0)} \underset{H_0}{\overset{H_1}{\gtrless}} \left(\frac{p}{q}\right)\left(\frac{C_{10}}{C_{01}}\right) \tag{5.71}$$

If the costs are equal, we get much simpler LR as

$$\frac{f(r|H_1)}{f(r|H_0)} \underset{H_0}{\overset{H_1}{\gtrless}} \left(\frac{p}{q}\right) \tag{5.72}$$

If the two hypotheses are equal ($p = q = 1/2$), the right-hand side of Eq. (5.72) equals 1, and by taking the logarithm, we get the simple expression for the likelihood ratio as

$$\log \frac{f(r|H_1)}{f(r|H_0)} \underset{H_0}{\overset{H_1}{\gtrless}} 0 \tag{5.73}$$

Equation (5.73) is identical to Eq. (3.206) presented in connection with the discussion of receiver operating characteristics.

5.3.1 ROC Analysis Revisited Through Bayes' Decision Theory

The ROC analysis undertaken in Chap. 3 and earlier in this chapter had not taken into account the actual a priori probabilities or costs associated with decisions in obtaining the optimal threshold. The optimum threshold was obtained on the basis of the simple approach that we wanted to get close to the top left corner of the ROC plot ignoring a priori probabilities or costs associated with decisions. In addition, the positive predictive value and the error rate were calculated through the use of a priori probabilities determined by the population counts of the data samples. However, in medical decision-making and in some machine vision experiments, the a priori probability associated with the prevalence of a disease or a faulty product may be very low. For example, with clinical diagnostic systems where the presence of cancer needs to be confirmed, the $p(H_1)$ may be very low ($\sim$ 1e-3 or 1e-5), and the number of patients involved in the study may be too low. This makes it necessary to revisit the concept of optimal threshold, positive predictive value, and, error rate because all these values are dependent on a priori probabilities. It should be noted that the probabilities of miss and false alarm used for the ROC plot are not influenced by the optimal threshold. In other words, area under the ROC curve is still valid regardless of the a priori probabilities.

Let us go back to the Bayes' decision theory discussion earlier and express the cost and a priori dependent term on the right-hand side of eq. (5.70) as

$$S = \left(\frac{p}{q}\right)\left(\frac{C_{10} - C_{00}}{C_{01} - C_{11}}\right) = \left(\frac{1 - \text{prev.}}{\text{prev.}}\right)\left(\frac{C_{10} - C_{00}}{C_{01} - C_{11}}\right) \qquad (5.74)$$

In Eq. (5.74), the a priori probability (q) associated with hypothesis H_1 is the **prevalence** (prev.) of the disease. The parameter S is identified as the slope of a line in P_F–P_D plane in the ROC plot and Eq. (5.74) is rewritten as (Zweig and Campbell 1993)

$$S = \left(\frac{FPc - TNc}{FNc - TPc}\right)\left(\frac{1 - \text{prev.}}{\text{prev.}}\right) \qquad (5.75)$$

In eq. (5.75), the cost terms are expressed directly as

$$\begin{aligned} FPc &: \text{cost of a false positive decision} \\ FNc &: \text{cost of a false negative decision} \\ TPc &: \text{cost of a true positive decision} \\ TNc &: \text{cost of a true negative decision} \end{aligned} \qquad (5.76)$$

If the decision making is cost-neutral, the slope becomes (Zweig and Campbell 1993)

$$S = \left(\frac{1 - \text{prev.}}{\text{prev.}} \right) \tag{5.77}$$

It should be noted that prevalence of a disease in the general population is always very low, and it is implied that the value of S will exceed unity ($S > 1$).

5.3.2 Optimal Operating Point (Revisited)

The optimum threshold for the decision maker occurs when the imaginary line with the slope S touches the point on the ROC plot and the probabilities of false alarm and detection at that point are the optimal values (Zweig and Campbell 1993). The corresponding threshold is the optimal threshold. In other words, the optimal threshold corresponds to the point on the P_F–P_D plot (ROC curve) where

$$[P_F, P_D]_{\text{optimum}} \rightarrow \left(P_F - \frac{P_D}{S} \right)_{\text{min}} \tag{5.78}$$

We will illustrate the inclusion of the prevalence in ROC analysis using an example. Consider the machine vision data collected with 70 samples of target absent and 60 samples of target present. The samples are given in Table 5.13.

The ROC plot is shown in Fig. 5.39.

We can now examine the different optimal threshold values and the corresponding probabilities of false alarm and detection.

Closest to the top left corner of the ROC plot

The threshold is 2.0065 and the corresponding $[P_F, P_D]$ are [0.2714, 0.8167]. The positive predictive value is 0.8167*6/(0.8167*6 + 0.2714*7) = 0.7206.

Based on the prevalence defined from the population count (prev. = 60/130).

In this case, the optimal threshold is obtained by starting from the top left corner with a line having the slope $S = 7/6$ and moving the line down in the ROC plot until it touches a point on the ROC plot. That point represents the optimal operating point.

In this case, the threshold is also 2.0065 and the corresponding $[P_F, P_D]$ are [0.2714, 0.8167]. The positive predictive value is 0.8167*6/(0.8167*6 + 0.2714*7) = 0.7206.

Based on arbitrary prevalence of 5% (prev. = 0.05)

In this case, the threshold is also 3.**8522** and the corresponding $[P_F, P_D]$ are [0.0, 0.3333]. The positive predictive value is 0.3333*0.05/(0.3333*0.05 + 0.0*0.95) = 1. It should be noted that even though the PPV is 100%, the sensitivity is too low to be acceptable!

Based on arbitrary prevalence of 20% (prev. = 0.2)

In this case, the threshold is also 3.4398 and the corresponding $[P_F, P_D]$ are [0.0143, 0.4167]. The positive predictive value is 0.4167*0.2/(0.4167*0.2 + 0.0143*0.8) = 0.8793. Once again, the low prevalence results in low sensitivity.

Table 5.13 Data set used

Target absent:

1.0650	1.4527	1.2136	0.5438	0.9607	1.4839	2.4966
1.8480	3.4215	0.9317	2.7098	1.4702	2.7796	1.2094
1.4237	1.9893	0.3166	2.9643	1.1842	0.6276	2.7992
1.0000	1.3539	0.8195	2.3391	3.3296	0.3461	2.3408
1.1051	1.1607	2.1160	3.1226	3.4398	1.5508	0.9616
1.4548	0.9880	1.9580	3.2232	0.9940	1.6618	3.8522
2.2179	2.6836	0.9415	0.5999	3.1532	1.2060	1.3110
0.1206	1.1292	1.0089	1.9378	2.4539	0.9694	0.7312
0.7583	1.5243	0.9880	2.2120	1.4416	0.8780	1.2381
0.5812	0.5502	0.3405	0.9917	0.4564	1.7705	2.0065

Target present:

2.3833	2.2629	4.2882	3.4470	0.9503	2.0914
3.8669	4.1612	2.7489	4.3879	2.7874	4.4084
3.1586	6.3640	2.0118	4.3661	7.1283	2.0643
2.8864	2.0157	2.3335	1.0294	6.0826	1.2949
4.2997	3.3567	6.8318	2.3034	1.6827	7.8573
3.3800	3.2767	3.8012	3.7993	0.9925	5.0523
3.0062	2.7570	3.1757	5.9732	3.1500	3.4679
2.4024	0.3027	3.3986	4.7591	2.6701	4.9292
4.2595	1.6926	2.3120	1.4589	1.2391	4.5932
4.0767	2.8594	1.9243	1.7278	6.2024	3.5736

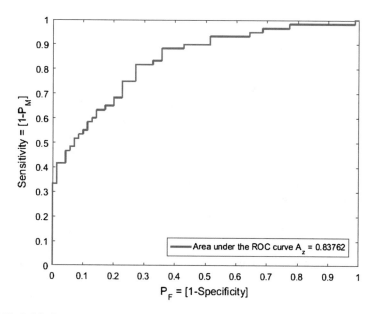

Fig. 5.39 ROC plot associated with data in Table 5.13

Another optimal threshold is defined on the basis of Youden's index (Youden 1950; Greiner et al. 2000; Hwang et al. 2018). In this case, the "optimal" threshold is the point on the ROC plot where the difference between P_D and P_F is the maximum. The maximum value is the Youden's index given as

$$J = [\text{sensitivity} + \text{specificity} - 1]_{\max} = [P_D - P_F]_{\max}. \tag{5.79}$$

Youden's index is also the maximum distance between the ROC curve and the diagonal line (Roupp et al. 2018). Comparing Eqs. (5.78) and (5.79), it is seen that Youden's index is calculated for the case of a disease prevalence of 50% or a priori probabilities being equal. Another interesting observation can also be made. Optimal threshold based on Bayes' risk in Eq. (5.78) may be interpreted as the optimal condition associated with a weighted Youden's index, weight determined by disease prevalence. Often actual costs associated with decisions are not known, and Youden's index-based threshold is a reasonable approach under those conditions. Note that the optimal threshold based on Neyman-Pearson criterion does not consider the disease prevalence at all. The importance of a priori probabilities or prevalence lies with the fact that the positive predictive values, error rates, and overall accuracy (same as [1- error rate]) require the knowledge of the prevalence.

Three optimal thresholds are explored with another set of data given in Table 5.14. Figure 5.40 shows the ROC curve, the three optimal thresholds (Neyman-Pearson, Bayes' risk based on the actual prevalence, and Youden's index). The Youden's index value is also given. As seen, depending on the prevalence, the optimal threshold values and the corresponding operating points $[P_F, P_D]$ will be different.

5.4 Bootstrapping

Often, with just a single experiment, we may not be able to obtain reliable measures of the sample mean. To be more precise, we are interested in obtaining the statistical attributes of the mean, namely, the standard deviation and the 95% confidence interval of the mean. The techniques we learned so far do not permit us to acquire these even though we may get the standard deviation of the data (Appendix A). There are instances where we only have the data collected from a single experiment and we need to ascertain the statistical attributes of any information gathered from the analysis of this single experiment. We explored use of the area under the ROC curve as a means to characterize the performance of a receiver, detector, or a diagnostic tool. We used the notion of standard deviation of the area in Eq. (5.4) to obtain a measure of the degree of uncertainty in the estimation of the area. Bootstrapping offers another means to obtain statistical attributes of the results from data collected in a single experiment whether the interest is simply the mean or the area under the ROC curve (Efron and Tibshirani 1986; Bertail et al. 2008).

Table 5.14 Data set used to compare the three optimal thresholds

Target absent							Target present					
3.7508	0.8729	2.0003	1.2117	0.1743	10.2165	1.7278	12.6786	15.6254	5.4019	8.4995	11.5195	9.3106
2.6961	3.0977	13.5529	8.7552	3.9492	5.5787	4.6197	14.634	5.7586	6.0107	11.2675	32.9631	2.2449
2.5725	1.4306	2.2185	1.6978	4.2054	1.7268	8.0876	6.4646	13.1632	18.4454	10.0753	7.2175	19.763
0.5345	4.6765	7.4689	3.513	7.6652	2.069	6.4688	12.8277	26.8269	14.404	24.4311	13.1363	14.0038
8.1225	4.8319	3.5925	3.8543	1.6243	8.9059	6.5962	9.4886	33.5173	17.9235	12.341	3.6827	8.9241
4.4795	24.8518	1.9869	0.7318	0.5151	9.7452	2.7263	20.2435	9.5191	7.8539	14.6728	8.892	20.8638
3.8151	11.4629	11.9305	4.979	4.4797	5.1983	3.9216	5.1332	22.6335	10.1706	11.7166	6.133	14.0148
18.7112	1.3521	5.314	7.8762	1.7524	0.7761	1.0716	10.1901	15.8181	4.0912	15.2988	9.3273	4.1789
1.9564	1.6111	9.4673	1.6061	1.6691	5.3408	0.4691	29.1289	13.0934	5.0242	10.4597	4.2245	10.5663
4.0739	1.1474	9.531	15.8469	1.0449	14.1081	5.9575	21.4942	14.7032	13.2955	6.5028	13.5853	10.3073

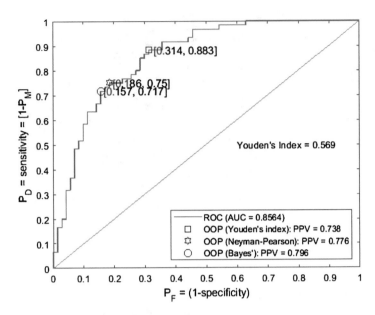

Fig. 5.40 The ROC curve, the three optimal thresholds (Neyman-Pearson, Bayes' risk based on the actual prevalence, and Youden's index), and the Youden's index (Data from Table 5.14)

Often we may be using two sensors in a machine vision experiment, and we are interested in determining which sensor performs better than the other. Similar situations arise in medical diagnostics where two competing imaging modalities may be available for patient screening and we need to establish the better of the two modalities based on ROC analysis. Since the subject cohorts studied using both modalities are the same, correlation may exist between the areas under the ROC curves obtained for both modalities (for example, magnetic resonance imaging versus CT scan). Bootstrapping offers a statistical means to explore the relationship between the two AUC values and provide quantitative measures of the comparison of the ability of the two modalities (Moise et al. 1985; Dorfman et al. 1995; Desgagné et al. 1998).

Examples are provided next to demonstrate bootstrapping using the basic concepts first before implementation using Matlab routine *bootstrp(.)*.

5.4.1 Bootstrapping of the Population Mean

We are given a data set of 20 samples ($m = 20$) collected in a single experiment. We are interested in obtaining the statistics of the mean. The bootstrapping procedure is described next (Shankar 2019d; Shankar 2020a).

The data set #1 consisted of $m = 20$ integer samples (numbers 1, 2, 3, 4, and 5) identified as vector, X (bottom row, Fig. 5.41). Bootstrapping involves resampling of data with replacement (Efron and Tibshirani 1986; Hall and Maiti 2006; Hu and Dannenberg 2006; Hesterberg 2011, 2015). This procedure means that we pick a number from the set and note its value. The number is returned back to the original set and another number is picked and its value is noted. This process of picking a number, noting its value, and returning it to the original set is repeated m times. The new set created constitutes the first bootstrap set. Since we are returning the number back to the original set, it can be understood that the numbers (1, 2, 3, 4 or 5) may be repeated. If the original m samples are set in a single column, bootstrapping implies choosing any row randomly each time. In other words, we are choosing numbers randomly between 1 and m (row indices). The row indices for resampling to create the first bootstrap set is the vector $\bar{v}$

$$\bar{v} = ceil(rand(1, m) * m) = randi(m, [m, 1]) \tag{5.80}$$

In Eq. (5.80), $rand(1,m)$ generates m uniform numbers between 0 and 1 and $ceil(.)$ provides the upper integer value of the uniform random number scaled by m. The command $randi(m)$ generates an integer between 1 and m. On the other hand, $randi(m,[m,1])$ generates m integer values between 1 and m. Equation (5.80) mimics the process of picking a ball from a box of balls numbered 1, 2, 3, 4, and 5 noting its value, putting the ball back, and repeating it m times. The newly created boot set is

$$Y = X(\bar{v}) \tag{5.81}$$

We repeat the process M times, creating M boot sets or boot samples. For each boot set, we may estimate the mean which gives us M boot samples of the mean. With M samples of the mean, we may calculate the statistics of the mean, specifically, the mean, standard deviation, and 95% confidence interval (CI) of the mean. We notice that the sample mean and sample variance may also be obtained directly from Section A.2 as

$$E(\overline{X}) = 3.2 \tag{5.82}$$

$$var(\overline{X}) = 0.0979 \tag{5.83}$$

$$std(\overline{X}) = 0.3129 \tag{5.84}$$

The process of bootstrapping is illustrated in Fig. 5.41. The original set and ten boot sets are shown. The column on the right provides the mean of the boot set.

The histogram of the boot samples ($M = 5000$) is shown in Fig. 5.42. It can be seen that the values of the mean and standard deviation in eqns. (5.82) and (5.84) very closely match those obtained through bootstrap.

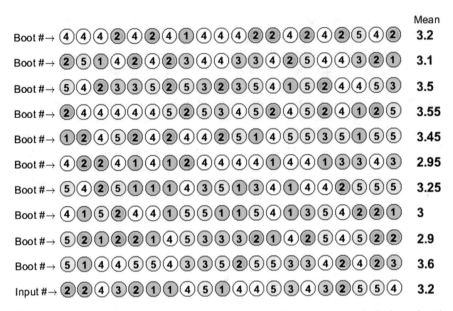

Fig. 5.41 Original data set of 20 integers (bottom row) and 10 boot sets (means in the last column)

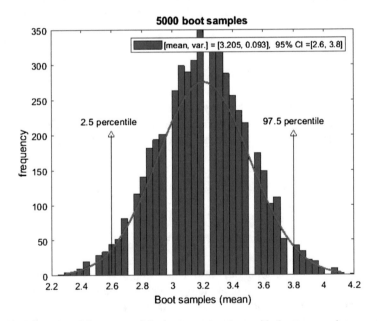

Fig. 5.42 Histogram of the means of the boot samples along with the mean, variance, and 95% confidence interval of the mean

5.4.2 Bootstrapping of Area Under the ROC Curve (AUC)

While bootstrapping of the mean of a data set undertaken in the previous section is simple and straightforward, the bootstrapping used in ROC analysis requires a little different approach. Consider a data set similar to the ones seen earlier. We have 70 samples of target absent and 60 samples of target present. The data set is given in Table 5.15.

As explained in Sect. 5.4.1, we first create a matrix consisting of data (first column) and the labels (second column), with "0" for target absent and "1" for target present. This means that we now have a $[N \times 2]$ matrix, with $N = 130$. This is shown in Table 5.16.

The bootstrapping is carried out using the index vector:

$$\bar{v} = ceil(rand(1, N) * N) = randi(N, [N, 1]) \tag{5.85}$$

Now, rows are sampled using the index in Eq. (5.85). This means that a single boot set will consist of resampled rows from the Table 5.16. Since we are resampling the rows and retaining the labels, the boot set will lead to cohorts (target absent and target present) different from 70 and 60 as in the original. We may have more of the target absent or target present samples. The area under the ROC curve is obtained with the newly created boot set. The process is repeated M times, thus generating M samples of the area under the ROC curve AUC (Az). With these M samples of the AUC, we can estimate the statistics of the AUC. Figure 5.43 shows the ROC curve, AUC, and, the standard deviation calculated using the formula (Hanley and McNeil 1982). Figure 5.44 shows the histogram of the boot samples of AUC. It also shows the Gaussian fit (line plot), mean, standard deviation, and the 95% confidence interval of the mean. The mean and standard deviation shown in Fig. 5.44 closely match values in Fig. 5.43.

5.4.3 Bootstrapping in Machine Vision and Medical Diagnostics: AUC Comparison

Another interesting application of bootstrapping in data analytics pertains to competing claims by two groups of researchers who come up with new techniques to detect a certain disease. Both groups worked on the same set of subjects in the study. In this case, it may be the same image of an organ to examine whether the image contains malignant regions. In radar and sonar applications, identical images are studied by two groups, each group has its own signal interpretation technique, and each group comes up with sets of values for the test images, one containing some target and the other one normal or containing no target. Each group analyzed the values and based on their analysis showed that the technique by group #1 resulted in an area under the ROC curve (AUC) of A_{z1}, while group #2 resulted an area under

Table 5.15 Data set used in bootstrapping of the AUC

Target absent							Target present					
0.042	0.498	2.466	0.186	0.31	1.617	1.605	3.851	3.253	0.455	3.204	3.484	2.654
0.609	0.488	1.491	0.317	0.834	0.636	0.744	1.118	0.271	1.174	10.82	3.507	0.648
0.914	0.387	0.168	0.16	0.6	0.839	4.026	3.25	0.415	4.903	1.037	2.323	12.08
0.072	0.728	0.271	0.066	1.802	0.781	0.38	5.705	3.43	1.485	3.627	7.731	2.602
0.411	1.256	0.669	0.339	0.967	0.772	0.893	4.086	4.246	3.562	2.684	5.309	1.457
1.92	3.112	0.423	1.658	0.879	0.144	0.332	5.766	5.536	3.963	2.3	2.761	5.801
2.121	3.093	0.648	0.451	0.328	0.418	0.957	1.952	4.347	1.599	4.95	2.662	1.674
1.987	1.164	0.189	1.475	1.244	0.543	0.174	6.657	5.436	0.45	5.609	0.966	2.167
0.31	1.243	0.653	1.212	2.944	1.026	0.426	3.196	5.663	2.05	1.808	1.752	2.147
0.316	0.079	0.324	0.137	0.277	3.04	1.603	6.064	2.936	2.721	9.323	4.05	1.623

Table 5.16 The 130-element composite data and the corresponding labels (second column)

0.0417	0	
0.6093	0	
0.9142	0	
0.0715	0	
0.4109	0	
1.9195	0	70
2.1212	0	
.	.	
.	.	
1.6026	0	
3.8512	1	
1.118	1	
3.2501	1	
5.7053	1	
4.0862	1	
.	.	60
.	.	
1.6281	1	

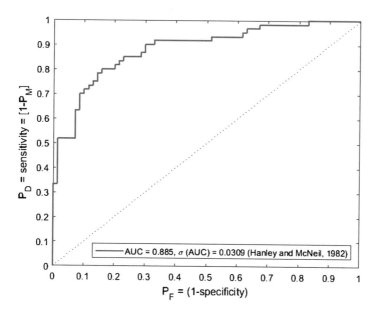

Fig. 5.43 ROC curve, AUC, and standard deviation of AUC of the data set in Table 5.16

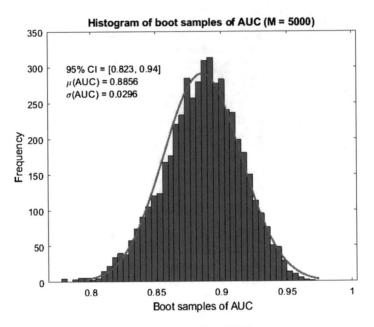

Fig. 5.44 Histogram and statistics of the boot samples of AUC

the curve (AUC) of A_{z2}, with $A_{z2} > A_{z1}$. The issue of interest is whether the difference in the areas is statistically significant, making the approach of group #2 better than that of group #1. Similar approaches are needed to see which of the two sensors in a machine vision system offers a better performance.

This case can be analyzed using the z-test described in Appendix A. But, this requires that we have reliable means of calculating the statistics of the two areas including any correlation that may exist between the areas. The correlation may exist because both groups used identical subjects (common set of images, or common target in machine vision). With only a single value of the area under the ROC curve from each group, it is not possible to estimate the correlation that may exist between the two areas. The bootstrapping is a convenient tool to accomplish this task. A simple z-test to check whether there is any statistical significance to the higher values will be based on the test statistic given in eq. A.43. Such cases are quite common in medical diagnostics and the z_{stat} can be expressed as (Hanley and McNeil 1983; Moise et. al 1985; Shankar 2019d)

$$z_{\text{stat}} = \frac{\text{mean}(A_{z1}) - \text{mean}(A_{z2})}{\sqrt{\sigma_{x1}^2 + \sigma_{x2}^2 - 2\rho\sigma_{x1}\sigma_{x2}}} \qquad (5.86)$$

In Eq. (5.86), σ_{x1}^2 is the variance of the mean of the area under the ROC curve (group #1), and σ_{x2}^2 is the variance of the mean of the area under the ROC curve (group #2). The subjects (or the images) of the study in both groups being identical, ρ is the correlation coefficient of the sample means. Bootstrapping technique may be used to estimate the mean of the areas and the related variances.

The two sets of data are given in Table 5.17.

A composite matrix of size $[N \times 3]$ created from these two data sets (paired: this means that first sample value from #1 sensor and first sample from #2 came from the same subject) is shown in Table 5.18. Bootstrapping is now implemented on this $[N \times 3]$ matrix, with index vector in eq. (5.85).

Each boot set will now be a $[N \times 3]$ matrix, retaining any and all association among the data from the two cohorts. With M boot sets, we will have 5000 samples of AUC from #1 and 5000 samples of AUC from #2. This means that z_{stat} in eq. (5.86) can be calculated. Figure 5.45 displays the two ROC curves corresponding to the two sensors showing that the AUC of sensor #2 is higher than that of sensor #1.

Figure 5.46 shows the histograms of the boot samples of the areas, mean and standard deviations. The statistics of the difference in areas are illustrated in Fig. 5.47. With the availability of boot samples of the areas, we can calculate the correlation coefficient of the areas. The summary analysis is tabulated in Fig. 5.48. If we are interested in seeing whether difference of two areas is statistically significant, we use a double sided z-test. But, in the present case, we are specifically interested in knowing whether the larger AUC of sensor #2 is statistically significant compared to the AUC from sensor #1. We expect the z_{stat} (same as z_{score} defined in Chap. 3) in Eq. (5.86) to be negative. We see the one-sided test clearly shows that the null hypothesis (increased area is not significant) is rejected at a significance level of 5% as seen by the p-value of 0.0317. On the other hand, the two-sided test clearly shows that the null hypothesis cannot be rejected at a significance level of 5% as seen by the p-value of 0.0634.

Another pair of data sets from two sensors is shown in Table 5.19. Table 5.20 shows the composite matrix used in bootstrapping. Figure 5.49 shows the two ROC curves. Figure 5.50 shows the histograms of the boot samples of the two areas. Figure 5.51 shows the statistics of the difference in areas. Figure 5.52 captures the summary analysis. It shows that in both tests (two sided and left sided), the null hypothesis is rejected in favor of the alternate hypothesis that the larger area from sensor #2 is statistically better than the area from sensor #1.

While ROC studies are important, other applications of engineering probability exist in fields of wireless communications, networking, and reliability of engineering systems, among a number of other areas. We will now look at the relevant issues in wireless communications first. We will start with a way to model the signal strength fluctuations seen in wireless channels using the statistical approaches developed.

Table 5.17 Two sets of measurements from the same machine vision set up

Set#1 sensor

Target absent							Target present					
0.653	0.083	0.188	0.886	2.989	0.189	1.248	1.514	5.683	0.897	1.113	3.968	1.102
0.566	0.406	0.194	0.277	0.128	1.055	0.694	0.269	7.566	2.629	0.815	1.105	2.464
0.067	1.736	1.449	1.267	0.714	0.032	2.103	1.873	3.644	1.019	1.374	1.85	4.537
1.54	0.786	0.763	1.226	1.345	2.9	0.474	1.715	2.766	3.611	2.28	0.381	2.609
1.004	2.167	1.554	0.117	0.633	1.813	0.448	2.955	0.983	1.234	1.188	2.155	3.289
0.661	0.655	1.174	1.411	1.564	1.333	0.266	2.308	3.846	2.648	2.643	1.388	2.837
0.583	0.169	1.337	1.808	0.375	0.122	0.826	0.43	0.394	1.032	5.247	1.672	5.104
0.294	1.341	0.643	0.027	1.046	1.071	0.337	1.344	4.453	3.962	1.754	2.337	3.669
0.465	0.327	0.032	0.654	0.03	0.114	0.523	2.265	3.313	3.035	2.003	1.401	1.327
2.013	1.047	1.551	0.257	0.309	2.365	0.367	6.327	1.638	2.006	2.032	6.364	3.514

Set #2 sensor

Target absent							Target present					
0.492	0.159	0.234	0.868	2.735	0.56	1.565	0.856	2.208	0.973	1.243	1.389	1.388
0.869	0.54	0.265	0.355	0.403	1.225	0.676	0.501	3.743	2.019	0.904	1.051	2.094
0.12	0.546	0.943	1.419	1.035	0.122	0.601	1.14	2.279	0.752	1.671	1.69	2.667
1.637	0.599	1.224	0.616	1.335	1.541	0.122	1.499	2.49	2.252	0.854	0.978	1.506
0.61	0.44	1.311	0.102	0.141	0.48	0.523	1.66	1.401	1.077	1.521	1.133	2.332
0.669	0.851	1.181	1.612	1.06	0.411	0.699	1.673	2.762	1.579	1.946	1.638	1.702
0.425	0.069	0.98	1.466	0.254	0.293	0.448	0.727	0.745	1.551	2.257	0.929	2.673
0.679	0.719	0.65	0.288	0.668	0.699	0.526	1.253	2.912	1.626	1.714	1.064	2.079
0.116	0.485	0.1	0.627	0.219	0.357	1.258	1.068	2.363	1.263	1.378	1.268	1.58
0.252	0.944	0.417	0.151	0.231	0.83	0.6	2.073	1.404	1.738	1.133	3.612	2.915

Table 5.18 Composite three-column data. First column has the composite data from sensor #1, and the second column has the composite data from sensor #2. The third column consists of labels common to both sensors

#1	#2	Labels	
0.653	0.492	0	
0.566	0.869	0	
0.067	0.12	0	
1.54	1.637	0	
.	.		
.	.	.	70
0.367	0.6	0	
1.514	0.856	1	
0.269	0.501	1	
1.873	1.14	1	
.	.		
.	.	.	60
3.514	2.915	1	

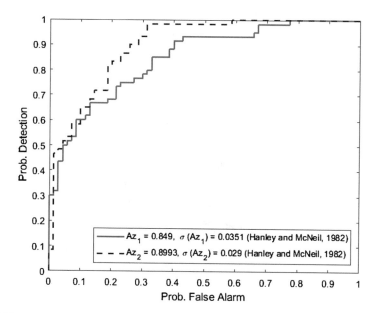

Fig. 5.45 ROC curves, respective AUC values, standard deviations

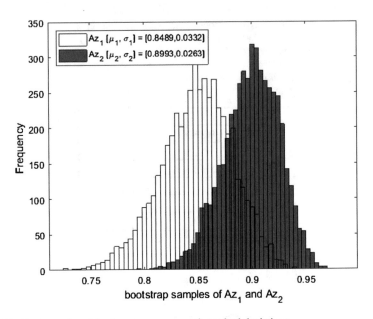

Fig. 5.46 Boot samples of the two areas, mean and standard deviations

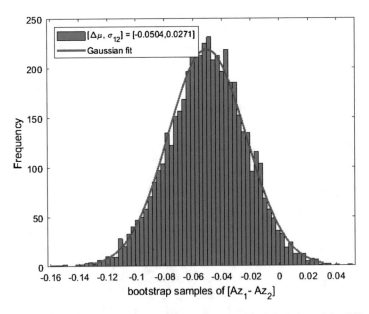

Fig. 5.47 Histogram of the difference in areas, mean and standard deviation of the difference in areas

Classifier # 1   Area #1, $<Az_1> = \mu_1 = 0.8489$, $\sigma(Az_1) = \sigma_1 = 0.0332$

Classifier # 2   Area #2, $<Az_2> = \mu_2 = 0.8993$, $\sigma(Az_2) = \sigma_2 = 0.0263$

correlation coefficient $([Az_1, Az_2]) = \rho = 0.6063$

$$z_{score} = \frac{mean(Az_1 - Az_2)}{std.\,dev(Az_1 - Az_2)} = \frac{\mu_1 - \mu_2}{\sqrt{\sigma_1^2 + \sigma_2^2 - 2\rho\sigma_1\sigma_2}} = \frac{\Delta\mu}{\sigma_{12}}$$

$$\Delta\mu = -0.0504 \qquad \sigma_{12} = 0.0271$$

$$z_{score} = -1.8566 \quad (z_{score} \text{ is negative !})$$

z-test (two-sided)   Alternate Hypothesis: Statistically significant difference between #1 and #2
Null hypothesis: no difference between #1 and #2
p(Null Hypothesis) = 0.0634

z-test (one-sided)   Alternate Hypothesis: #2 better than #1
[left tail, $z_{score} < 0$]   Null Hypothesis: no difference between # 2 and #1
p(Null Hypothesis) = 0.0317

Fig. 5.48 Summary analysis of the bootstrapping of areas from two sensors

5.5 Applications in Telecommunications

Concepts developed in all the previous chapters find applications in telecommunications. Wireless communications is one area where the notion of one, two, or more random variables finds particular applications. These include the modeling of statistical fluctuations of signal strengths observed in wireless channels and techniques used to mitigate the effects of signal strength fluctuations.

5.5.1 Wireless Channel Models

A number of concepts developed in Chaps. 2, 3, and 4 can be used to model the statistical fluctuations observed in wireless channels. As described in previous sections, simple models of these include Rayleigh, Rician, Nakagami, lognormal, etc. (Nakagami 1960; Suzuki 1977; Holm and Alouini 2004; Shankar 2004, 2015, 2016, 2017). While Rayleigh and Rician models can easily be developed by invoking the central limit theorem and lognormal model may be developed by invoking the central limit theorem for products, the more commonly used Nakagami model and some of the other models related to the Nakagami ones require a slightly different approach.

The transmission of signals takes place in the channel that comprises of all types of obstructions such as buildings, structures, particulate matter in the atmosphere,

Table 5.19 Second set of data for analysis

Set#1 sensor

Target absent							Target present					
0.972	0.836	0.918	1.173	1.262	1.628	0.379	2.334	1.294	3.228	1.567	1.336	1.784
1.103	0.639	1.555	1.042	1.078	1.48	0.845	2.285	1.815	2.716	1.734	1.392	1.333
1.262	0.987	1.755	0.732	1.908	2.004	1.502	1.357	1.047	2.104	1.258	1.349	1.058
3.052	0.578	0.591	0.476	1.143	1.681	2.465	1.443	2.352	1.599	2.539	1.065	2.404
0.957	1.038	1.478	1.904	1.548	1.699	1.432	2.387	2.792	2.413	1.908	2.72	1.025
1.279	1.036	1.965	1.084	1.118	1.62	0.701	1.485	1.617	1.468	1.444	1.781	1.801
1.149	0.965	0.273	1.435	0.623	1.518	1.29	2.213	0.992	0.861	1.975	0.915	1.347
1.67	1.054	0.956	0.283	0.405	1.032	0.595	1.692	1.316	2.113	0.639	1.133	2.245
1.757	1.86	0.792	0.61	1.434	0.801	0.697	1.887	3.227	1.729	1.679	1.729	2.319
0.914	1.621	0.851	0.774	1.844	1.216	0.402	1.356	0.969	1.17	1.412	1.664	0.931

Set #2 sensor

Target absent							Target present					
1.863	1.208	1.877	2.006	2.167	1.688	1.203	5.366	3.538	4.241	2.373	2.18	2.401
1.974	1.271	2.251	0.985	0.623	2.567	1.887	5.436	3.355	1.186	3.053	4.069	2.708
1.2	2.239	2.556	1.237	3.137	2.469	1.363	3.013	3.03	3.671	2.003	3.245	2.105
2.818	1.502	2.206	1.023	2.048	3.443	2.047	2.862	4.073	3.021	3.979	2.28	5.533
1.655	1.662	2.232	3.166	1.29	2.815	2.575	5.519	4.187	3.967	4.186	5.253	1.961
2.828	2.291	2.617	2.279	3.132	2.828	1.696	2.706	2.888	2.058	3.304	4.436	5.807
0.933	2.622	0.735	2.133	1.301	2.06	1.162	5.456	2.249	1.933	4.461	2.669	3.523
1.679	2.124	1.687	0.857	1.38	2.691	1.152	4.867	3.666	2.643	2.986	3.123	1.146
0.791	3.885	1.732	0.868	2.777	0.741	1.677	2.934	3.353	4.193	3.469	3.57	2.794
1.685	2.332	1.714	1.358	3.592	1.992	1.518	3.437	2.191	1.128	2.68	4.037	1.974

Table 5.20 Composite data for bootstrapping

#1	#2	Labels	
0.972	1.863	0	
1.103	1.974	0	
1.262	1.2	0	
3.052	2.818	0	
.	.	.	
.	.	.	70
0.492	1.518	0	
2.334	5.366	1	
2.285	5.4366	1	
1.137	3.013	1	
.	.		
.	.	.	60
0.931	1.974	1	

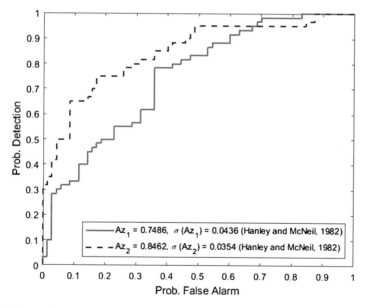

Fig. 5.49 The two ROC curves

etc. Instead of treating each of these obstructions as a single element, a more appropriate way to model the channel is treat the channel to be comprising of several clusters of scatterers. Such an approach even allows us to treat a single obstruction (i.e., a building) as a cluster. The concept of the clusters is illustrated in Fig. 5.53.

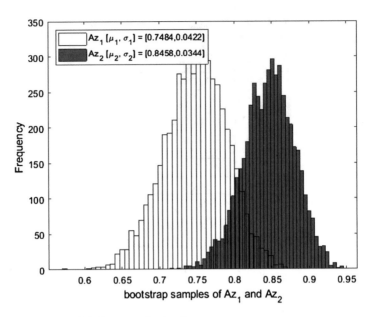

Fig. 5.50 Histograms of the boot samples (area)

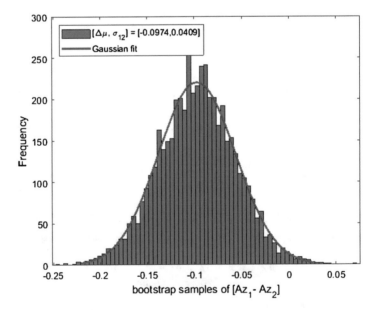

Fig. 5.51 Statistics of the difference in areas

Each cluster is assumed to have enough scatterers so that the signal generated by the cluster can be modeled as Gaussian by invoking the central limit theorem for sums. If the signal amplitude is designated by the amplitude X, the density of X will be Gaussian distributed with zero mean, and a variance determined by the number of

Classifier # 1   Area #1, $<Az_1> = \mu_1 = 0.7484$, $\sigma(Az_1) = \sigma_1 = 0.0422$
Classifier # 2   Area #2, $<Az_2> = \mu_2 = 0.8458$, $\sigma(Az_2) = \sigma_2 = 0.0344$
correlation coefficient $([Az_1, Az_2]) = \rho = 0.4456$

$$z_{score} = \frac{mean(Az_1 - Az_2)}{std.\,dev(Az_1 - Az_2)} = \frac{\mu_1 - \mu_2}{\sqrt{\sigma_1^2 + \sigma_2^2 - 2\rho\sigma_1\sigma_2}} = \frac{\Delta\mu}{\sigma_{12}}$$

$\Delta\mu = -0.0974$ $\sigma_{12} = 0.0409$

$z_{score} = -2.3824$ (z_{score} is negative !)

z-test (two-sided) Alternate Hypothesis: Statistically significant difference between #1 and #2
Null hypothesis: no difference between #1 and #2
p(Null Hypothesis) = 0.0172

z-test (one-sided) Alternate Hypothesis: #2 better than #1
[left tail, $z_{score} < 0$] Null Hypothesis: no difference between # 2 and #1
p(Null Hypothesis) = < 0.01

Fig. 5.52 Summary statistical analysis

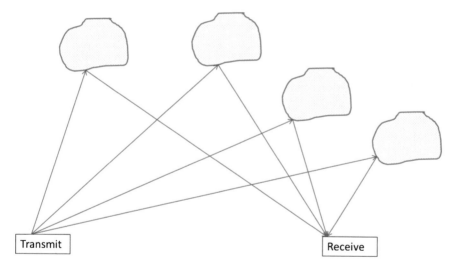

Fig. 5.53 Concept of clustered scattering

scatterers in the cluster and their scattering strengths. Initially, let us assume that there are n such clusters. These clusters act independently and therefore, the power of the received signal will be (Shankar 2015, 2017)

$$Z = \sum_{i=1}^{n} X_i^2 \qquad (5.87)$$

We also assume that the clusters are identical implying that amplitudes X_i, $i = 1,2,\ldots,n$ are identically distributed. The density of the sum of n independent and identically distributed squares of Gaussian variables is the chi-square density given by (see Chap. 4)

$$f_Z(z) = \frac{z^{\left(\frac{n}{2}\right)-1}}{Z_n^{\left(\frac{n}{2}\right)}\Gamma\left(\frac{n}{2}\right)} \exp\left(-\frac{z}{Z_n}\right)U(z) \tag{5.88}$$

In Eq. (5.88), Z_n is related to the variance of X_i, $i = 1,2,\ldots,n$. We may replace $n/2$ with m

$$m = \frac{n}{2} \tag{5.89}$$

The average power

$$Z_0 = mZ_n \tag{5.90}$$

The density in eq. (5.88) now becomes

$$f_Z(z) = \left(\frac{m}{Z_0}\right)\frac{z^{m-1}}{\Gamma(m)} \exp\left(-\frac{m}{Z_0}z\right)U(z) \tag{5.91}$$

Without any loss of generality, we can treat m to be a non-integer, and it is possible to identify the density in Eq. (5.91) as the gamma density. However, if we insist that we must have at least one cluster, Eq. (5.91) can be expressed with a limit imposed on m as

$$f_Z(z) = \left(\frac{m}{Z_0}\right)\frac{z^{m-1}}{\Gamma(m)} \exp\left(-\frac{m}{Z_0}z\right)U(z), \quad m > \frac{1}{2} \tag{5.92}$$

With the limitation on m, Eq. (5.92) bears resemblance to the density of the signal-to-noise ratio in a Nakagami faded channel. Note that the density of $\sqrt{Z}$ will be the traditional Nakagami density for the amplitude (Shankar 2017).

We will now use the concept of clustering to arrive at other models for fading. Let us assume that we now only have two clusters, and both of them are identical. The power received will be

$$Z = X_1^2 + X_2^2 \tag{5.93}$$

Since X_1 and X_2 are $N(0,\sigma^2)$, the density of Z will be

$$f_Z(z) = \frac{1}{2\sigma^2} \exp\left(-\frac{z}{2\sigma^2}\right) \tag{5.94}$$

The amplitude A and its density will be

$$A = \sqrt{Z} \tag{5.95}$$

$$f_Z(a) = \frac{a}{\sigma^2} \exp\left(-\frac{a^2}{2\sigma^2}\right) \tag{5.96}$$

This leads to the Rayleigh fading channel with a power that is exponentially distributed in Eq. (5.94) and amplitude Rayleigh distributed in Eq. (5.96). Note that the average power is

$$Z_0 = 2\sigma^2 \tag{5.97}$$

Let us consider two other special cases where the two clusters are not identical with (1) both having means of zero but have unequal variances (2) both having identical variances and different means, one with a mean of zero and the other having a non-zero mean. While unequal variances can easily be understood from having varying number of elements in each cluster, Gaussian variables with non-zero mean requires an explanation that will be provided when the results of case (2) are discussed.

Case 1: The two Gaussian variables have variances that are unequal,

$$\begin{aligned} X_1 &\Rightarrow N\left(0, \sigma_1^2\right) \\ X_2 &\Rightarrow N\left(0, \sigma_2^2\right) \end{aligned} \tag{5.98}$$

We can use the results of exercise 4. 22. The density of the power will be (Nakagami 1960)

$$f_z(z) = \frac{1}{\sqrt{\sigma_1^2 \sigma_1^2}} \exp\left(-\frac{z}{2}\left(\frac{1}{2\sigma_1^2} + \frac{1}{2\sigma_2^2}\right)\right) I_0\left(\frac{z}{2}\left[\frac{1}{2\sigma_1^2} - \frac{1}{2\sigma_2^2}\right]\right) U(z) \tag{5.99}$$

Equation (5.99) can be simplified in terms of the average power and the ratio of the variances.

$$Z_0 = \sigma_1^2 + \sigma_2^2 \tag{5.100}$$

$$q^2 = \frac{\sigma_1^2}{\sigma_2^2}, \quad 0 < q^2 < 1 \tag{5.101}$$

Note that in Eq. (5.101), the choice of σ_2^2 being larger than σ_1^2 is purely arbitrary and the definition assures that the ratio is always less than unity. Using Eqs. (5.100) and (5.101), the density in Eq. (5.99) becomes

Fig. 5.54 Scattering within
a single cluster

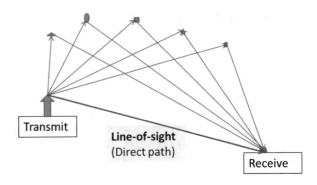

$$f(z) = \frac{(1+q^2)}{2qZ_0} \exp\left[-\frac{(1+q^2)^2}{4q^2Z_0}z\right] I_0\left[\frac{(1-q^4)}{4q^2Z_0}z\right] \qquad (5.102)$$

Eq. (5.102) represents the density of the power in a Nakagami-Hoyt fading channels (Shankar 2017). It can be easily seen that when $q \to 1$, Eq. (5.102) becomes the exponential density associated with the density of the power in a Rayleigh fading channel.

Case 2 We have

$$\begin{aligned} X_1 &\Rightarrow N(0, \sigma^2) \\ X_2 &\Rightarrow N(\mu, \sigma^2) \end{aligned} \qquad (5.103)$$

In Eq. (5.103), μ is the non-zero mean. To understand why there is a possibility that one or more clusters will have a non-zero mean, we will examine the physics of scattering within a cluster. Consider the case of a single cluster with several multiple scattering centers shown in Fig. 5.54.

While scattering takes place from multiple scattering centers, depending on the location, it is possible to have a direct path between the transmitter and receiver. In the absence of any direct (or line-of-sight) path, the vectorial sum of scattering leads to a Gaussian density for the amplitude with zero mean. When the direct path is present, the strength of the direct path adds a constant value leading to a non-zero mean Gaussian. Therefore, it is possible to have one cluster having a direct path, while the other does not. It should be noted that while vectorial addition takes place within each cluster (resulting in Gaussian densities), the powers from the clusters add as seen in Eq. (5.87).

The density function of the power was derived in Chap. 4, Eq. 4.91 and the density of the power is

$$f_Z(z) = \frac{1}{2\sigma^2} \exp\left(-\frac{z+\mu^2}{2\sigma^2}\right) I_0\left(\frac{\mu}{\sigma^2}\sqrt{z}\right) U(z) \qquad (5.104)$$

Note that the average power in this case will be

$$Z_0 = 2\sigma^2 + \mu^2 \tag{5.105}$$

Equation (5.104) represents the density of the power in a Rician fading channel (also referred to as the Nakagami-Rice channel). It can be easily seen that when $\mu \to 0$, the density in Eq. (5.104) becomes the exponential density associated with the Rayleigh fading channel.

All these densities are obtained on the basis of the additive modeling of the channel (additive in amplitude if scatterers are considered and additive in power if cluster are considered). That approach may be inadequate to describe fluctuations in signal strengths observed under all circumstances. For example, if the channel contains structures (very high density of tall buildings and other scattering centers made of widely diverse materials such as concrete, brick, stucco, aluminum, steel, wood, glass, plants, trees, etc.) making it impossible for the signals to reach the receiver after a single bounce, multiple scattering must be considered as a means of signal reaching the receiver. This scenario leads to the multiple scattering model (shadowing) and using central limit theorem for products, the density of the power will be lognormal. Since power in communications systems is described in decibels, the density of the power in mW is expressed as (Suzuki 1977; Shankar 2004)

$$f_L(z) = \left[\frac{10}{\log_e(10)}\right] \frac{1}{\sqrt{2\pi z^2 \sigma_{dB}^2}} \exp\left[-\frac{(10\log_{10} z - \mu)^2}{2\sigma_{dB}^2}\right] U(z) \tag{5.106}$$

In Eq. (5.106), the average power measured is μdBm and σ_{dB} is the shadowing level. The channel conditions get worse as shadowing levels go up.

While it is possible to treat fading and shadowing as two separate effects, they occur simultaneously. This means that the average power observed in a fading channel itself is varying due to shadowing (this approach is similar to the case of two stage experiments presented in Chap. 4). Thus Z_0 can be modeled in Eq. (5.91) as a random variable Y taking a value y such that (5.91) becomes

$$f_{Z|y}(z|y) = \left(\frac{m}{y}\right) \frac{z^{m-1}}{\Gamma(m)} \exp\left(-\frac{m}{y} z\right) U(z) \tag{5.107}$$

In Eq. (5.107), y represents the shadowing component with a density

$$f_Y(y) = \left[\frac{10}{\log_e(10)}\right] \frac{1}{\sqrt{2\pi y^2 \sigma_{dB}^2}} \exp\left[-\frac{(10\log_{10} y - \mu)^2}{2\sigma_{dB}^2}\right] U(y) \tag{5.108}$$

The density of the SNR in a shadowed fading channel now becomes

$$f_Z(z) = \int_0^\infty f_{Z|y}(z|y) f_Y(y) dy \tag{5.109}$$

Substituting the expressions,

$$f_Z(z) = \int_0^\infty \left(\frac{m}{y}\right) \frac{z^{m-1}}{\Gamma(m)} \exp\left(-\frac{m}{y} z\right) \left[\frac{10}{\log_e(10)}\right]$$

$$\times \frac{1}{\sqrt{2\pi y^2 \sigma_{dB}^2}} \exp\left[-\frac{(10\log_{10} y - \mu)^2}{2\sigma_{dB}^2}\right] dy \tag{5.110}$$

In arriving at Eq. (5.109), we made use of the properties of the conditional densities seen in Chap. 4. Equation (5.110) represents the density of the SNR in a Nakagami-lognormal channel. This approach is also considered as a product model because the density in Eq. (5.110) may also be obtained by treating the signal-to-noise ratio as the product of two random variables, one having a density in Eq. (5.91) and the other one having a density in Eq. (5.108).

The Nakagami-lognormal model does not lead to a closed form solution to the density. This can be overcome by approximating the lognormal density in Eq. (5.108) to a gamma density [ref]. In this case, the density of the SNR in a shadowed fading channel (referred to as a gamma-gamma channel, Nakagami-gamma channel or generalized K channel) is written as (Shankar 2004, 2017)

$$f_Z(z) = \frac{2}{\Gamma(m)\Gamma(c)} \left(\sqrt{\frac{mc}{Z_0}}\right)^{m+c} z^{\left(\frac{m+c}{2}\right)-1} K_{m-c}\left(2\sqrt{\frac{mc}{Z_0}} z\right) U(z). \tag{5.111}$$

In Eq. (5.111), c is the order of the gamma density representing the shadowing in place of the lognormal density. The relationship is

$$\sigma_{dB}^2 = \left[\frac{10}{\log_e(10)}\right]^2 \psi'(c) \tag{5.112}$$

In Eq. (5.112), $\psi'(.)$ is the trigamma function (Gradshteyn and Ryzhik 2000).

As described above, it is possible to utilize the concepts of random variables and their properties to develop models for describing the statistical fluctuations of signal strengths. Other models such as the cascaded one can also be developed using those concepts.

Once we have the models for the statistical fluctuations of the signal strengths, we can now examine the effects of fading, shadowing, and shadowed fading in wireless channels.

5.5.1.1 Error Rates and Outage Probabilities

In Chap. 3, the concept of error rates was introduced. While we will not explore the specific aspects of error rates, we examine how the error rates vary based on the models described on the concept of conditioning of variables. Consider the example of a simple coherent binary phase shift keying (BPSK). The error rates have been estimated as (Shankar 2017)

$$p(er) = \frac{1}{2} erfc\left(\sqrt{Z}\right) \tag{5.113}$$

The SNR is Z. The issue that arises in wireless systems is the effect of fading, shadowing, or shadowed fading. When these phenomena exist, the receiver signal power fluctuates randomly. This is modeled by making the error rate a random variable by expressing Eq. (5.113) as

$$p_e(er|Z = z) = p_e(er|z) = \frac{1}{2} erfc(\sqrt{z}) \tag{5.114}$$

The average probability of error in a fading channel can now be expressed as

$$\overline{p}(er) = \int p_e(er|z) f_Z(z) dz = \int_0^\infty \frac{1}{2} erfc(\sqrt{z}) f_Z(z) dz \tag{5.115}$$

If the fading channel is modeled as a Rayleigh fading channel, the density function of the SNR becomes

$$f_Z(z) = \frac{1}{Z_0} \exp\left(-\frac{z}{Z_0}\right) U(z) \tag{5.116}$$

It can be seen that the pdf of the SNR is exponentially distributed. The average probability of error now becomes

$$\overline{p}(er) = \int_0^\infty \frac{1}{2} erfc(\sqrt{z}) \frac{1}{Z_0} \exp\left(-\frac{z}{Z_0}\right) dz = \frac{1}{2}\left[1 - \sqrt{\frac{Z_0}{1+Z_0}}\right]. \tag{5.117}$$

For the sake of uniformity of notations, we can express the probability of error in an ideal Gaussian channel in Eq. (5.113) as

$$p(er) = \frac{1}{2} erfc\left(\sqrt{Z_0}\right) \tag{5.118}$$

Note that Eq. (5.115) can be used with any density of the SNR that describes the statistical fluctuations.

It is also possible to see how the error rates deteriorate in a Nakagami fading channel. The average error rate in a Nakagami channel can be expressed from Eq. (5.115) as

$$\bar{p}(er) = \int\limits_0^\infty \frac{1}{2} erfc(\sqrt{z}) \left(\frac{m}{Z_0}\right)^m \frac{z^{m-1}}{\Gamma(m)} \exp\left(-\frac{m}{Z_0}z\right) dz \tag{5.119}$$

An analytical expression for the error rates in a Nakagami fading channel can be obtained from Eq. (5.119) as (Shankar 2017)

$$\bar{p}(er) = \left(\frac{1}{2\sqrt{\pi}}\right) \frac{\Gamma\left(m+\frac{1}{2}\right)}{m\Gamma(m)} \left(\frac{m}{Z_0}\right)^m {}_2F_1\left(\left[m, \frac{1}{2}+m\right], [1+m], -\frac{m}{Z_0}\right) \tag{5.120}$$

In Eq. (5.120), $_2F_1(.)$ is the hypergeometric function (Gradshteyn and Ryzhik 2000).

The effect of fading on error rates can be seen from Fig. 5.55.

As the value of the Nakagami parameter m increases, the error rates move toward the error rates in an ideal channel (additive white Gaussian channel). It is also possible to understand the concept of power penalty in fading. This is the value of the additional SNR needed (in dB) to reach a certain error rate in an ideal channel. In

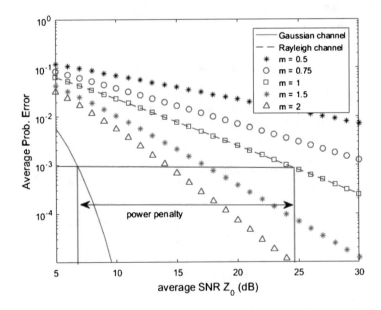

Fig. 5.55 Error rates in a Gaussian channel compared to error rates in fading channels

Fig. 5.55, it can be seen that for the case of a Rayleigh channel, the power penalty will be about 18 dB at an error rate of 1e-3.

While error rates provide one way to quantify the performance of systems in wireless channels, signal strength fluctuations also lead to outage because the instantaneous SNR may not reach the threshold SNR needed to achieve a certain level of performance. For example, if the maximum error rate tolerated is 1e-4, the minimum SNR required is obtained from eq. (5.118) as

$$Z_{thr} = erfcinv(2 * e - 4) \tag{5.121}$$

The outage probability in a fading, shadowing or shadowed fading channel will be

$$P_{out} = \int_0^{Z_{thr}} f_Z(z)dz = F_Z(Z_{thr}) \tag{5.122}$$

5.5.2 Diversity Techniques

We have discussed methods of finding the density of the sum, maximum, product, or in general functions of several random variables. We also explored improving the ROC performance through creation of multiple data sets. It is also known that measurement errors decrease if we take multiple independent measurements and use the average. The techniques which rely on the use of multiple observations belong to a class of techniques generally referred to as "**diversity techniques**" implying the creation and use of "**diverse**" versions of data collected on the same experiment (Brennan 1959; Shankar 2017). An example of diversity is the method of locating multiple receivers in a geographical region to collect wireless, acoustic, or infrared signals from the same source (space diversity). It is also possible to envision an experiment where the same information is transmitted over multiple frequency bands and the receiver collecting all these signals coming over multiple frequency bands (frequency diversity). In another example, it is possible to transmit the same information repeatedly (time diversity). Yet another example will be the transmission of information using at both X and Y polarized waves (polarization diversity). It is also possible to have multiple transmitters and multiple receivers offering a more enhanced means of creating diverse signals (Brennan 1959; Shankar 2017).

Generation of diverse signals is only the first step in reaping the benefits of such a procedure. Post-processing of the diverse signals is the next step. In this connection, we will concentrate on three methods that are typically implemented in wireless systems. The pre-processing of the signals to generate the amplitude of the received signal is beyond the scope of this book, and we will examine the post-processing based on the amplitude or the power of the signal noting that the power is the square of the amplitude. Post-processing involves content from Chap. 4, while pre-processing involves systems theory concepts not presented in this book.

The simple technique of processing the diversity signals is selection combining (SC) algorithm. If we have M independent and identically distributed signals (these represent the diverse signals), the SC algorithm can be expressed as

$$Z_{SC} = \max (Z_1, Z_2, \cdots, Z_M) \tag{5.123}$$

In Eq. (5.123), Z_k, $k = 1,2,..,M$ are the powers of the M-diverse signals. The other processing algorithm is maximal ratio combining (MRC). This implies that the each complex amplitude is weighted by itself and MRC leads to an output as

$$Z_{MRC} = \sum_{i=1}^{M} Z_i \tag{5.124}$$

While the SC algorithm is implemented easily, it wastes resources because we are only using a single output. MRC is complicated to implement because it requires perfect capture of the complex amplitude. A compromise is the algorithm termed in generalized selection combining (GSC). In GSC, the outputs are sorted and instead of simply applying the SC algorithm, the top k outputs are used and combined further using MRC. Thus, GSC only requires MRC implementation on a smaller subset of diversity outputs. GSC(M,k) implies that the k strongest components of M diversity elements are combined using the MRC algorithm and GSC($M,1$) implies SC, while GSC(M,M) implies MRC.

It should be noted that the density functions of the power or the signal-to-noise ratio (SNR) will have identical characteristics because the primary noise is additive white Gaussian noise and the noise power is not random. Therefore, power and signal-to-noise ratio may be used interchangeably in wireless. This means that density function such as the one in Eq. (5.91) may be used to describe the pdf of the SNR or power with Z_0 representing either the average SNR or the average power.

For the case of a Nakagami fading channel, using the concept of the maximum of a set of variables, the density of the output of the SC algorithm will be

$$f_{SC}(z) = M \left(\frac{m}{Z_0}\right)^m \frac{z^{m-1}}{\Gamma(m)} \exp\left(-\frac{m}{Z_0}z\right) \left[\gamma\left(m, \frac{mz}{Z_0}\right)[\Gamma(m)]^{-1}\right]^{M-1} U(z) \tag{5.125}$$

Similarly, the density of the SNR at the output of the MRC algorithm will be

$$f_{MRC}(z) = \left(\frac{m}{Z_0}\right)^{mM} \frac{z^{mM-1}}{\Gamma(mM)} \exp\left(-\frac{m}{Z_0}z\right) U(z) \tag{5.126}$$

A simple expression for the output of the GSC algorithm in a Nakagami fading channel does not exist.

These techniques can now be examined to enhance the performance of communication systems in the presence of fading channels. In addition to error rates and outage probabilities, two additional measures could be used to quantify the performance following diversity. One is the improvement in average SNR. If we look at the average SNR in a single channel is Z_0, using results from Chap. 4, the SNR following MRC diversity will be $M Z_0$. The other measure is the amount of fading (*AF*) defined as

$$AF = \frac{\langle Z^2 \rangle}{\langle Z \rangle^2} - 1 = \frac{\text{var}(Z)}{\langle Z \rangle^2}. \tag{5.127}$$

5.6   Applications to Reliability of Interconnected Systems

Often large scale systems are assembled by interconnecting several systems. We can examine three types of interconnected of systems, series, parallel, and standby modes. These are shown in Fig. 5.56.

For the system operating in a series mode, it is clear that the system reliability will be dictated by the units that has the smallest lifetime. Therefore, the lifetime of the system in series mode is the minimum. If Z represents the lifetime of the complete system, when operating in a series mode

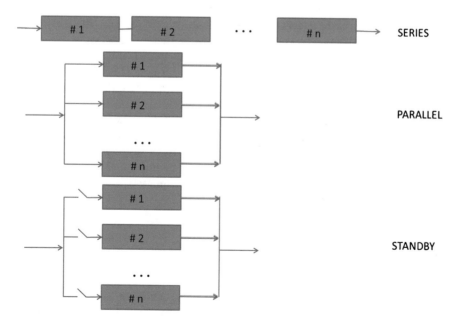

Fig. 5.56 Interconnected systems (series, parallel, and standby)

$$Z = \min (X_1, X_2, \cdots, X_n) \qquad (5.128)$$

In Eq. (5.128), X's represent the lifetimes of the individual units.

For the system operating in a parallel mode, the reliability will be determined by the longest lifetime of the group. In other words, the lifetime Z in the parallel mode is

$$Z = \max (X_1, X_2, \cdots, X_n) \qquad (5.129)$$

In the standby mode, at any given time only unit is working. In the event, one fails, the next unit becomes active and so on. Therefore the lifetime Z in the standby mode is

$$Z = \sum_{k=1}^{n} X_k \qquad (5.130)$$

5.7 Summary

In this chapter provided an overview of applications of probability to engineering problems, concentrating mostly on machine vision. Starting with the simple premise of identifying the origins of the data (whether it represented the case when a target is present or absent), the analysis relied on the topics covered in three previous chapters. Additional materials needed to complete the data analytics required modeling of the statistics of the data and testing of the statistics of the data. These two steps involved point estimation and chi-square testing. In addition, the receiver operating characteristic analysis was examined in light of the models of the data and signal processing steps were explored to improve the performance. The ROC analysis was also reexamined on the basis of Bayes' risk. The technique of bootstrapping was demonstrated through examples.

While the ROC analysis was the cornerstone of the chapter, an illustrative example of statistical modeling was presented to describe the statistical fluctuations in wireless channels. This was followed with analysis of ways to mitigate the problems caused by these fluctuations in terms of diversity and associated signal processing algorithms. The statistical aspects of reliability of engineering systems were also examined in this chapter.

Exercises

Problems 5.1–5.10 These problems are on chi square testing to establish the statistical validity of transformations seen in Chaps. 3 and 4. Most of these can be completed using the built-in function *chi2gof*(.) function in Matlab. Two of them require additional steps since these need the use of a chi-square test to establish the hypothesis of a uniform density. In every problem, generate a set or sets (as required) of 400 random numbers. In each case, undertake the simulation at least 10 times.

1. If X is $U(0,1)$, show that $Y = \sqrt{-\log_e(X)}$ is Rayleigh distributed.
2. If X is $N(0,4)$ and Y is $N(0,4)$ and they are independent, show that $Z = X^2 + Y^2$ is gamma distributed.
3. If X is Nakagami distributed, show that $Y = X^2$ is gamma distributed.
4. If R is Rayleigh distributed and α is $U(0,2\pi)$, show that $X = R \cos(\alpha)$ and $Y = R \sin(\alpha)$ are each Gaussian with zero mean.
5. If X and Y are independent and each is $N(0,1)$, show that $W = \tan^{-1}\left(\frac{Y}{X}\right)$ is uniform in $(0,2\pi)$.
6. If X and Y are independent and exponentially distributed, show that $Z = X + Y$ is gamma distributed.
7. If X_k, $k = 1,2,3,4$ are independent and identically distributed and each is $N(0,1)$, $Z = \sum_{k=1}^{4} X_k^2$ has a chi square distribution.
8. If X is Rayleigh distributed with a parameter b, show that $Y = 2\pi\left[1 - \exp\left(-\frac{X^2}{2b^2}\right)\right]$ is uniform in $(0,2\pi)$.
9. If X is exponentially distributed with a parameter a, show that $Y = 2\pi\left[1 - \exp\left(-\frac{X}{a}\right)\right]$ is uniform in $(0,2\pi)$.
10. If X is exponentially distributed with a parameter a, show that $Y = X^{\frac{1}{a}}, \alpha > 0$ has a Weibull density.

Problems 5.11–5.20 require both estimation and chi-square testing. While it may be possible to use built in Matlab function "*mle*" to estimate the parameters, chi-square testing must be carried out separately since the densities being tested are not 'native' to Matlab.

11. Establish the statistics of given data set as the one resulting from the product of two independent and identically distributed exponential variables. (*K* distribution)
12. Establish the statistics of given data set as one resulting from the square root of the sum of the squares two independent and Gaussian variables, $N(0,a^2)$ and $N(0, b^2)$. (Hoyt density)

$$Z = \sqrt{X^2 + Y^2}, \quad X = N(0, a^2), \quad Y = N(0, b^2), \quad a \neq b.$$

13. Establish the statistics of the given data set as the one resulting from taking the maximum of three independent and identically distributed Rayleigh variables.
14. Establish the statistics of given data set as the one resulting from taking the minimum of three independent and identically distributed exponential variables.

15. Establish the statistics of given data set as the one resulting from the ratio of two independent Rayleigh variables.

16. Establish the statistics of the given data set as the one resulting from the geometric mean of two independent and non-identical Rayleigh variables.

17. Establish the statistics of the given data set as the one resulting from the geometric mean of two independent and identically distributed gamma variables.

18. Establish the statistics of given data set as the one resulting from the arithmetic mean of two independent and identically distributed gamma variables.

19. Establish the statistics of the given data set as the one resulting from the maximum of two independent and identically distributed gamma variables.

20. Establish the statistics of the given data set as one resulting from the minimum of two independent and identically distributed gamma variables.

5.21–5.120 For the given data, test for the appropriate statistical fit based on the mean square error (MSE): Normal, Laplacian, Nakagami, gamma, Weibull. 100 data sets are given, each with 200 samples.

The Laplacian has the following properties:

$$f_X(x) = \frac{1}{2b} \exp\left(-\frac{|x-a|}{b}\right) \qquad F_X(x) = \begin{cases} \dfrac{1}{2} \exp\left(\dfrac{x-a}{b}\right), x < a \\ 1 - \dfrac{1}{2} \exp\left(-\dfrac{x-a}{b}\right), x > a \end{cases}$$

$$E(X) = a \quad \text{var}(X) = 2b^2$$

Note that you should test the data to see if negative values exist. In this case, testing will be limited to Laplacian and normal. If the data contains only positive values, tests will be limited to Nakagami, gamma, and Weibull.

Problem 5.121–5.220 You are given 200 samples of data. Undertake a chi-square test (number of bins: 8, fixed bin width) to determine the closest fit among Rayleigh, Rician, Nakagami, Weibull, and gamma to the data based on the highest p-value.

Problem 5.221–5.320 You are given 200 samples of data (100 sets). Undertake a chi-square test (number of bins: 8, fixed bin count) to determine the closest fit among Rayleigh, Rician, Nakagami, Weibull, and gamma to the data. Compare the results to those from the fixed bin width case.

Problem 5.321–5.420 You are given 200 samples of data (100 sets). Choose 100 samples out of the given 200 through random permutation. Undertake a chi-square test (number of bins: 8, fixed bin count and fixed count) to determine the closest fit among Rayleigh, Rician, Nakagami, Weibull, and gamma to the data. Repeat the choice of 100 samples 5 times and examine whether the hypothesis tests lead to identical conclusions (100 data sets).

Problems 5.421–5.520 Analyze the data (70 samples of target absent and 60 samples of target present), and undertake the ROC analysis similar to part 1 described (Sect. 5.2).

Problems 5.521–5.620 Analyze the data (70 samples of target absent and 60 samples of target present), and undertake the ROC analysis similar to part 2 described (Sect. 5.2).

Problems 5.621–5.720 Data collected from a machine vision lab to determine the efficiency of an object recognition system is given. The 200 data points given are collected as follows: Two receivers mounted on the vehicle are receiving the backscattered signal from the observation region. The first set of 100 points is from receiver # 1 and the second set of 100 points is from receiver # 2. In this experiment, the interest is to see how the performance of the receiver could be improved. To accomplish this, three algorithms are explored, namely, the arithmetic mean, geometric mean, and the maximum. The performance is characterized by the performance index

$$\eta = \frac{\text{mean}}{\text{standard deviation}}$$

Obtain the performance indices for the raw data, arithmetic mean, geometric mean, and the maximum.

If X and Y represent the two outputs,

$$V = \text{arithmetic mean} = \frac{X + Y}{2}$$
$$W = \text{geometric mean} = \sqrt{XY}$$
$$Z = \text{Maximum} = \max(X, Y)$$

In each case, conduct a chi-square test also to determine the best fit.

Problems 5.721–5.820 Analyze the data (70 samples of target absent and 60 samples of target present), and undertake the ROC analysis similar to part 3 described (Sect. 5.2).

Problems 5.821–5.920 Analyze the data (70 samples of target absent and 60 samples of target present), and undertake the ROC analysis similar to part 4 described (Sect. 5.2).

Problems 5.921–5.1020 Analyze the data (70 samples of target absent and 60 samples of target present). Obtain the ROC. Undertake bootstrapping of the area under the ROC and obtain the 95% confidence interval of the mean AUC and estimate the standard deviation of AUC. Compare this standard deviation of AUC to the standard obtained using Hanley and McNeil (1982).

Problems 5.1021–5.1080 You are given machine vision data from a pair of sensors. Undertake statistical testing (requires bootstrapping) to see whether the null hypothesis true (there is no difference in the AUC values of the two sensors).

Problems on Reliability

Problem 5.1081 Ten switches are connected (1) in parallel (2) in series and (3) in a stand-by mode (if one fails, the other one come on). If the lifetimes of the switches are independent and identically distributed, Rayleigh random variables with a mean lifetime of 5 years, estimate the mean life time of the three arrangements. Verify results through random number simulation.

Problem 5.1082 Generate 5 (diversity M) sets of gamma random variables to mimic a Nakagami channel with $m = 1.8$. With 50,000 in each set, obtain the densities of the SC(5), MRC(5), GSC(5,2), GSC(5,3), and GSC(5,4) estimate the improvement in performance in terms of (1) average SNR (2) amount of fading.

Problem 5.1083 Use the data sets from Problem 272 and obtain the outage probabilities as a function of the average SNR in a channel for SC(5), MRC(5), GSC (5,2) GSC(5,3), and GSC(5,4). First estimate the threshold needed to maintain an average error rate of 1e-4.

References

Abramowitz M, Segun IA (1972) Handbook of mathematical functions with formulas, graphs, and mathematical tables. Dover Publications, New York

Andrae R, Schulze-Hartung T, Melchoir P (2010) Dos and don'ts of reduced chi-squared. arXiv:1012.3754. https://arxiv.org/pdf/1012.3754

Baggerly KA (2001) Probability binning and testing agreement between multivariate immunofluorescence histograms: extending the chi-squared test. Cytometry 45:141–150

Barton DE (1956) Neyman's test of goodness of fit when the null hypothesis is composite. Scand Actuar J 1956(2):216–245

Bertail P, Clémençcon SJ, Vayatis N (2008) On bootstrapping the ROC curve. In: Proceedings of the 21st international conference on neural information processing systems (NIPS 2008), pp 137–144

Bradley AP (1997) The use of the area under the ROC curve in the evaluation of machine learning algorithms. Pattern Recogn 30(7):1145–1159

Brennan DG (1959) Linear diversity combining techniques. Proc IRE 47(6):1075–1102

Cochran WG (1954) Some methods for strengthening the common $\chi 2$ tests. Biometrics 10:417–451

Desgagné A, Castilloux A-M, Angers J-F, LeLorier J (1998) The use of the bootstrap statistical method for the pharmacoeconomic cost analysis of skewed data. PharmacoEconomics 13:487–497

Dorfman DD, Bernbaum KS, Lenth RV (1995) Multireader, multicase receiver operating characteristic methodology: a bootstrap analysis. Acad Radiol 2:626–633

Efron B, Tibshirani R (1986) Bootstrap methods for standard errors, confidence intervals, and other measures of statistical accuracy. Stat Sci 1:54–75

Gradshteyn IS, Ryzhik IM (2000) Table of integrals, series, and products, 6th edn. Academic, New York

Greiner M, Pfeiffer D, Smith RD (2000) Principles and practical application of the receiver-operating characteristic analysis for diagnostic tests. Prev Vet Med 45(1–2):23–41

Hall P, Maiti T (2006) On parametric bootstrap methods for small area prediction. J R Stat Soc Series B Stat Methodol 68(2):221–238

Hand DJ, Till RJ (2001) A simple generalization of the area under the ROC curve for multiple class classification problems. Mach Learn 45:171–186

Hanley JA, McNeil B (1982) The meaning and use of the area under a receiver operating characteristic (ROC) curve. Radiology 143:29–36

Hanley JA, McNeil BJ (1983) A method for comparing areas under the receiver operating characteristic curves derived from the same cases. Radiology 148:839–843

Helstrom CW (1968) Statistical theory of signal detection. Pergamon Press, Oxford/New York

Hesterberg T (2011) Bootstrap. Wiley Interdiscip Rev Comput Stat 3(6):497–526

Hesterberg TC (2015) What teachers should know about the bootstrap: resampling in the undergraduate statistics curriculum. Am Stat 69(4):371–386

Holm H, Alouini M-S (2004) Sum and difference of two squared correlated Nakagami variates in connection with the McKay distribution. IEEE Trans Commun 52:1367–1376

Hooper D, Coughlan J, Mullen M (2008) Structural equation modelling: guidelines for determining model fit. Electron J Bus Res Methods 6:53–60

Hu N, Dannenberg RB (2006) Bootstrap learning for accurate onset detection. Mach Learn 65 (2–3):457–471

Hwang Y-T, Hung Y-H, Wang CC, Terg H-J (2018) Finding the optimal threshold of a parametric ROC curve under a continuous diagnostic measurement. Revstat Stat J 16(1):23–24

Kester AD, Buntinx F (2000) Meta-analysis of ROC curves. Med Decis Mak 20(4):430–439

Liu X (2012) Classification accuracy and cut point selection. Stat Med 31(23):2676–2686

McClish DK (1989) Analyzing a portion of the ROC curve. Med Decis Mak 9:190–195

Metz CE (2006) Receiver operating characteristic analysis: a tool for the quantitative evaluation of observer performance and imaging systems. J Am Coll Radiol 3(6):413–422

Moise A, Clément B, Ducimetière P, Bourassa MG (1985) Comparison of receiver operating curves derived from the same population: a bootstrapping approach. Comput Biomed Res 18:125–131

Nakagami M (1960) The m-distribution—a general formula of intensity distribution of rapid fading. In: Hoffman WC (ed) Statistical methods in radio wave propagation. Pergamon, Elmsford

Obuchowski NA (2005) Fundamentals of clinical research for radiologists. Am J Roentgenol 184 (2):364–372. https://doi.org/10.2214/ajr.184.2.01840364

Obuchowski NA, Lieber ML, Wians FH (2004) ROC curves in clinical chemistry: uses, misuses, and possible solutions. Clin Chem 50(7):1118–1125. https://doi.org/10.1373/clinchem.2004. 031823

Papoulis A, Pillai U (2002) Probability, random variables, and stochastic processes. McGraw-Hill, New York

Rohatgi VK, Saleh AKME (2001) An introduction to probability and statistics. Wiley, New York

Rota M, Antolini L (2014) Finding the optimal cut-point for Gaussian and Gamma distributed biomarkers. Comput Stat Data Anal 69:1):1–1)14. https://doi.org/10.1016/j.csda.2013.07.015

Roupp MD, Perkins NJ, Whitcomb BW, Schisterman EF (2018) Youden index and optimal cut-point estimated from observations affected by a lower limit of detection. Biom J 50 (3):419–430

Shankar PM (2004) Error rates in generalized shadowed fading channels. Wirel Pers Commun 28 (3):233–238

Shankar PM (2015) A composite shadowed fading model based on the McKay distribution and Meijer G functions. Wirel Pers Commun 81(3):1017–1030

Shankar PM (2016) Performance of cognitive radio in N*Nakagami cascaded channels. Wirel Pers Commun 88(3):657–667

Shankar PM (2017) Fading and shadowing in wireless systems, 2nd edn. Springer, Cham

Shankar PM (2019a) Pedagogy of Bayes' rule, confusion matrix, transition matrix, and receiver operating characteristics. Comput Appl Eng Educ 27(2):510–518. https://doi.org/10.1002/cae. 22093

Shankar PM (2019b) Pedagogy of chi square goodness of fit test for continuous distributions. Comput Appl Eng Educ 27(3):679–689

Shankar PM (2019c) Pedagogy of diversity and data analytics: theory to practice. Comput Appl Eng Educ 27(5):1277–1285

Shankar PM (2019d) Pedagogy of bootstrapping. WSEAS Trans. Adv Eng Educ 16:18–27

Shankar PM (2020a) Tutorial overview of simple, stratified, and parametric bootstrapping. Eng Rep 2(1):1–11. https://doi.org/10.1002/eng2.12096

Shankar PM (2020b) Introduction of data analytics in the engineering probability course: Implementation and lessons learnt. Comput Appl Eng Educ 28(5):1072–1082

Suzuki H (1977) A statistical model for urban radio propagation. IEEE Trans Commun 25:673–680

Swets JA (1979) ROC analysis applied to the evaluation of medical imaging techniques. Investig Radiol 14:109–121

Swets JA (1988) Measuring the accuracy of diagnostic systems. Science 240(4857):1285–1293

Timischl F (2015) The contrast-to-noise ratio for image quality evaluation in scanning electron microscopy. Scanning 37:54–62

Van Trees HL (1968) Detection, estimation, and modulation theory, part I. Wiley, New York

Watson GS (1958) On chi-square goodness-of-fit tests for continuous distributions. J R Stat Soc Series B Methodol 20:44–72

White LF, Bonetti M, Pagano M (2009) The choice of the number of bins for the M statistic. Comput Stat Data Anal 53:3640–3649

Youden WJ (1950) Index for rating diagnostic tests. Cancer 3(1):32–35

Zweig MH, Campbell G (1993) Receiver-operating characteristic (ROC) plots: a fundamental evaluation tool in clinical medicine. Clin Chem 39(4):561–577

Appendix A: t-Tests, z-Tests, and p-Values

Introduction

Data analytics involve modeling, analysis, and interpretation of the data collected from experiments. While the first step is the collection of data, it is usually followed by the estimation of the mean and variance of the data besides advanced techniques of extraction of any other useful information. Depending on the need, we may be interested in finding out whether the estimated mean is close to a known mean value and if it is not close, whether the difference between them is statistically significant. We may also have collected two sets of data, and we are interested in determining whether one set has a mean higher than the other. All these steps require statistical testing to establish or reject the null hypothesis. The null hypothesis assumes that the estimated mean is not different from the known mean or that the two sets of data are not statistically different. The validation of the null hypothesis is undertaken using the t-tests or z-tests, and results are typically expressed in terms of p-values (probability values). A low value of p supports the *rejection* of the null hypothesis in favor of an alternate hypothesis. A high value of p informs the user that the null hypothesis *cannot be rejected*.

In this Appendix, we undertake the study of the statistics of the mean and parametric hypothesis testing through t-tests and z-tests.

Data, Samples, Sample Mean, and Sample Variance

While we may consider samples of the data to be deterministic, experiments and measurements introduce randomness, and therefore samples must be treated as random. If we have n samples (representing n observations) from an experiment, the randomness implies that the sample set $\{X_1, X_2, \ldots, X_n\}$ constitute an n-dimensional set of random variables. It is safe to assume that all these samples or

© Springer Nature Switzerland AG 2021
P. M. Shankar, *Probability, Random Variables, and Data Analytics with Engineering Applications*, https://doi.org/10.1007/978-3-030-56259-5

n-random variables are identically distributed and if the experiment or measurement has been carefully crafted, these are also independent. Based on these assumptions, we can obtain the relationship between the arithmetic mean of the data and the sample mean. We represent the sample mean by $\overline{X}$.

$$\overline{X} = \frac{X_1 + X_2 + \cdots + X_n}{n}. \tag{A.1}$$

The right hand side of Eq. (A.1) is also the population mean or the arithmetic mean. X's are independent and identically distributed random variables with each mean η and variance σ^2. In Eq. (A.1), the sample mean $\overline{X}$ itself is a random variable (X's are random). We may therefore obtain the mean and variance of the sample mean. Taking the statistical average of both sides of Eq. (A.1), we have

$$E(\overline{X}) = E\left(\frac{X_1 + X_2 + \cdots + X_n}{n}\right) = \frac{\eta + \eta + \cdots + \eta}{n} = \eta. \tag{A.2}$$

Equation (A.2) suggests that the *mean* of the sample mean possesses the same mean as the original random variable X (note that $X_1, X_2, \ldots, X_n$ are samples of X). If we now calculate the variance of the sample mean,

$$\text{var}(\overline{X}) = \sigma_x^2 = \text{var}\left(\frac{X_1 + X_2 + \cdots + X_n}{n}\right) = \frac{\sigma^2 + \sigma^2 + \cdots + \sigma^2}{n^2} = \frac{n\sigma^2}{n^2}$$

$$= \frac{\sigma^2}{n}. \tag{A.3}$$

Equations (A.2) and (A.3) indicate that while the mean of the *sample mean* and the *population mean* of X are identical, the *variance of the sample mean* is a scaled down value of the *population variance* σ^2.

In arriving at Eqs. (A.2) and (A.3), we assumed that we know the mean and variance of X namely η and σ^2, respectively. If we do not know these, we may estimate the mean ($\widehat{\eta}$) and variance ($\widehat{\sigma}^2$) using the maximum likelihood approach described in Chap. 5. Given a set of data values $X_1, X_2, \ldots, X_n$ constituting the samples of X, we have the estimate of the mean η represented as

$$\widehat{\eta} = \frac{1}{n} \sum_{i=1}^{n} X_i. \tag{A.4}$$

The estimate of the variance σ^2 will be

$$\widehat{\sigma}^2 = \frac{1}{n} \sum_{i=1}^{n} (X_i - \widehat{\eta})^2 = \frac{1}{n} \sum_{i=1}^{n} X_i^2 - \widehat{\eta}^2. \tag{A.5}$$

While we have these estimates, we may examine if they are *unbiased*. An estimator is *unbiased*, if the expected value of the estimate equals the actual value (estimates are random variables). Let us examine whether estimates of the mean and variance are unbiased. Taking the expectation of Eq. (A.4), we have

$$E(\hat{\eta}) = \frac{1}{n} \sum_{i=1}^{n} E(X_i) = \frac{1}{n} n \; \eta = \eta. \tag{A.6}$$

Equation (A.6) shows that the estimate of the mean given in Eq. (A.4) is *unbiased*. Taking the expectation of Eq. (A.5), we have

$$E(\hat{\sigma}^2) = E\left(\left[\frac{1}{n}\sum_{i=1}^{n}X_i^2\right] - \overline{X}^2\right) = \frac{1}{n}\sum_{i=1}^{n} E(X_i^2) - E(\overline{X}^2) \tag{A.7}$$

Rewriting Eq. (A.7)

$$E(\hat{\sigma}^2) = \frac{1}{n} \sum_{i=1}^{n} (\sigma^2 + \eta^2)_i - E(\overline{X}^2) \tag{A.8}$$

Noting that $\overline{X}$ has a mean of η and a variance of $\frac{\sigma^2}{n}$, we have

$$E(\overline{X}^2) = \frac{\sigma^2}{n} + \eta^2. \tag{A.9}$$

Equation (A.8) can now be rewritten using Eq. (A.9) as

$$E(\hat{\sigma}^2) = \frac{1}{n}(n\sigma^2 + n\eta^2) - \frac{1}{n}\sigma^2 - \eta^2 = \sigma^2 + \eta^2 - \frac{1}{n}\sigma^2 - \eta^2 \tag{A.10}$$

Rearranging, we have

$$E(\hat{\sigma}^2) = \frac{n-1}{n}\sigma^2 \neq \sigma^2 \tag{A.11}$$

Equation (A.11) shows that the estimate of the variance in Eq. (A.5) is *not unbiased*. The estimate of the variance becomes *unbiased* if we use the following expression (Papoulis and Pillai 2002; Rohatgi and Saleh 2001):

$$\hat{\sigma}^2 \Big|_{unbiased} = \frac{1}{n-1} \sum_{i=1}^{n} (X_i - \hat{\eta})^2 \tag{A.12}$$

At large values of n, there is little difference between using n or $n - 1$ in the denominator in Eq. (A.5) or (A.12) to obtain the estimate of the variance. For the remainder of this section, we will use the unbiased estimators of the mean and variance of X as (we also drop the top hat symbol)

$$\eta = \frac{1}{n} \sum_{i=1}^{n} X_i \tag{A.13}$$

$$\sigma^2 = \frac{1}{n-1} \sum_{i=1}^{n} (X_i - \overline{X})^2. \tag{A.14}$$

Using central limit theorem (Chap. 4), we may conclude that the sample mean $\overline{X}$ expressed in Eq. (A.1) is normally distributed with a mean of η and variance $\frac{\sigma^2}{n}$ respectively. The Gaussian density associated with $\overline{X}$ is $N\left(\eta, \frac{\sigma^2}{n}\right)$.

Confidence Intervals for the Estimation of Mean

Once we have collected the data, we may use the sample mean to gain an under-standing on the sample size of the data to estimate the mean. Using the z_{score} defined in Chap. 3, for the sample mean $\overline{X}$, we may define the z_{score} as

$$z_{score} = \frac{\overline{X} - \eta}{\left(\frac{\sigma}{\sqrt{n}}\right)}. \tag{A.15}$$

It can be easily seen from the standard normal probability table (Chap. 3) that

$$\text{Prob}(-1.96 < z_{score} < 1.96) = 0.95. \tag{A.16}$$

Equation (A.16) means that

$$P\left(-1.96 < \frac{\overline{X} - \eta}{\left(\frac{\sigma}{\sqrt{n}}\right)} < 1.96\right) = P\left[-1.96\left(\frac{\sigma}{\sqrt{n}}\right) < \overline{X} - \eta < 1.96\left(\frac{\sigma}{\sqrt{n}}\right)\right]$$

$$= 0.95. \tag{A.17}$$

Equation (A.17) can be reworked to get

$$P\left[\overline{X} - 1.96\left(\frac{\sigma}{\sqrt{n}}\right) < \eta < \overline{X} + 1.96\left(\frac{\sigma}{\sqrt{n}}\right)\right] = 0.95. \tag{A.18}$$

Fig. A.1 The 95% confidence interval

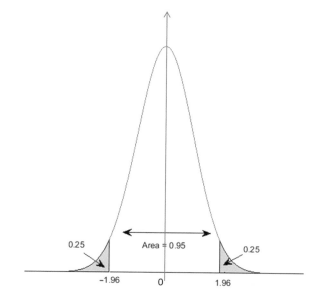

Equation (A.18) can be interpreted to infer that the 95% confidence interval for the mean η is given by

$$\left[\bar{x} - 1.96\left(\frac{\sigma}{\sqrt{n}}\right), \ \bar{x} + 1.96\left(\frac{\sigma}{\sqrt{n}}\right)\right] \Rightarrow 95\% \text{ Confidence interval for the mean } \eta.$$

$$(A.19)$$

In Eq. (A.19), we have replaced $\overline{X}$ with $\bar{x}$. The confidence interval is shown in Fig. A.1.

Expressing Eq. (A.19) in a slightly different way, the 95% confidence interval (C.I.) for the mean will be

$$
\begin{aligned}
\text{C.I.} &= \left[\bar{x} + 1.96\left(\frac{\sigma}{\sqrt{n}}\right)\right] - \left[\bar{x} - 1.96\left(\frac{\sigma}{\sqrt{n}}\right)\right] = (2)(1.96)\left(\frac{\sigma}{\sqrt{n}}\right) \\
&= 3.92\left(\frac{\sigma}{\sqrt{n}}\right).
\end{aligned}
$$

$$(A.20)$$

Equation (A.20) allows us to estimate the number of samples or the number of times an experiment needs to be undertaken or repeated to ensure that the measured mean will fall within the confidence interval. In other words, Eq. (A.20) provides a method to determine the minimum number of samples that must be collected to satisfy the 95% confidence interval.

We may generalize Eq. (A.20) to arbitrary levels of confidence intervals. We can see that a $100(1 - \alpha)\%$ confidence interval for the average of a Gaussian population with a known standard deviation σ is

Fig. A.2 Confidence level
is shown

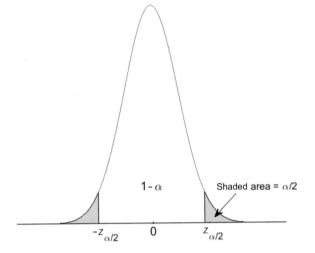

$1-\alpha$ Shaded area = $\alpha/2$

$-z_{\alpha/2}$ 0 $z_{\alpha/2}$

$$\left[\bar{x} - z_{\frac{\alpha}{2}}\left(\frac{\sigma}{\sqrt{n}}\right), \ \bar{x} + z_{\frac{\alpha}{2}}\left(\frac{\sigma}{\sqrt{n}}\right)\right] \Rightarrow 100(1-\alpha)\% \ \text{Confidence interval for the mean } \eta$$

$$(A.21)$$

Equation (A.21) implies that if W is the confidence interval,

$$W = 2z_{\frac{\alpha}{2}}\frac{\sigma}{\sqrt{n}}. \qquad (A.22)$$

It should be obvious that z_{score} in Eq. (A.15) itself is a random variable and when the number of samples is large, z_{score} is Gaussian distributed (Central Limit Theorem). Figures A.1 and A.2 represent this Gaussian approximation to the sample mean. We may also refer to the z_{score} in Eq. (A.15) as the Z-statistic.

Example A.1: A manufacturer of resistors (high value of Ω) has determined that standard deviation of the resistors is 10 Ω. If the manufacturer expects a 95% confidence interval of 5 Ω, how many resistors must be tested?

Solution: We have

$$\text{C.I.} = 3.92\left(\frac{\sigma}{\sqrt{n}}\right) = 5$$

This means that

$$\sqrt{n} = 3.92 * \frac{10}{5} = 7.84$$

This corresponds to $n = 61.4$ or 62 resistors need to be tested.

If the number of samples is not large, the distribution of the sample mean is best described using the Student's t-distribution (Rohatgi and Saleh 2001). In this case, Eq. (A.15) is identified as the T-statistic (instead of the Z-statistic) given by (Rohatgi and Saleh 2001)

$$T = \frac{\overline{X} - \eta}{\left(\frac{\sigma}{\sqrt{n}}\right)}. \tag{A.23}$$

The t-distribution is said to have ν degrees of freedom and its density is given as

$$f_T(t) = \frac{\Gamma\left(\frac{\nu+1}{2}\right)}{\sqrt{\pi\nu}\,\Gamma\left(\frac{\nu}{2}\right)}\left(1 + \frac{t^2}{\nu}\right)^{-\frac{\nu+1}{2}}, \quad \nu \geq 1. \tag{A.24}$$

The degrees of freedom ν is related to n as

$$\nu = n - 1. \tag{A.25}$$

The density in Eq. (A.24) is symmetric about zero and approaches a standard normal density when ν becomes large. The mean of the t-distribution is given by

$$E(T) = 0. \tag{A.26}$$

The variance of the t-distribution is given by

$$\mathrm{var}(T) = \frac{\nu}{\nu - 2}, \quad \nu > 2. \tag{A.27}$$

The Student's t-distribution is shown in Fig. A.3. As ν increases, the density approaches the standard normal density, $N(0,1)$. Figure A.4 displays the cumulative distribution functions of the t-density.

Interpretation of the Test Statistic

The test statistic given in Eq. (A.15) or Eq. (A.23) is useful in hypothesis testing. We limit our discussion to examining two sets of data where two competing claims are being made:

- Two Internet providers claim that one of them provides higher speeds consistently.

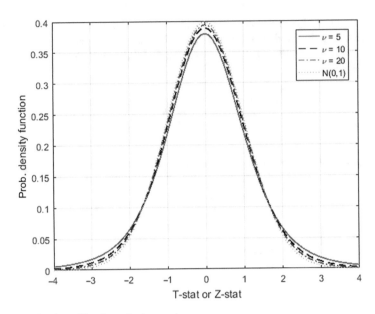

Fig. A.3 Student's t-pdf and standard normal

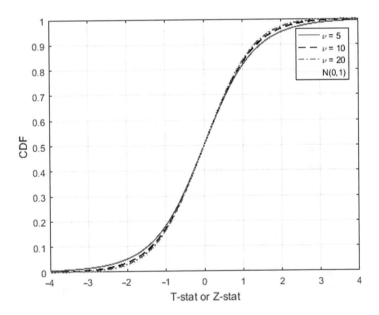

Fig. A.4 Student's t-cdf and standard normal cdf

- Two drug manufacturers make competing claims of the efficacy of the new drug in lowering blood pressure.
- Two schools make claims of having accepted students with higher SAT scores, etc.

All these claims can be analyzed using hypothesis testing to determine whether the claim is statistically valid. In Chap. 5, we explored hypothesis testing to see whether the data fits a specific probability density function. In the present case, the hypothesis testing is modeled with two possibilities, either **to reject** the null hypothesis (the difference is **not** statistically significant) in favor of the alternate hypothesis (the possibility that a difference exists between the two entities) or **not to reject** the null hypothesis. We may examine the two data sets further and determine whether the mean of the first set is higher than the mean of the second set or vice versa. These three situations are studied through what are termed as "two-tailed" or "single-tailed" tests.

The first step in hypothesis testing is to modify the test statistic Eq. (A.15) or Eq. (A.23):

$$Z_{\text{stat}} = T_{\text{stat}} = \frac{\overline{X} - \overline{Y}}{\sigma\sqrt{\dfrac{2}{n}}}. \tag{A.28}$$

In Eq. (A.28), $\overline{X}$ and $\overline{Y}$ are the sample means of the data sets from the two entities represented by random variables X and Y, respectively. The number of samples in each case is n. The appearance of σ in Eq. (A.28) indicates that the variance is known and the variances of the data from the two entities are equal. We have also assumed that X and Y are independent and this leads to the addition of variance and hence the presence of 2 in the denominator. If the number of samples is large, we use the Z_{stat} and perform a z-test (use the Normal distribution). Otherwise, we use the T_{stat} and perform a t-test (use the t-distribution).

Null Hypothesis: The two sets of data are not statistically different.

We may do a *single-tailed* test in which case the null hypothesis *is* rejected at a significance level of 100α % if

$$Z_{\text{stat}} \geq Z_{\alpha}$$

or   (A.29)

$$T_{\text{stat}} \geq T_{\alpha,\nu} \quad (\nu = n - 1).$$

Equation (A.29) the threshold values Z_α and $T_{\alpha,\nu}$ are given by

$$\alpha = 1 - N(Z_\alpha, 0, 1)$$
$$\alpha = 1 - \text{tcdf}(T_\alpha, \nu). \tag{A.30}$$

In Eq. (A.30), $N(Z_\alpha, 0, 1)$ is the cumulative distribution function of the normal distribution of mean zero, and standard deviation of unity and $\text{tcdf}(T_\alpha, \nu)$ is the

Fig. A.5 Right-tailed, left-tailed, and two-tailed tests

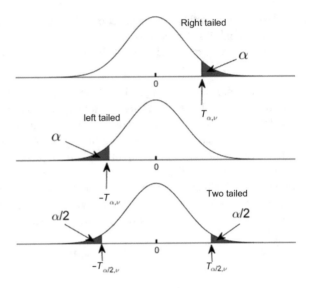

cumulative distribution function of the Student's t-distribution of order ν. The test in Eq. (A.29) is the single-tailed (*upper or right*) test. The details are shown in Fig. A.5. The single tailed (*left or lower*) test will correspond to

$$Z_{stat} \leq -Z_\alpha$$

or   (A.31)

$$T_{stat} \leq -T_{\alpha,\nu} \quad (\nu = n - 1)$$

Equation (A.31) leads to the rejection of the null hypothesis in favor of the alternate hypothesis that data set Y has higher mean than the set X.

For the case of a *two-tailed test*, null hypothesis is rejected at a significance level of 100α % if

$$\text{Either } Z_{stat} \geq Z_{\alpha/2} \text{ or } Z_{stat} \leq -Z_{\alpha/2}.$$ (A.32)

If t-test is used, the null hypothesis is rejected at a significance level of 100α % when

$$\text{Either } T_{stat} \geq T_{\alpha/2,\nu} \text{ or } T_{stat} \leq -T_{\alpha/2,\nu} \quad (\nu = n - 1).$$ (A.33)

In Eq. (A.28) we assume that the variance is known. It also assumes that the two data sets are paired with equal number of samples. Therefore, Eq. (A.28) represents the case of the test statistic associated with a *paired t-test* or *z-test* (Zimmerman 1997, 2012; Yue and Pilon 2004; Kim 2015).

Often, we encounter two issues. First, the sample size in the initial experiment may be different from the sample size in the latter experiment. Second, we may not know the variance and variances may be unequal. These situations require the modification of the Z- or T-statistic in Eq. (A.28). The derivation of the T-statistic under these conditions is beyond the scope of this book even though its origins can be easily traced by expanding the denominator in Eq. (A.28) to the case when the variances are not equal. The test statistic is (Rohatgi and Saleh 2001)

$$Z_{stat} = T_{stat} = \frac{\overline{X} - \overline{Y}}{\sqrt{\frac{S_x^2}{m} + \frac{S_y^2}{n}}}. \tag{A.34}$$

In Eq. (A.34), m is the number of samples of X and n is the number of samples of Y. The variance of the data set X is S_x and variance of the data set Y is S_y. The unbiased estimates of these variances are given by

$$S_x^2 = \frac{1}{m-1} \sum_{i=1}^{m} \left(X_i - \overline{X} \right)^2 \tag{A.35}$$

$$S_y^2 = \frac{1}{n-1} \sum_{i=1}^{n} \left(Y_i - \overline{Y} \right)^2. \tag{A.36}$$

These conditions also lead to a change in the degrees of freedom ν given as (Rohatgi and Saleh 2001)

$$\nu = \frac{(1+R)^2}{\left(\frac{R^2}{m-1} + \frac{1}{n-1} \right)}. \tag{A.37}$$

In Eq. (A.37), R is given by

$$R = \frac{\left(\frac{S_x^2}{m} \right)}{\left(\frac{S_y^2}{n} \right)}. \tag{A.38}$$

If ν in Eq. (A.37) is not an integer, we round it off to the nearest integer. With the values of the Z_{stat}, T_{stat}, in Eq. (A.34) and ν in Eq. (A.37), the hypothesis testing follows the same pattern described earlier in Eqs. (A.29), (A.31) and (A.32).

It is also possible that while variances are unknown, they may be very close. Under these conditions, additional modifications are needed for the test statistic. It becomes (Moser and Stevens 1992; Rohatgi and Saleh 2001)

$$Z_{stat} = T_{stat} = \frac{\overline{X} - \overline{Y}}{S_P \sqrt{\frac{1}{m} + \frac{1}{n}}}. \tag{A.39}$$

In Eq. (A.39), S_p^2 is identified as the pooled variance given by

$$S_p^2 = \frac{(m-1)S_x^2 + (n-1)S_y^2}{m+n-2}.$$ (A.40)

The number of degrees of freedom ν in this case becomes

$$\nu = m + n - 2.$$ (A.41)

Note that if $m = n$ and the sample variances are very approximately equal, R in Eq. (A.38) will be unity and the number of degrees of freedom in Eq. (A.37) becomes

$$\nu = 2n - 2.$$ (A.42)

If the number of degrees of freedom exceeds 30 or so (or we have a large number of samples), we use the z-test. The underlying assumption in z-test is that we have samples with statistics that matches a Gaussian density. Even though this is a loosely stated one, it assumes that there are no outliers and we may use the t-test even when samples do not follow the Gaussian and use the z-test when the number of samples is large.

The previous description of hypothesis testing did not specifically examine cases of correlated sets, X and Y (equal size). For example, we conduct an experiment and collect the data on how a system is working. We are not happy with the results. We make adjustments to the system and take another set of measurements. Another example pertains to medical diagnostics. We collect the statistics on hypertension (blood pressure measurements) in patients. Once the patients are put on medications and waiting for a certain number of days, blood pressure measurements are taken again. In both cases (medical and non-medical), we are interested in finding out whether the second set of measurements shows any improvement over the first one. Clearly, in these setting, it is likely that correlation may exist and we need to examine paired t-test or z-test. Both sets have equal samples, each equal to n. In another example, a number of high resolution images (brain scans) have been collected. These images have been analyzed by two modalities to assess whether a specific location is abnormal. We are interested to know which modality is better and since the same images were used, we expect the data to be correlated. We may now go back to Eq. (A.28) and rewrite it as

$$Z_{\text{stat}} = T_{\text{stat}} = \frac{\overline{X - Y}}{\sqrt{\frac{var(X-Y)}{n}}}.$$ (A.43)

In Eq. (A.43),

$$\overline{X - Y} = \frac{1}{n} \sum_{i=1}^{n} [X_i - Y_i] \tag{A.44}$$

$$\mathrm{var}(X - Y) = \mathrm{var}(X) + \mathrm{var}(Y) - 2\rho\sqrt{\mathrm{var}(X)\mathrm{var}(Y)}. \tag{A.45}$$

In Eq. (A.45), ρ is the correlation between the two data sets. We are using paired observations and this means that the degrees of freedom will be $(n - 1)$.

If the observations are independent, we do not have a paired t-test and Eq. (A.43) becomes

$$Z_{\mathrm{stat}} = T_{\mathrm{stat}} = \frac{\overline{X} - \overline{Y}}{\sqrt{\frac{\mathrm{var}(X) + \mathrm{var}(Y)}{n}}}. \tag{A.46}$$

The degrees of freedom will now be $2n - 2$ instead of being $(n - 1)$ when the observations are paired. It should be noted that the numerator in Eqs. (A.43) and (A.46) are equal because expectation (first moment) is a linear operation. When the variances are equal, Eq. (A.46) matches Eq. (A.28). The degrees of freedom associated with the unpaired test, $2n - 2$ matches the one given in Eq. (A.41) when the numbers of samples of X and Y are equal.

p-Value of a Test, z-Tests, and t-Tests

Every test is an examination of whether the null hypothesis is true or not. The p-value of test is the probability that *the null hypothesis **cannot be** rejected*. This means that a very high p-value implies that the null hypothesis cannot be rejected. If the p-value is low, we **reject** the null hypothesis, and we **may not reject** the alternate hypothesis. The p-value lies between 0 and 1, and an accepted p-value is $\alpha = 0.05$, stated as a level of significance of 5% (Rohatgi and Saleh 2001). If the p-value is below 0.05, we reject the null hypothesis (5% level of significance). Note that we may set our threshold p-values to be higher or lower than 0.05.

The p-values can be calculated from the test statistic and the corresponding t-densities with relevant degrees of freedom or the normalized Gaussian densities. The p-values for the three cases, the upper or right tailed, lower or left tailed, and two-tailed, are shown in Fig. A.6.

It can be stated that if the calculated p-value is smaller than the level of significance α, the null hypothesis is rejected. When the sample size is large, we use the Z-stat and the results shown in Fig. A.5 still have the same meaning with the t-pdf replaced by the standard Gaussian. Note that there is no notion of degrees of freedom associated with the standard normal.

Let us look at a few examples z- or t-tests as the case may be. All examples are treated in the most general case as variance unknown and unequal:

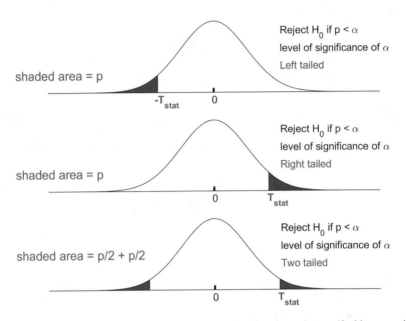

Fig. A.6 *p*-values for t-tests. In each case, the density function shown is a t-pdf with appropriate degrees of freedom

Example A.2: For the two data sets (data sets are provided in the Matlab script), determine whether there is any difference between the two sets.

Solution: Since we are interested in finding out whether there is any statistically significant difference between the two sets, we may obtain the absolute value of the T_{stat} and undertake a two-tailed t-test. We have two data sets of unequal lengths and we are treating this case as the one with unequal variances. This means that the T_{stat} is

$$T_{\text{stat}} = \frac{|\overline{X} - \overline{Y}|}{\sqrt{\frac{S_x^2}{m} + \frac{S_y^2}{n}}}.$$

The number of samples of data (*x*) *m* is 40 and the number of samples of the data (*y*) *n* is 30. The rest of the computation is directly done in Matlab. The degrees of freedom is given by Eq. (A.37).

```
clear;close all;clc
x=[55.7753    53.3593    48.6845    51.0092    59.1362    50.4814    56.9805,...
   37.0497    54.7183    56.0814 57.7628    53.5403    49.7233    47.2292,...
   55.2403    54.2985    58.2826    53.1765    56.7235    49.8549 59.1168,...
   48.3325 51.1845 51.1950 57.6732 44.2669 55.1225 53.8251 52.0994 47.4036,...
   47.5005    57.6242    40.8114    52.1842    57.6400    55.8348    40.3077,...
   54.6059    52.3211    55.9291];
y=[ 55.1879    45.6122    58.7621    56.0610    47.6339    49.0548    61.9124,...
    54.5625    52.1237    51.4540 56.8316    60.7656    51.9112    44.9606,...
    59.1641    54.6062    47.6636    56.2763    46.4618    54.9929,...
    54.5219    60.1287    58.3339    47.6224    48.8323    53.3526    51.6347,...
    55.3572    55.7476    63.8340];
X_bar=mean(x);
Y_bar=mean(y);
Sxsq=var(x);
Sysq=var(y);
m=length(x);
n=length(y);
Sed=sqrt(Sxsq/m+Sysq/n);
R=(Sxsq/m)/(Sysq/n);
nu=(1+R)^2/(R^2/(m-1)+1/(n-1));
nu=round(nu);
Tstat=abs((X_bar-Y_bar))/Sed;
pvalue1=tcdf(-Tstat,nu-1)+(1-tcdf(Tstat,nu-1));
[h,pvalueMatlab,ci,statsm]=ttest2(x,y,'Vartype','unequal'); %
disp('          Results based on the formulas')
disp(['T-stat = ',num2str(Tstat)])
disp(['Degrees of fredom nu = ',num2str(nu)])
disp(['p-value (two tailed) = ',num2str(pvalue1)])
disp('          Results based on Matlab')
disp(['T-stat = ',num2str(statsm.tstat)])
disp(['Degrees of fredom nu = ',num2str(statsm.df)])
disp(['p-value (two tailed) using ttest2 in Matlab = ',num2str(pvalueMatlab)])
disp('the p-value is larger than 0.05. This means that we cannot reject the')
disp('null hypothesis that there is no statistical difference between the two sets')
```

```
              Results based on the formulas
T-stat = 1.2058
Degrees of fredom nu = 64
p-value (two tailed) = 0.2324
              Results based on Matlab
T-stat = -1.2058
Degrees of fredom nu = 63.8379
p-value (two tailed) using ttest2 in Matlab = 0.23234
the p-value is larger than 0.05. This means that we cannot reject the
null hypothesis that there is no statistical difference between the two sets.
```

Example A.3: We are given two data sets (same as in the previous example) and the set (y) has a higher mean than the first set (x). Does the higher mean of (y) imply that the difference is statistically significant?

Solution: In this case, we expect the difference in the mean to be negative based on the information given that the second set has a higher mean. This means that the absolute value sign used in Example A.2 should be removed and the T_{stat} now becomes

$$T_{\text{stat}} = \frac{\overline{X} - \overline{Y}}{\sqrt{\frac{S_x^2}{m} + \frac{S_y^2}{n}}}.$$

In this case, we undertake the ***lower or left tailed*** t-test. The null hypothesis is that the means are not different (statistically).

```
clear;close all;clc
x=[0.6782   2.4994    5.0133    0.9601    1.9857    7.0013    4.6306,...
   3.5208   2.6695    2.0998, 1.4203   6.2639    0.5779    2.9961,...
   0.7497   4.9269    2.1044    3.9763    0.5118    4.4869  3.0951,...
   9.4297  0.8753  0.7169  4.9819 2.2060 6.0575 0.6585 0.0919   3.2614];
y=[4.2328   3.1844    3.7383    3.3633    4.6515    4.6940    3.0327,...
   2.5691   3.7927    3.6475 4.5745    3.7023    4.3430    3.2149,...
   4.6759   4.4065    3.0364    4.7841    3.9250    4.5707  4.9887,...
   3.5325   4.5031    4.2113    4.6538    1.8056    4.7860    4.4255,...
   4.6557   4.3322  6.0199    3.9407    4.1991    3.3570    3.5627];
X_bar=mean(x);
Y_bar=mean(y);
Sxsq=var(x);
Sysq=var(y);
m=length(x);
n=length(y);
Sed=sqrt(Sxsq/m+Sysq/n);
R=(Sxsq/m)/(Sysq/n);
nu=(1+R)^2/(R^2/(m-1)+1/(n-1));
nu=round(nu);
Tstat=(X_bar-Y_bar)/Sed;
pvalue1=tcdf(Tstat,nu-1);
[h,pvalueMatlab,ci,stats]=ttest2(x,y,'Vartype','unequal','tail','left'); %
disp(['Tstat (calculated) = ',num2str(Tstat)])
disp(['Degrees of fredom nu (calculated) = ',num2str(nu)])
disp(['p-value (left tailed) directly = ',num2str(pvalue1)])
disp(['Degrees of fredom nu (from Matlab) = ',num2str(round(stats.df))])
disp(['Tstat (from Matlab) = ',num2str(stats.tstat)])
disp(['p-value (left tailed) using ttest2 in Matlab = ',num2str(pvalueMatlab)])
disp('the p-value is smaller than 0.05. This means that we cannot reject the')
disp('alternate hypothesis that mean of the second set is higher')
disp('and the difference is statistically significant')
```

```
Tstat (calculated) = -2.3174
Degrees of fredom nu (calculated) = 35
p-value (left tailed) directly = 0.013317
Degrees of fredom nu (from Matlab) = 35
Tstat (from Matlab) = -2.3174
p-value (left tailed) using ttest2 in Matlab = 0.013221
the p-value is smaller than 0.05. This means that we cannot reject the
alternate hypothesis that mean of the second set is higher
and the difference is statistically significant.
```

Example A.4: You are given the scores of students who took the test in a subject within the first 2 weeks of classes (INITIAL) and took the test in the same subject in another 2 weeks (FINAL). Undertake a hypothesis test to determine whether the performance of the students improved in a statistically significant way.

Solution: The data sets are given in the Matlab document below. The solution is prepared in Matlab directly by estimating the t-test statistic and then performing the test in Matlab. Several steps and results are shown.

Paired t-test: This means that the correlation between the two sets (if it exists) is obtained and t-test statistic calculated. Results are compared to those obtained using ttest(.) in Matlab.

Unpaired t-test: In this case, the two sets are treated as independent and there are two ways of looking at the data. One is to assume that variances are known and in this case, the degrees of freedom will be $2n - 2$ as given in Eq. (A.41). If variances are treated as unequal and unknown, the degrees of freedom will be lower as given in Eq. (A.37). These results are also verified using ttest2(.) in Matlab.

The importance of including the correlation (that is the realistic case) can be seen in the lower p-value with the paired t-test. This shows that correlation should not be ignored and leads to lower p-value even with a lower number of degrees of freedom.

The analysis repeated with z-testing as well since we have 23 samples. Results indicate that the alternate hypothesis that the performance improved cannot be rejected at a significance level of 5%. In all the calculations, the p-values are smaller than 0.05.

paired vs. unpaired t-test

```
clear;clc;
% demonstrates the difference between paired and unpaired samples
% these are the scores of students in a topic 2 weeks into the course
% and 4-weeks into the course. The student cohort being the same, we expect
% correlation between the two sets of scores. Unpaired implies independent
% sets.
%
% initial scores (maximum of 20)
score_iniT=[15.5,17.0,16.5,16.0,15.5,14.0,16.0,13.0,16.5,12.0,13.0,13.,...
12.0,12.,15.,14.,11.,14.,13.5,15.5,17.0,12.5,11.0];
% final scores (maximum of 20)
score_finL=[20,13,20,20,20,18,19,18,10,16,17,20,12,11,16,20,19,13,17,12,...
 15,15,10]; % final scores (max 20)
n=length(score_iniT);
% we are looking to see if there is any improvement in scores from Initial
% to  Final and if the improvement is statistically significant.
%
% We expect improvement and therefore, the mean(score_finL) nust be larger
% than mean(score_iniT). In other words, we expect the difference
% mean(score_iniT)-mean(score_finL) to be negative.
%
% Therefore, we use a tailed t-test (left or lower tailed t-test).
%
% while there is no difference between the numretor of the test statistic
% for the paired and unpaired tests, the denominator will be impacted by
% the presence of any correlation as far as the denominator is concerned.
%
% In addition, the pairing leads to (n-1) degrees of freedom while unpaired
% t-test will be undertaken at n-1+n-1 or 2n-2 degrees of freedom. unpaired
% implies that we have two sets of samples that are independent.
%
% start the analysis of the data
score_diff=score_iniT-score_finL; % difference in scores: vector
Meanscrore_diff=mean(score_diff); %sample mean: paired as well as unpaired
r=corrcoef(score_iniT,score_finL);% correlation matrix
% paired t-test
varscore_diff=var(score_diff);% this automatically includes the correlation
varscore_diff_sample=varscore_diff/n;
T_paired=Meanscrore_diff/sqrt(varscore_diff_sample);
p_valdirect1=tcdf(T_paired,n-1);
[h1,p1_matlab,ci,stats]=ttest(score_iniT,score_finL,'tail','left');
disp('----###############################################------')
disp('------------ paired t-test(direct calculations)------------')
disp(['correlation between the initial and final scores = ',...
    num2str(r(1,2))])
disp(['test statistic = ',num2str(T_paired)])
disp(['number of degrees of freedom (n-1) = ',num2str(n-1)])
disp(['p-value calculated using (n-1) degrees of freedom = ',...
    num2str(p_valdirect1)])
disp('------------paired t-test(ttest in Matlab)----------------')
disp(['test statistic = ',num2str(stats.tstat)])
disp(['number of degrees of freedom from Matlab = ',num2str(stats.df)])
disp(['p-value calculated using degrees of freedom above = ',...
    num2str(p1_matlab)])
% unpaired t-test; known variance approach
varsumSample=(var(score_iniT)+var(score_finL))/n;
T_unpaired=Meanscrore_diff/sqrt(varsumSample);
p_valdirect2=tcdf(T_unpaired,2*n-2);
[h2,p2_matlab,ci,stats2]=ttest2(score_iniT,score_finL,'tail','left');
```

Conduct a z-test since we have 23 samples

We get the paired difference in scores and determine whether its mean is 0 and test whether the data consisting of the difference in scores comes from a population of standard normal distribution.

```
% unpaired t-test; known variance approach; use equation from notes
R=(var(score_iniT)/n)/(var(score_finL)/n);
nu=round((1+R)^2/(R^2/(n-1)+1/(n-1)));
p_valdirect3=tcdf(T_unpaired,nu);
[h2,p3_matlab,ci,stats3]=ttest2(score_iniT,score_finL,...
    'Vartype','unequal','tail','left');
disp('----#####################################################------')
disp('----Unpaired t-test(direct calculations): Variance known---')
disp(['test statistic = ',num2str(T_unpaired)])
disp(['number of degrees of freedom (2n-2) = ',num2str(2*n-2)])
disp(['p-value calculated using (2n-2) degrees of freedom = ',...
    num2str(p_valdirect2)])
disp('--------- Unpaired t-test(ttest in Matlab): Variance known------')
disp(['test statistic = ',num2str(stats2.tstat)])
disp(['number of degrees of freedom from Matlab = ',num2str(stats2.df)])
disp(['p-value calculated using degrees of freedom above = ',...
    num2str(p2_matlab)])
disp('----#####################################################------')
disp('-------- Unpaired t-test(direct calculations): Variance unknown--')
disp(['test statistic = ',num2str(T_unpaired)])
disp(['number of degrees of freedom nu = ',num2str(nu)])
disp(['p-value calculated using ',num2str(nu),' degrees of freedom = ',...
    num2str(p_valdirect3)])
disp('-----Unpaired t-test(ttest in Matlab): Variance unknown-----------')
disp(['test statistic = ',num2str(stats3.tstat)])
disp(['number of degrees of freedom from Matlab = ',...
    num2str(round(stats3.df))])
disp(['p-value calculated using degrees of freedom above = ',...
    num2str(p3_matlab)])


----#####################################################------
------------ paired t-test(direct calculations)------------
correlation between the initial and final scores = 0.15823
test statistic = -2.5408
number of degrees of freedom (n-1) = 22
p-value calculated using (n-1) degrees of freedom = 0.0093125
------------paired t-test(ttest in Matlab)----------------
test statistic = -2.5408
number of degrees of freedom from Matlab = 22
p-value calculated using degrees of freedom above = 0.0093125
----#####################################################------
----Unpaired t-test(direct calculations): Variance known---
test statistic = -2.3657
number of degrees of freedom (2n-2) = 44
p-value calculated using (2n-2) degrees of freedom = 0.011233
--------- Unpaired t-test(ttest in Matlab): Variance known------
test statistic = -2.3657
number of degrees of freedom from Matlab = 44
p-value calculated using degrees of freedom above = 0.011233
----#####################################################------
-------- Unpaired t-test(direct calculations): Variance unknown--
test statistic = -2.3657
number of degrees of freedom nu = 34
p-value calculated using 34 degrees of freedom = 0.011919
-----Unpaired t-test(ttest in Matlab): Variance unknown-----------
test statistic = -2.3657
number of degrees of freedom from Matlab = 34
p-value calculated using degrees of freedom above = 0.011915
```

We get the paired difference in scores and determine whether its mean is 0 and test whether the data consisting of the difference in scores comes from a population of standard normal distribution.

```
disp('Results of z-test')
z_value=T_paired; % T_paired is identical to z_score)
[hz,pz1,cz,statz]=ztest(T_paired,0,1,'tail','left');
disp(['p-value calculated from the z-test = ',num2str(pz1)])
% note that there is no direct way to use the z-test in Matlab when we have
% paired data. One way is to directly use the test statstic as used above.
```

```
             Results of z-test
p-value calculated from the z-test = 0.0055296
```

Input the score difference divided by the std. deviation of the score diff

```
testdat=score_diff/sqrt(varscore_diff);
[he,pe,ci,zvalue]=ztest(testdat,0,1,'tail','left');
disp(['z-value from the test = ',num2str(zvalue)])
disp(['p-value calculated from the z-test = ',num2str(pe)])
```

```
z-value from the test = -2.5408
p-value calculated from the z-test = 0.0055296
```

Published with MATLAB® R2019a

On the basis of the p-values, it is seen that we reject the null hypothesis in favor of the alternate hypothesis (improvement in performance in the second test).

Appendix B

appendix_book_shankar

Illustrates summation series expansions, simplifications of known forms, solution of equations, differentiation

    char(.) converts the symbolic quantity in the brackets to text or string. double(.) converts the symbolic quantity in the brackets into numbers. num2str(.) converts the number in the bracket to text or string. simplify(.) simplifies the mathematical symbolic expression. Conversion of symbolic expressions to inline functions use of LaTex to create equations.

Find the Sum of a Symbolic Expression I

```
clear; close all; clc
```

Find the Sum of a Symbolic Expression II

```
syms x y xs yy k
xs=symsum(1/k^3,1,15);%
disp(xs)%
disp('This does not look good! Convert to regular numbers')
xs=double(xs);%converts to numbers from text
disp(['sum of inverse of the 3rd power of Nos: 1 to 15',' --->',num2str(xs)])
```

```
56154295334575853/46796108014656000

This does not look good! Convert to regular numbers
sum of inverse of the 3rd power of Nos: 1 to 15 --->1.2
```

© Springer Nature Switzerland AG 2021
P. M. Shankar, *Probability, Random Variables, and Data Analytics with Engineering Applications*, https://doi.org/10.1007/978-3-030-56259-5

Find Sum of Infinite Series in Symbolic Form I

```
xs=symsum(k*(k+1)^2,1,10);
xs=double(xs);%converts to numbers from text
disp(['sum of k(k+1)^2, k= 1 : 15',' --->',num2str(xs)])
disp('--------------------------')
```

```
sum of k(k+1)^2, k= 1 : 15 --->3850
--------------------------
```

Find Sum of Infinite Series in Symbolic Form I

```
clear;
syms x y xs yy
syms k n integer
xs=symsum((1/k)*(-1)^(k+1),k,1,Inf); % summation from 1 to infinity
disp(['sum of [(-1)^(k+1)]/k, k= 1 : inf',' --->',char(xs)])
```

```
sum of [(-1)^(k+1)]/k, k= 1 : inf --->log(2)
```

Find Sum of Infinite Series in Symbolic Form II

```
xs=symsum(1/k^2,k,1,Inf);% summation from 1 to infinity
disp(['sum of [(-1)^(k+1)]/k, k= 1 : inf',' --->',char(xs)])
```

```
sum of [(-1)^(k+1)]/k, k= 1 : inf --->pi^2/6
```

Find Sum of Infinite Series in Symbolic Form III

```
xs=symsum(k,k,1,n); % summation from 1 to n
disp('--------------------------')
disp(['sum of natural numbers from 1:n is','--------->',char(xs)])
```

```
--------------------------
sum of natural numbers from 1:n is--------->(n*(n + 1))/2
```

Find Sum of Infinite Series in Symbolic Form IV

```
xs=symsum(k^2,k,1,n);
disp('--------------------------')
disp(['sum of squares of natural numbers from 1:n is','------->',char(xs)])
```

```
--------------------------
sum of squares of natural numbers from 1:n is------->(n*(2*n + 1)*(n + 1))/6
```

Series Expansions: $\log(1 + x)$, $-\log(1 - x)$, $\sin(x)$, $\cos(x)$, Euler's Identity

```
xs=symsum(2*k-1,k,1,n);
xs=simplify(xs);
disp('--------------------------')
disp(['sum (2k-1),k= 1:n is','------->',char(xs)])
```

```
--------------------------
sum (2k-1),k= 1:n is------->n^2
```

Differentiation

```
clear
syms x a
f=log(1+x);
y=taylor(f,x,'order',8);
disp(['log(1+x)=',char(y)])
ff=-log(1-x);
yy=taylor(ff,x,'order',8);
disp('----------------------------------------')
disp(['-log(1-x)=',char(yy)])
disp('----------------------------------------')
fs=sin(x);
ys=taylor(fs,x,'order',8);
disp('----------------------------------------')
disp(['sin(x)=',char(ys)])
disp('----------------------------------------')
fc=cos(x);
yc=taylor(fc,x,'order',8);
disp('----------------------------------------')
disp(['cos(x)=',char(yc)])
disp('----------------------------------------')
f1=(1/2)*(exp(i*x)+exp(-i*x));
yc1=taylor(f1,x,'order',8);
disp('----------------------------------------')
disp(['(1/2)*[exp(jx)+exp(-jx)] = ',char(yc1),' = cos(x)'])
disp('----------------------------------------')
u=simplify(cosh(x)+sinh(x));
disp('-------------------------')
disp('----------------------------------------')
disp(['cosh(x)+sinh(x) is','----->',char(u)])
disp('----------------------------------------')
u=simplify(exp(x*log(a)));
disp('-------------------------')
disp(['exp(x.log(a)) is','-------->',char(u)])
disp('----------------------------------------')
u=simplify((1/2)*(exp(i*x)+exp(-i*x)));
disp('-------------------------')
disp(['(1/2)*[exp(i*x)+exp(-i*x)] is','----->',char(u)])
u=simplify((1/(2*i))*(exp(i*x)-exp(-i*x)));
disp('-------------------------')
disp(['(1/2i)*[exp(i*x)-exp(-i*x)] is','----->',char(u)])
clear
```

```
log(1+x)=x - x^2/2 + x^3/3 - x^4/4 + x^5/5 - x^6/6 + x^7/7
----------------------------------------
-log(1-x)=x + x^2/2 + x^3/3 + x^4/4 + x^5/5 + x^6/6 + x^7/7
----------------------------------------
----------------------------------------
sin(x)=x - x^3/6 + x^5/120 - x^7/5040
----------------------------------------
----------------------------------------
cos(x)=x^4/24 - x^2/2 - x^6/720 + 1
----------------------------------------
----------------------------------------
(1/2)*[exp(jx)+exp(-jx)] = x^4/24 - x^2/2 - x^6/720 + 1 = cos(x)
----------------------------------------
-------------------------
-------------------------
cosh(x)+sinh(x) is----->exp(x)
----------------------------------------
-------------------------
exp(x.log(a)) is-------->a^x
----------------------------------------
-------------------------
(1/2)*[exp(i*x)+exp(-i*x)] is----->cos(x)
-------------------------
(1/2i)*[exp(i*x)-exp(-i*x)] is----->sin(x)
```

Implicit Differentiation

Implicit differentiation of equations in more than one variable with respect to one variable f=x^2+y^2+2*x*y w.r.t. x. The result will be in terms of the derivative of y w.r.t. x.

```
syms x a
f=sin(x);dif=diff(f,x);
disp(['Derivative of sinx(x)',' --->',char(dif)])
disp('--------------------------')
f=sin(x)*log(x); dif=diff(f,x);
disp('--------------------------')
disp(['Derivative of sin(x)log(x)',' --->',char(dif)])
f=exp(-x)*x^2*sin(x);dif=diff(f,x);
disp('--------------------------')
disp(['Derivative of x^2.sin(x)exp(-x) ',' --->',char(dif)])
disp('--------------------------')
f=a^x;dif=diff(f,x);
disp(['Derivative of a^x ',' --->',char(dif)])
disp('--------------------------')
clear
```

```
Derivative of sinx(x) --->cos(x)
--------------------------
--------------------------
Derivative of sin(x)log(x) --->cos(x)*log(x) + sin(x)/x
--------------------------
Derivative of x^2.sin(x)exp(-x)  --->x^2*exp(-x)*cos(x) - x^2*exp(-x)*sin(x) + 2*x*exp(-x)*sin(x)
--------------------------
Derivative of a^x --->a^x*log(a)
--------------------------
```

Integration

```
clear
syms t y(t) a b c z w zz z1 z2
w=a*t+b*t^2*y+y^3;
z=diff(w,t);
zz=diff(z,t);
z1=simplify(z);z2=simplify(zz);
disp('--------------------------')
disp(['d[a*t+b*t^2*y+y^3]/dt = ',char(z)])
disp(['d^2[a*t+b*t^2*y+y^3]/dt^2 = ',char(zz)])
```

```
--------------------------
d[a*t+b*t^2*y+y^3]/dt = a + 3*y(t)^2*diff(y(t), t) + 2*b*t*y(t) + b*t^2*diff(y(t), t)
d^2[a*t+b*t^2*y+y^3]/dt^2 = 6*y(t)*diff(y(t), t)^2 + 2*b*y(t) + 3*y(t)^2*diff(y(t), t, t) +
b*t^2*diff(y(t), t, t) + 4*b*t*diff(y(t), t)
```

Definite Integrals

```
syms x y f a b
f=log(x);intf=int(f,x);
disp('-------------------------')
disp(['Integral of log(x) ',' --->',char(intf)])
f=sin(x)*x^2*exp(-x);intf=int(f,x);
disp('-------------------------')
disp(['Integral of sin(x).x^2.exp(-x)',' --->',char(intf)])
f=sin(x)^4/cos(x)^4;intf=int(f,x);
disp('-------------------------')
disp(['Integral of sin(x)^4/cos(x)^4 is',' --->',char(intf)])
f=log(x+1);intf=int(f,x);
disp('-------------------------')
disp(['Integral of log(x+1) is',' --->',char(intf)])
f=1/(1+x^2);intf=int(f,x);
disp('-------------------------')
disp(['Integral of 1/(1+x^2) is',' --->',char(intf)])
```

```
-------------------------
Integral of log(x)  --->x*(log(x) - 1)
-------------------------
Integral of sin(x).x^2.exp(-x) --->-(exp(-x)*(x + 1)*(cos(x) - sin(x) + x*cos(x) + x*sin(x)))/2
-------------------------
Integral of sin(x)^4/cos(x)^4 is --->x - tan(x) + tan(x)^3/3
-------------------------
Integral of log(x+1) is --->(log(x + 1) - 1)*(x + 1)
-------------------------
Integral of 1/(1+x^2) is --->atan(x)
```

Definite Integration with More Specifications

```
syms x
y=x*exp(-x);
F=int(y,x,0,inf); % integrate y for x=0 to inf
disp(['Integral of x exp(-x): x o to infinity is ',char(F)])
```

```
Integral of x exp(-x): x o to infinity is 1
```

Symbolic Substitution

Replace *x* with *y*.

```
clear
% declare x a b as positive symbolic variables
syms x a
assumeAlso(a>0)
f=(x/a^2)*exp(-x^2/(2*a^2));
disp('-------------------------')
intf=int(f,x,0,inf);
disp(['integral of (x/a^2)*exp(-x/(2*a^2)) is',' --->',char(intf)])
```

```
-------------------------
integral of (x/a^2)*exp(-x/(2*a^2)) is --->1
```

sin(pi*x)/(pi*x) in the limit x → 0

```
syms x y a positive
fx=(x/a^2)*exp(-x^2/(2*a^2));
fy=subs(fx,x,y);
disp(fy)
F=int(fy,y,0,x);%
disp(['integrate fy, y= 0 to x = ',char(F)])
```

```
(y*exp(-y^2/(2*a^2)))/a^2

integrate fy, y= 0 to x = 1 - exp(-x^2/(2*a^2))
```

Substitution, Differentiation, Integration, Inline Functions, LaTeX Use

```
clear
syms x
y=sin(pi*x)/(sym(pi)*x);%
F_limit=limit(y,x,0)% express as limit
```

```
F_limit = 1
```

```
clear;
syms x y a positive
fx=(x/a^2)*exp(-x^2/(2*a^2)); % this the Rayleigh density
fy=subs(fx,x,y);

disp(['replace x with y ',char(fy)])
Fx=int(fy,y,0,x);% Rayleigh CDF
syms f_X(x)  F_X(x)
f1=[f_X(x)==fx];
F1=[F_X(x)==Fx];
% create an inline function
ff=matlabFunction(fx)
FF=matlabFunction(Fx)
% chose a value of a as 2 and plot for values of x
x1=0:.2:10;
ff1=ff(2,x1); % values for plotting
FF1=FF(2,x1); % values for plotting
figure,plot(x1,ff1)
xlabel('x'),ylabel('pdf')
% now display the title as a formal equation
title(['$' latex(f1) '$'],'interpreter','latex','fontsize',16,'color','b')
figure,plot(x1,FF1)
xlabel('x'),ylabel('CDF')
% now display the title as a formal equation
title(['$' latex(F1) '$'],'interpreter','latex','fontsize',16,'color','b')
```

```
replace x with y (y*exp(-y^2/(2*a^2)))/a^2

ff = @(a,x)1.0./a.^2.*x.*exp(1.0./a.^2.*x.^2.*(-1.0./2.0))

FF = @(a,x)-exp(1.0./a.^2.*x.^2.*(-1.0./2.0))+1.0
```

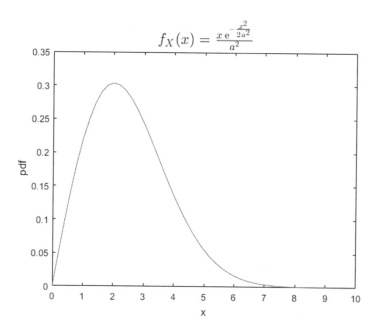

$$f_X(x) = \frac{x\, e^{-\frac{x^2}{2a^2}}}{a^2}$$

Gamma Density: CDF the Ratio of Two Gamma Functions

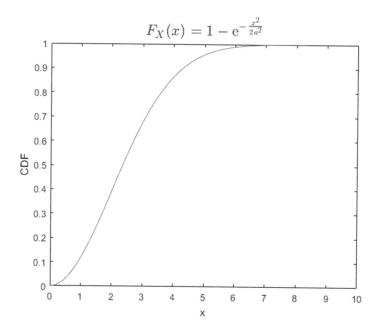

$$F_X(x) = 1 - e^{-\frac{x^2}{2a^2}}$$

```
clear;
syms x y a b positive
fx=x^(a-1)*exp(-x/b)/(b^a*gamma(a));
fy=subs(fx,x,y);
disp(fy)
Fx=int(fy,y,0,x);%
syms f_X(x)  F_X(x)
f1=[f_X(x)==fx];
F1=[F_X(x)==Fx];
% create an inline function; note the order of the variables (in
% alphabetical order  a, b, ..x, y, z
ff=matlabFunction(fx)
FF=matlabFunction(Fx)
% chose a value of a as 2 and b as .5 and plot for values of x
a=2;
b=0.5;
x1=0:.2:10;
ff1=ff(a,b,x1); % values for plotting
FF1=FF(a,b, x1); % values for plotting
figure,plot(x1,ff1)
xlabel('x'),ylabel('pdf')
% now display the title as a formal equation
title(['$' latex(f1) '$'],'interpreter','latex','fontsize',16,'color','b')
figure,plot(x1,FF1)
xlabel('x'),ylabel('CDF')
% create title by directly typing in LaTex for mat
CDFF='$$ F_X(x)=1-\frac{\Gamma\left(a,\frac{x}{b}\right)}{\Gamma(a)}$$';
title(CDFF,'interpreter','latex','fontsize',16,'color','b')
```

```
(y^(a - 1)*exp(-y/b))/(b^a*gamma(a))

ff = @(a,b,x)(b.^(-a).*x.^(a-1.0).*exp(-x./b))./gamma(a)

FF = @(a,b,x)-igamma(a,x./b)./gamma(a)+1.0
```

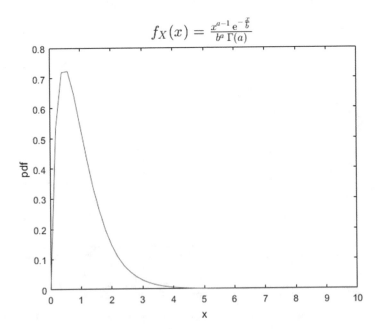

$$f_X(x) = \frac{x^{a-1}\,e^{-\frac{x}{b}}}{b^a\,\Gamma(a)}$$

Conditional Density Modeling and Analysis

The conditional density Example 3.23. Additionally, the conditional mean is also estimated for part 1.

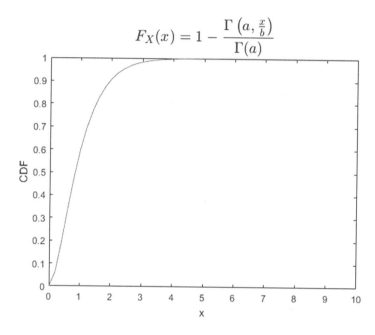

$$F_X(x) = 1 - \frac{\Gamma\left(a, \frac{x}{b}\right)}{\Gamma(a)}$$

```
clear
% Rayleigh parameter b corresponds to average speed of 60 MPH
b=60*sqrt(2/pi);
x=raylrnd(b,1,1e7);
HS=75; % higher speed
%y=x.*(x>HS); % values less than HS will be zeros
%yy=y(y~=0); % values that are not zeros
[fi,x1]=ksdensity(x);% original pdf;
yy=x(x>HS); % get the conditional random number (numbers exceeding 75)
w=HS:1:250;
[f]=ksdensity(yy,w);
ff=raylpdf(w,b)/(1-raylcdf(HS,b));
figure, plot(x1,fi,'b--',w,ff,'r-',w,f,'k-.*')
xlabel('speed mph'),ylabel(' pdf')
xlim([0,max(w)]),ylim([0,.05]),legend('f(x)','f(x|x>75)- theory','f(x|x>75)- simulation')
str1='$$ f(x|X\ge 75)=\frac{(x/48^2)e^{-x^2/2(48^2)}}{e^{-75^2/2(48^2)}}U(x-75) $$';
text(90,.036,str1,'interpreter','latex','color','b','fontweight','bold')
% get the conditional mean
ME=mean(yy);% mean of the speed|speed exceeds HS MPH
syms y % define y as the speed
f=y*exp(-y*y/(2*b^2))/b^2;
ff=exp(-HS^2/(2*b^2));
mes=double(int(y*f/ff,y,HS,inf));
  text(100,.03,['E[X|X>75] = ',num2str(round(ME*1000)/1000),'(sim), ',...
      num2str(round(mes*100)/100),' (integral) '],'color','b','fontweight','bold')
text(200,.01, 'PM Shankar','color','g','fontsize',8)
% lower limit of 40 MPH
clear yy f w ff ME mes y
LS=40;
yy=x(x<LS); % get the conditional random number (numbers below LS)
w=0:1:LS;
[f]=ksdensity(yy,w);
ff=raylpdf(w,b)/raylcdf(LS,b);
figure, plot(x1,fi,'b--',w,ff,'r-',w,f,'k-.*')
xlabel('speed mph'),ylabel(' pdf')
xlim([0,200]),ylim([0,.05]),legend('f(x)','f(x|X<40)- theory','f(x|X<40)- simulation')
str11='$$ f(x|X\le 40)=\frac{(x/48^2)e^{-x^2/2(48^2)}}{1-e^{-40^2/2(48^2)}}\left[U(x)-U(x-
40)\right] $$';
text(50,.036,str11,'interpreter','latex','color','b','fontweight','bold')

ME=mean(yy); % mean from random number generation
syms y
f1=y*exp(-y*y/(2*b^2))/b^2;
f2=1-exp(-40^2/(2*b^2));
mes=double(int(y*f1/f2,y,0,LS)); % mean using the density
text(50,.03,['E[X|X<40] = ',num2str(round(ME*1000)/1000),'(sim), ',...
      num2str(round(mes*100)/100),' (integral) '],'color','b','fontweight','bold')
% speed between 40 and 75
clear yy f w ff
yy=x(x>LS & x <HS); % get the conditional random number (numbers below LS)
w=LS:1:HS;
[f]=ksdensity(yy,w);
ff=raylpdf(w,b)/(raylcdf(HS,b)-raylcdf(LS,b));
ME=mean(yy); % mean from simulation
syms y
f1=y*exp(-y*y/(2*b^2))/b^2;
f2=exp(-LS^2/(2*b^2))-exp(-HS^2/(2*b^2));
mes=double(int(y*f1/f2,y,LS,HS)); % mean from density
figure, plot(x1,fi,'b--',w,ff,'r-',w,f,'k-.*')
xlabel('speed mph'),ylabel(' pdf')
xlim([0,200]),ylim([0,.05]),legend('f(x)','f(x|40<X<75)- theory','f(x|40<X<75)- simulation')
str11='$$ f(x|40\le X \le 75)=\frac{(x/48^2)e^{-x^2/2(48^2)}}{\left[e^{-40^2/2(48^2)}-e^{-
75^2/2(48^2)}\right]}\left[U(x-40)-U(x-75)\right] $$';
text(2,.036,str11,'interpreter','latex','color','b','fontweight','bold')
text(80,.03,['E[X|40<X<75] = ',num2str(round(ME*100)/100),'(sim), ',...
      num2str(round(mes*100)/100),' (integral) '],'color','b','fontweight','bold')
```

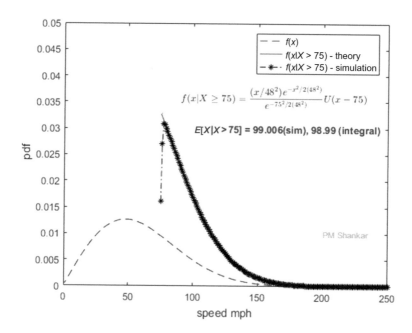

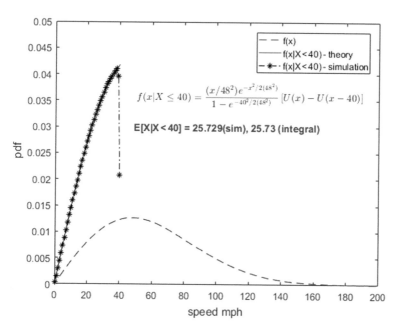

Data Analysis, Extra Data Creation Using Random (.) or Bootstrapping

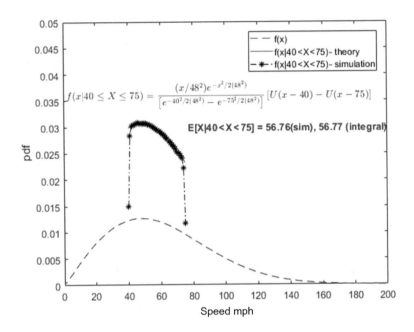

```
clear x
% use the data given below
x=[3.7899    4.9786    11.8411    10.6441    5.8730
   10.3001    3.0888     6.0906     0.4442    5.2253
    5.3235    8.8603     3.1639     7.8682    6.2912
    3.0422    6.3404     5.5841     8.5316   10.8773
    2.4165   13.0514    14.8414     7.0778    8.6197
    6.7786    4.8938     3.8280    11.1229    9.5716
    2.0945    5.0428     3.0181     8.5509    2.5235
   10.8716    4.9474    26.9505     4.7526    7.7246
    2.4514    2.6551     7.7272    12.3863    8.9034
    5.0004    4.0185     2.9678    14.0975    2.7562];
x=x(:);
% how to use ksdensity(.)
 xval=0:.1:1.5*max(x); %x-values for finding the pdf
 fx=ksdensity(x,xval); % estimated pdf of the data
 Fx=ksdensity(x,xval,'function','cdf'); % estimated CDF of the data
 % now test whether the data fits gamma, Rayleigh, etc.
 % to do this, there is a need to estimate the parameters
 pd=fitdist(x,'gamma');
 a1=pd.a;
 b1=pd.b;
 [h,p,statsG]=chi2gof(x,'CDF',pd);
 disp(h)
 disp(statsG)
 disp('shows that h = 0: Null hypothesis of GAMMA cannot be rejected')
 % now get a plot of the theoretical fit. This requires the parameters
 fxth=pdf('gamma',xval,a1,b1); % theoretical pdf
 Fxth=cdf('gamma',xval,a1,b1);% theoretical pdf
figure, plot(xval,fxth,'r-',xval,fx,'k-')
legend('theoretical fit','estimated data fit')
xlabel('values'),ylabel('estimated pdf')
figure, plot(xval,Fxth,'r-',xval,Fx,'k-')
legend('theoretical fit','estimated data fit')
xlabel('values'),ylabel('estimated CDF')
% how to find the area under a curve
% verify that theoretical density integrates to unity
area1=polyarea(xval,fxth) % use polyarea  command:  this must be unity
area2=trapz(fx)*(xval(2)-xval(1)) % use trapz;
% now generate a new set of samples having same statistics as the given data
xnew=random('gamma',a1,b1,50,1);%
y=[x,xnew];
figure,hist(y)
legend('original data','simulated data')
xlabel('values')
ylabel('Relative Frequency')
pd2=fitdist(xnew,'gamma');
 [h,p,statsG]=chi2gof(xnew,'CDF',pd2)
 disp('shows that h = 0: Null hypothesis of GAMMA cannot be rejected')
%generate one new set having the same statistics using bootstrap
[bootstat,bootsam] = bootstrp(1,[],x);
newdat= x(bootsam);
figure, hist([x,newdat]),legend('original data','recreated data')
xlabel('values'),ylabel('frequency')
title('Comparison: orginal and recreated data (bootstrapping) sets','color','b')
text(max(x)-2,5,'P M Shankar','color','g','fontsize',9)
pd3=fitdist(newdat,'gamma');
 [h,p,statsG]=chi2gof(newdat,'CDF',pd3)
 disp('shows that h = 0: Null hypothesis of GAMMA cannot be rejected')
```

```
           0
      chi2stat: 2.6009
            df: 2
         edges: [0.4442 3.0948 5.7455 8.3961 11.0467 26.9505]
             O: [11 13 9 10 7]
             E: [8.4610 14.4879 11.7974 7.3840 7.8697]
shows that h = 0: Null hypothesis of GAMMA cannot be rejected
area1 = 0.9999
area2 = 0.9682
h = 0
p = 0.2496
statsG =
      chi2stat: 2.7759
            df: 2
         edges: [2.1259 4.5612 6.9966 9.4319 11.8673 26.4795]
             O: [7 18 10 7 8]
             E: [9.5749 13.0929 11.4667 7.6086 8.2569]
shows that h = 0: Null hypothesis of GAMMA cannot be rejected
h = 0
p = 0.3076

statsG =
      chi2stat: 3.6030
            df: 3
         edges: [0.4442 2.9656 4.2264 5.4871 6.7478 8.0085 13.0514]
             O: [8 8 11 5 3 15]
             E: [9.6864 8.2684 7.9851 6.7572 5.2668 12.0362]

shows that h = 0: Null hypothesis of GAMMA cannot be rejected
```

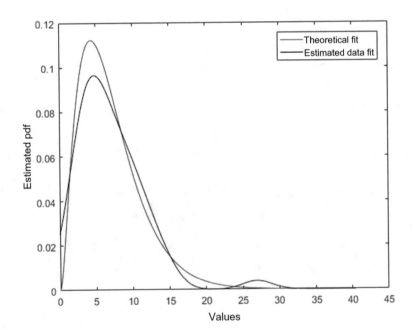

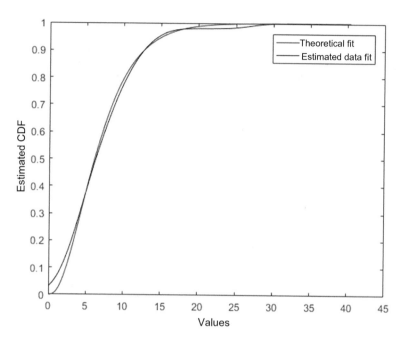

symbolic_MeijerG-use

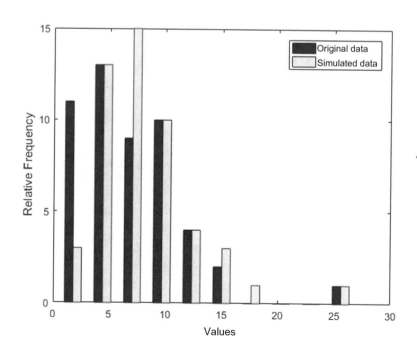

Use of Meijer G Function Numerically

This requires slight modification from the previous segment.

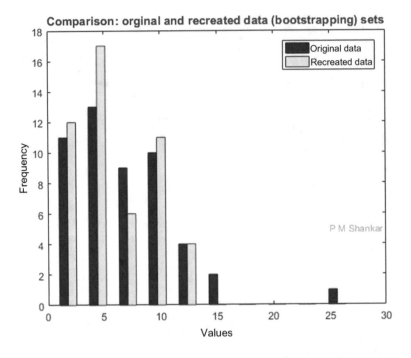

```
syms x Z m n positive
%following expression is the pdf of the product of 4 i.i.d. gamma variables
f=evalin(symengine,'meijerG([[], []], [[m, m, m, m], []], m^4*x/Z)/(x*gamma(m)^4)');
% check whether this is a valid pdf
verf1=int(f,x,0,inf)
% get the CDF
F=int(f,x,0,x)
% get the pdf back by differentiation
f1=diff(F,x)
% check for the mean; it must be Z
Me=int(f*x,x,0,inf)
% express gamma density as a Meijer G function
fg=(m/Z)^m*x^(m-1)*exp(-m*x/Z)/gamma(m); % this is a gamma pdf
fgM=evalin(symengine,'meijerG([[], []], [[m], []], m*x/Z)/(x*gamma(m))');
% verify that the representation is correct
verf4=simplify(fgM/fg)% this value should be unity
.% product of two independent gamma variables for different orders m and n
fk=2*(m*n*x/Z)^(m/2+n/2)*besselk(m-n,2*sqrt(m*n*x/Z))/(x*gamma(m)*gamma(n));
fkM=evalin(symengine,'meijerG([[], []], [[m, n], []], m*n*x/Z)/(x*gamma(m)*gamma(n))');
% verify that the representation is correct
verf5=simplify(fkM/fk)% this value should be unity
```

```
verf1 = 1

F = meijerG([[1], []], [[m, m, m, m], [0]], (m^4*x)/Z)/gamma(m)^4

f1 = meijerG([[], []], [[m, m, m, m], []], (m^4*x)/Z)/(x*gamma(m)^4)

Me = Z

verf4 = 1

verf5 = 1
```

MLE and Chi-Square Testing: Generalized Gamma pdf, Not a Built-in pdf

As an example, a three-parameter gamma distribution is used. It is also known as the Stacy distribution. It is obtained by scaling a gamma variable of parameters a and b to the power k. In other words, if G(a,b) is a gamma variable, [G(a,b)]^k will be a generalized gamma variable.

```
clear
% plot the pdf of the product of 3 i.i.d. gamma variables
m=1.7;   gm=gamma(m);
Z=15;
xval=0.01:.1:8;
for k=1:length(xval)
    x=xval(k);   xZ=x/Z;
  pd(k)=(1/x)*double(evalin(symengine,sprintf('meijerG([[ ], [ ]], [[%e,%e,%e], []],
%e)',m,m,m,(m^3)*xz)))/gm^3;
end;
figure,plot(xval,pd,'linewidth',2,'color','k')
xlabel('x '),ylabel('pdf')
title('meijerG([[ ], [ ]], [[m,m,m], []], (Z*m^3/x))/(\Gamma^3(m)x)')
legend(['m = ',num2str(m),', E(X) = Z = ',num2str(Z)])
```

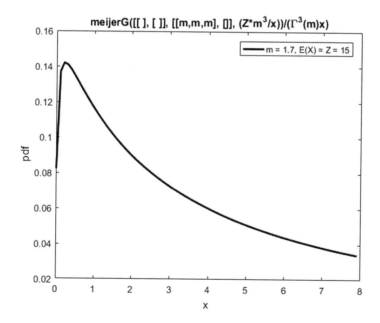

Characteristic Function, Laplace and Inverse Laplace, Mellin

```
syms x y z k a b positive
fx=(x^(a/k-1))*exp(-x^(1/k)/b)/(k*gamma(a)*b^a);
verf=simplify(int(fx,x,0,inf),'steps',10);
F=1-igamma(a,x^(1/k)/b)/gamma(a);%CDF for the three parameter gamma
% % verify
% pdf=diff(F,x);
% simplify(pdf-fx) % must be 0
x1=0:1:200;
% plot the generalized gamma pdf for a=2.5, b=4 and k=1.8;
gf=subs(fx,[a,b,k],[2.5,4,1.8]);
gf1=matlabFunction(gf);% create the in-line function for plotting
figure,plot(x1,gf1(x1))
xlabel('x'),ylabel('pdf')
title('Generalized gamma density')
syms f_X(x)
syms F_X(x)
text(100,.0125,['$' latex([f_X(x)==fx]) '$'],'interpreter','latex',...
    'fontsize',16,'color','b')
text(100,.01,['$' latex([F_X(x)==F]) '$'],'interpreter','latex',...
    'fontsize',16,'color','b')
text(100,.008,'[a = 2.5, b = 4, k = 1.8]')

dat= [
    21.5017     3.1831   154.2424   205.1274    23.8291   102.9688     4.1345    21.3768     7.9424
 37.7401
    24.5300    49.3259    16.5008     5.0045    53.9964    92.7681    77.3781    75.6681    23.4622
267.7394
     8.5155     8.3571    12.7686     4.0325   123.5047    28.3785    24.6613     0.7295    15.7628
  2.3886
    38.2985    12.8360   183.9903    84.7612    28.9807   788.4190    38.3543     2.3606    67.8712
 38.7587
   135.8796    15.1251    16.1129    20.5816   248.1222   402.0427   398.4408   171.2651    22.8848
 63.3869
    19.9117    20.6321    24.5119   290.5381    23.6142    44.6659    26.5654   281.3627    54.6354
  3.6391
    58.5116    58.6039     0.0331    19.1041    53.3146    15.6900    13.7187    23.9311    80.4561
 20.1473
   107.5165    13.2864    17.3573    22.7856    23.6173    27.1641    75.1911    16.4866   117.2523
 59.4483
    32.2764    64.3589     9.3189    22.6501   119.1552    98.1383   129.0120    48.6907    72.0877
 25.2814
    65.7074     1.9527    15.8387    46.3445    41.4054     1.1819    29.5380   316.5675     5.9639
254.8011
    78.5312    31.9672     6.3450    89.9654     5.1491     7.7568    13.6582     7.3568    43.2329
143.6805
    21.7340    42.2495    23.0352   124.1584     4.4812    26.9991    87.4799   100.7563    10.2423
102.3278
     0.3771     2.3581     3.0748    74.7991     2.0464    24.7604   230.7695    14.1470   684.7978
148.4075
   140.1095    27.6369    57.3045     5.2766    84.6171     5.1914    49.3461     5.5886   282.1673
 12.8769
    51.7287   157.5918    41.8577     1.6998     6.9076     0.5960    62.0397    13.1748    14.9686
  9.2509
     8.6514    40.4503    90.1721    34.7709    29.1528    77.6155     7.2551   109.2082    25.0322
 71.7511
    16.5679    92.7065   415.8947     1.5984    17.4258    64.0209    53.2793   163.2720   165.6885
 28.7031
    25.4071    53.6847    56.4375    23.4761    26.9256   111.9488    19.1382    46.3235   309.8903
 36.4922
     2.2496    24.5133     8.4166     5.3026     7.0829    12.1289    45.1385    73.5977    15.1796
  4.1832
    24.8185     9.5451   308.3797    23.7986     3.6129    36.0475     4.8318   115.9825    49.9432
 25.2293
```

23.4000 400.3217 31.8927 26.2441 17.6319 180.7773 20.2014 79.2005 139.9371
112.1410

67.7981 67.1331 23.3264 19.7641 145.3293 38.3604 1.5981 52.1649 20.3011
155.9270

40.9387 296.9857 19.4667 50.2296 32.2110 64.7544 7.0334 1.8839 84.0826
9.0247

9.2731 27.9282 7.8564 17.7832 4.6630 28.4369 250.2817 16.2042 130.8028
168.6469

49.4548 33.5197 14.5750 3.0315 108.2028 38.8936 9.7545 42.2249 61.7707
28.4053

33.1550 26.1103 46.0439 72.4981 26.5686 84.9266 293.3571 35.2434 170.1173
2.5256

7.9485 46.6906 27.8529 87.0898 97.3859 8.2694 233.6620 111.5403 33.8085
34.8324

125.1440 4.9181 148.5131 54.9283 43.9878 7.3198 36.3928 146.3520 0.2837
21.0407

34.4713 113.6689 35.3086 193.8376 27.4139 51.6083 228.3258 171.3916 49.1588
58.7153

63.5229 74.1801 62.0739 58.4276 20.1359 125.7690 13.2848 8.1573 55.5109
215.2074

206.9716 44.7609 359.7120 4.4991 48.4783 16.2541 8.8177 70.9870 29.3731
109.5124

47.7950 33.4492 102.1773 51.6183 24.9483 40.7769 195.4396 6.6902 9.2987
37.7287

0.4762 105.3725 15.4321 61.5094 2.5402 0.6823 10.7875 52.6776 30.4445
138.8773

54.2213 20.5583 257.7243 38.6089 0.0119 2.3613 11.9520 8.4368 178.3801
158.0252

84.7105 31.5135 13.1950 34.8784 7.2193 101.5790 87.7741 33.2396 7.3877
6.2022

15.3452 18.0064 41.1349 33.3757 114.2155 19.7129 34.6006 8.6649 65.5095
84.3449

108.5346 43.7198 66.1866 29.8392 17.9513 39.8294 148.4294 80.3846 301.3913
52.1937

274.8825 37.0637 7.9066 16.7368 24.1421 162.9452 26.6983 38.6883 68.5915
41.4498

19.7770 30.0968 4.7592 108.2322 2.5500 42.2280 28.8682 7.6278 20.9222
5.1486

0.8900 181.3892 54.7741 7.8880 88.8691 56.6457 13.5902 20.1034 129.5821
39.3024

37.4886 28.8423 8.1145 46.1850 78.0746 101.3665 58.4793 43.7070 284.1930
7.3176

90.1152 85.8045 226.4074 34.9214 8.7540 64.2749 48.4746 15.8436 29.0345
98.8780

90.8786 60.2256 53.4164 40.7115 58.9418 19.7676 112.2353 335.5730 125.4176
30.0916

31.6973 30.3235 42.1029 12.4763 70.2300 49.8567 30.9060 12.8144 30.0453
5.9456

4.3720 12.8607 3.3972 108.9482 5.0576 72.5695 45.3753 24.1934 17.0549
90.0682

18.3225 5.6384 454.4624 174.0267 14.4527 13.4843 21.1183 258.6243 92.5277
13.7211

250.3684 2.5522 0.6541 14.1319 2.3368 5.2543 179.7539 5.2427 65.9451
37.2182

17.8949 27.3989 86.5518 172.8248 42.5940 12.4831 318.2036 2.9824 9.2143
18.6378

195.8066 26.2461 27.1318 44.3612 191.4820 10.4533 34.8251 26.2636 47.3690
79.5238

42.1747 6.5236 6.3123 35.2652 75.7189 9.1662 211.7725 53.9872 172.6013
257.7750

9.9136 3.7105 1.9863 8.3839 13.3175 51.6917 319.4972 77.5506 17.9177
139.0883

```
  29.2867  179.7881   47.7536  170.8965    8.9839    5.3404   25.9397   25.1154   32.3995
  32.0645
  58.3568    0.4327   14.9363    9.3531  149.6148    4.1097    9.8516   60.7674    6.5107
 369.5840
  96.8080   31.0478  134.2403   23.9150    1.5682   41.9756  231.9028    6.4258    3.5760
   8.3529
   5.7837  111.3143   46.4706   10.0178  120.6690    6.0627   12.1046   40.1180   72.6083
  33.7976
  94.5946   18.1863    9.7176   61.4225  247.2045   27.7172   82.2373   32.4192  639.9252
  21.7377
  20.5322   42.8500   52.9598   51.6337   10.9088   50.3957   74.3257   71.6871   56.9992
   8.3445
   9.0860   36.4154   26.1428    3.2405  134.6399   24.5427   28.2058   91.3756   32.6607
 111.8270
  25.3081  145.0382    8.6796    0.8054   57.8104  178.3289   66.7308   33.4875    6.6251
  13.5139
  16.4303   17.4354  170.1776  284.8113   14.6214    2.0843   12.1546    8.1750    0.4211
  21.5807
  23.1030   54.8018   64.2207   33.4588    4.1718  131.7946   41.9234    7.6157   59.3584
 163.6121
  16.2272   52.4461   44.0463   66.1231   67.7311   41.9912   40.0780   25.5734   54.1528
  14.4987
  56.7341   11.3308   13.7092   34.4540   75.5485   26.1206   28.7662   53.1423   67.8216
  39.7014
  11.2665  249.5792    9.5248    9.8115   20.4359   14.2771   32.0731  104.4305    1.8486
   1.9050
  98.1121   25.7005    5.7428   10.9962   36.4391    8.9120   99.3256   21.2700   13.5613
   4.0944
   3.6400  102.5711   90.1788   19.8800  128.8499   20.7388    9.3949   49.2305   64.3041
  88.1682
  13.6333   37.2413   32.8503  160.9963  108.5602  318.4472  138.1711   31.6410   23.6467
 197.8721
   3.1108   78.6077  253.2203   16.5910   29.1233    6.2372   12.7413   59.7340  224.7555
  96.6066
  29.3350    0.5414   26.3971   17.0500   65.9152   15.0480   38.6704   36.0979   25.9818
 147.6274
   9.4215  109.0098  441.2865   62.2729   96.7869  136.9699   11.1689   50.4865  191.7741
  92.2198
 705.2326  101.5650   14.9092    9.7315  190.5665  112.0488   41.4094   25.6819  126.2224
  18.7504
  20.2924  147.0423   89.9365   11.7887  140.4205   10.2014   51.1718   29.3203   14.3885
   3.7826
  53.7904    9.5761   10.0896  123.7467   89.5866    3.6215  158.4973   56.1279    9.0378
  53.5599
  14.6751   13.0760   33.7084  252.4395  197.4033   71.9664    2.2816    4.0304   20.4060
  37.2025
   1.4441  143.9641   39.9757   76.1361   44.3124    3.1361   14.2802   73.7958   41.7295
  18.1915
  98.3298   22.5931    4.1137  158.7003    5.2871   34.6017   45.1059    1.7635  129.2142
  92.3368
   2.6244   78.5428    2.2819   34.9014   52.0327   98.7417    0.0366   12.8311    5.4965
   6.0786
  41.9453   27.4784   21.7983   13.8748   40.5257   76.5167   20.8097    4.0740   62.1200
  76.7911
  56.8537  105.1812   15.4542   53.7731   46.4908  224.6305   57.9247   46.4082   18.7129
 283.8315
 357.3417   63.0152  119.8564   40.2873  130.0084   38.6664   78.2401   61.4684   32.6907
  15.2723
  22.3273   35.2498    1.1146   33.4297   39.8653   31.0540    7.2155   24.5120   16.0755
  55.0203
   0.9555   30.1257   95.7360   17.3554   24.1734    4.6328   61.9824   62.4317   15.9151
  12.3017
   0.6769   15.7582   11.8610   20.6944    8.8117   11.2436    8.5956   58.1168   70.1233
   7.5838
  22.9604  259.4303   40.5751   10.0531   97.2628  213.7740   50.8140   13.2999  117.4895
  60.3513
```

```
   15.8999    30.7648     6.8609    36.2809   120.3734     4.0148    93.5674   199.2389    78.6964
10.3329
   33.3775    17.0640   107.6621    48.7242   115.2965     1.7359    24.2983    42.6980    85.6573
77.8984
   18.3995    31.3889     5.1581    83.6393    31.6590    50.4362   293.7642    25.0514    94.1023
7.7524
  140.7602   148.1534     9.4727     3.8089   177.9784   121.5244    37.8256    51.9444    73.6025
28.2503
   11.7751    36.3642    62.1944    64.1061    53.6294    30.5467    63.8995   124.6098     7.4905
47.0393
   54.3453     1.4011    57.7024     7.9203    22.1534   319.7976    46.2160    18.0042     3.9895
2.8696
   73.7992     6.3909     7.7852    35.3120   213.8119    24.1367   104.1098    31.1877   127.3026
143.3549
    6.3916    47.4092     9.7238    48.3512    68.2875    16.2000   167.7536    59.3393    63.5888
18.2684
   57.3838     9.4696     6.2025   194.0325    60.7661    29.1813     1.0313     9.4525   141.5523
83.8502
    0.9405   181.7966   110.7760     1.5620    46.9796    41.0018     8.1548    90.0577    49.5288
104.5554
  149.0506     7.2051    27.5313     8.0819    64.5152    14.0691    20.3496    76.2621   132.1737
31.7600
   14.6757    74.3074    13.7705    14.2369    19.2466   107.6819    46.9279   100.9093   128.5390
13.3706
    6.8823     9.8036    47.1678     9.9776    29.9345    76.2844    12.1821   202.1878    34.2299
19.4109
   47.0296    44.3930    14.9779   107.9039   480.2757    25.8550    89.9825    39.1286    98.0472
24.1647
  103.5698     5.6874    95.8114    79.0958     6.2294    54.0607     3.4698    19.1553    66.8851
7.3736
   76.9730    34.5633    24.1626    27.7188    31.5231    10.9157    44.2254   140.9606     1.0107
34.7985
    ];
dat=dat(:);%
% create an in-line function: the variable must be first followed by
% the parameters to do this change x to X
syms X
ffx=subs(fx,x,X); % change the variable x to X
 pdff=matlabFunction(ffx) ;
% either the function above or the inline function below may
% be used. Results will match
% pdff=@(y,a1,b1,k1) (y.^(a1/k1-1)).*exp(-y.^(1/k1)/b1)/(k1*gamma(a1)*b1^a1);
 opt = statset('mlecustom');
 opt = statset(opt,'FunValCheck','off','MaxIter',4000,'TolX',1e-4,'TolFun',1e-5,'TolBnd',1e-
5','MaxFunEvals', 4000);
 pp = mle(dat,'pdf',pdff,'start',[1.6,4,2.3], 'options',opt); %;
 a_est=pp(1);
 b_est=pp(2);
 k_est=pp(3);
 disp(['MLE estimates: [ a_est = ',num2str(a_est),', b_est = ',num2str(b_est),...
     ', k_est = ',num2str(k_est),']'])
% now conduct a chi square test
 nbin=7; % number of bins
[O,edg]=histcounts(dat,nbin); % there will be nbin+1 edges
FW=subs(F,[a,b,k],[a_est,b_est,k_est]); % substitute a, b, and p estimates
Fy=double(subs(FW,x,edg));% get the CDF at the edges of the bins
p=diff(Fy); % probability in the subinterval; this will be of size nbin
N=length(dat);
np=p.*N;
fac=((O-np).^2)*1./np;
q=sum(fac); % chi2stat
df=nbin-1-3;% 3 is the number of parameters calculated from the DATA
Qthr=chi2inv(.95, df);
```

```
if q<=Qthr
    h=0;
    disp('------------------------------------------------------')
    disp('test statistics less than the threshold value: CANNOT..REJECT')
        disp('------------------------------------------------------')
else
    disp('FAIL....Reject')
    h=1;
end;
COUNTS=0
chi2_thr=Qthr
chi2stat=q
DegFreedom=df
```

MLE estimates: [a_est = 2.3935, b_est = 3.1834, k_est = 1.896]
----
test statistics less than the threshold value: CANNOT..REJECT

COUNTS = 852   102   34   7   1   3   1

chi2_thr = 7.8147

chi2stat = 7.3326

DegFreedom = 3

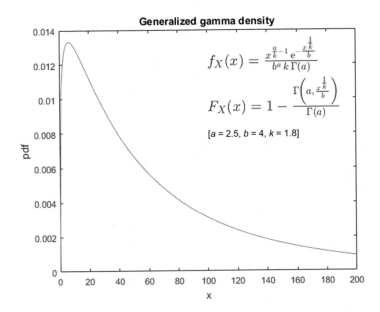

characteristic function, Laplace and inverse Laplace, Mellin

characteristic function, Laplace and inverse Laplace, Mellin

```
syms x u m sig
syms sig real
assumeAlso(sig>0)
fn=exp(-(x-m)^2/(2*sig*sig))/sqrt(2*sig*sig*sym(pi));
Fn=fourier(fn,x,u)

syms a b x m u s sig real, syms a b sig positive
fe=(1/a)*exp(-x/a);
Fe=simplify(int(fe*exp(i*u*x),0,inf),'steps',10)
fg=x^(a-1)*exp(-x/b)/(gamma(a)*b^a);
Fg=simplify(int(fg*exp(i*u*x),0,inf),'steps',10)
% Laplace and inverse laplace
Fle=laplace(fe,x,s)
Flg=laplace(fg,x,s)
feback=ilaplace(Fle,s,x)
fgback=ilaplace(Flg,s,x)
```

```
Fn = exp(-(sig^2*(u + (m*1i)/sig^2)^2)/2 -m^2/(2*sig^2))
Fe = -1/(-1 + a*u*1i)
Fg = 1/(b^a*(1/b -u*1i)^a)
Fle = 1/(a*(s + 1/a))
Flg = 1/(b^a*(s + 1/b)^a)
feback = exp(-x/a)/a
fgback = (x^(a - 1)*exp(-x/b))/(b^a*gamma(a))
```

Published with MATLAB® R2016a.

Bibliography

Baumer B (2015) A data science course for undergraduates: thinking with data. Am Stat 69 (4):334–342

Bertsekas D, Tsitsiklis J (2008) Introduction to probability. Athena Scientific, Boston

Carnell LJ (2008) The effect of a student-designed data collection project on attitudes toward statistics. J Stat Educ 16:1–15

de Winter JC, Dodou D (2012) Five-pPoint Likert items: t test versus Mann-Whitney-Wilcoxon. Practical assessment. Res Eval 15:1–6

Du Q (2006) Design of application-oriented computer projects in a probability and random processes course for Electrical Engineering majors. In: Proceedings of the American Society for Engineering Education, pp 11.409.1–11.409.1

Garfield J, Ahlgren A (1988) Difficulties in learning basic concepts in probability and statistics: implications for research. J Res Math Educ 19:44–63

Hagman J, Johnson E, Fosdick BK (2017) Factors contributing to students and instructors experiencing a lack of time in college calculus. Int J STEM Educ 4(1):12–15

Han KK, Kagan HP, Kass RD (2001) Simple demonstration of the central limit theorem using mass measurements. Am J Phys 69:1014–1019

Hardin J, Hoerl R, Horton NJ, Nolan D, Baumer B, Hall-Holt O, Murrell P, Peng R, Roback P, Temple Lang D, Ward MD (2015) Data science in statistics curricula: preparing students to "think with data". Am Stat 69(4):343–353

Kim TK (2015) T test as a parametric statistic. Korean J Anesthesiol 68:540–546

Koehler KJ, Larntz K (1980) An empirical investigation of goodness-of-fit statistics for sparse multinomials. J Am Stat Assoc 75:336–344

Krause S, Middleton J, Johnson E, Chen Y-C (2015) Factors impacting retention and success of undergraduate engineering students. In: Proceedings of the 122nd ASEE Annual Conference, June 14–17

Kvam PH (2006) The effect of active learning method on student retention in engineering statistics. Am Stat 54:136–140

Liu H, Li G, Cumberland WG, Wu T (2015) Testing statistical significance of the area under a receiving operating characteristics curve for repeated measures design with bootstrapping. J Data Sci 3(3):257–278

Mecatti F (2000) Bootstrapping unequal probability samples. Stat Appl 12(1):67–77

Miao Y, Xu X (2019) A note on the almost sure central limit theorem for the product of partial sums of m-dependent random variables. Commun Stat Theory Methods 48(9):2102–2112. Metz CE, Pan X (1999) "Proper" binormal ROC curves: theory and maximum-likelihood estimation. J Math Psychol 43(1):1–33

© Springer Nature Switzerland AG 2021
P. M. Shankar, *Probability, Random Variables, and Data Analytics with Engineering Applications*, https://doi.org/10.1007/978-3-030-56259-5

Moser BM, Stevens GR (1992) Homogeneity of variance in the two-sample means test. Am Stat 46:19–21

Munoz-Repiso AG, Tejedor FJ (2012) The incorporation of ICT in higher education. The contribution of ROC curves in the graphic visualization of differences in the analysis of the variables. Br J Educ Technol 43:901–919

Papoulis A, Pillai U (2002) Probability, random variables, and stochastic processes. McGraw-Hill, New York

Rohatgi VK, Saleh AKME (2001) An introduction to probability and statistics. Wiley, New York

Roscoe JT, Byars JA (1971) An investigation of the restraints with respect to sample size commonly imposed on the use of the chi-square statistic. J Am Stat Assoc 66:755–759

Smith G (1998) Learning statistics by doing statistics. J Stat Educ 6:1–12

Watkins JC (2010) On a calculus-based statistics course for life science students. CBE Life Sci Educ 9(3):298–310

Wehrens R, Putter H, Buydens LM (2000) The bootstrap: a tutorial. Chemom Intell Lab Syst 54 (1):35–52

Welch BL (1947) The generalization of student's problems when several different population variances are involved. Biometrika 34:28–35

Yue S, Pilon P (2004) A comparison of the power of the t-test, Mann-Kendall and bootstrap tests for trend detection. Hydrol Sci J 49:21–37

Zimmerman DW (1997) A note on interpretation of the paired-samples t test. J Educ Behav Stat 22:349–360

Zimmerman DW (2012) Correcting two-sample z and t tests for correlation: an alternative to one-sample tests on difference scores. Psicológica 33:391–418

Index